Principles of Regenerative Biology

Principles of Regenerative Biology

Bruce M. Carlson, M.D., Ph.D.

UNIVERSITY OF MICHIGAN
ANN ARBOR, MICHIGAN

AMSTERDAM • BOSTON • HEIDELBERG • LONDON
NEW YORK • OXFORD • PARIS • SAN DIEGO
SAN FRANCISCO • SINGAPORE • SYDNEY • TOKYO

Academic Press is an imprint of Elsevier

Academic Press is an imprint of Elsevier
30 Corporate Drive, Suite 400, Burlington, MA 01803, USA
525 B Street, Suite 1900, San Diego, California 92101-4495, USA
84 Theobald's Road, London WC1X 8RR, UK

This book is printed on acid-free paper. ♾

Library of Congress Cataloging-in-Publication Data
Application Submitted

British Library Cataloguing-in-Publication Data
A catalogue record for this book is available from the British Library.

ISBN 13: 978-0-12-369439-3
ISBN 10: 0-12-369439-6

For information on all Academic Press publications
visit our Web site at www.books.elsevier.com

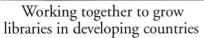

To the memory of Richard J. Goss, who stimulated me* and many others to probe into the mysteries of regeneration.

---

*As a student about to graduate from a small midwestern college, I was wandering through the library and ran across the first issue of the *Journal of Experimental Zoology* that the library had subscribed to. In it was an article by Richard Goss and Martha Stagg (Regeneration in lower jaws of newts after excision of the intermandibular regions. *J Exp Zool* 137:1–12, 1958). Even though I had already signed up to do graduate work on fish taxonomy at Cornell, that article affected me almost like a religious conversion, and I knew instantly that regeneration would be my ultimate field of endeavor. Although my interest in fish has not abated, I have never regretted the decision to switch to regeneration research, and it has led me down some fascinating pathways.

Table of Contents

Preface

Regeneration is one of the most fascinating phenomena in biology, but it is also one of the most complex. Virtually all species, from protozoa to humans, have the capacity to regenerate, but the extent of their regenerative ability varies greatly. Planaria, starfish and some worms can regenerate most of their body, whereas many other species are able to regenerate only parts of specific tissues. Among the vertebrates, urodele amphibians are the best adapted for regeneration; they can regenerate limbs, tails, jaws, eyes and a variety of internal structures.

For much of the twentieth century, mammals, including humans, were considered to have a poor capacity for regeneration. This was particularly true up to the late 1950s, when I first began to explore the literature on regeneration. A research visit to the Soviet Union during 1965–1966 exposed me to many new ways of looking at mammalian regeneration. Perhaps the most striking to me was the emphasis on *uslovie* ("conditions") of regeneration that abounded in the Russian regeneration literature of the 1960s. The essence of this concept was that the success of regeneration often is a function of the environment in which the regenerative process is taking place. I was initially inclined to attribute this emphasis to the waning influence of Lysenkoism, which still permeated the Russian biology of the time. This element was certainly present, but during my several decades of research on regeneration since that time, I have come to appreciate how important the environment is in supporting or failing to support regenerative processes in mammals.

With the explosion of knowledge from molecular biology and the burgeoning interest in generating or regenerating tissues or organs through various tissue engineering or stem cell approaches, many scientists and students have shown a renewed interest in the phenomenon of regeneration. Because relatively few have had the luxury of being able to approach the phenomenon of regeneration from a broad biologic perspective, I thought that it would be useful to write a short book that outlines some of the fundamental biologic principles of regeneration. As the book has evolved, the contents have focused principally on regeneration in vertebrate systems, but when certain points are best illustrated by examples taken from the very diverse universe of invertebrate regeneration, they are included, as well. In order to manipulate regenerative processes, it is important to understand the underlying principles of regeneration. Laying these out is what this book is all about.

It is often said that science progresses by stepping on the shoulders of one's predecessors. My hope is that this book will place in focus enough intellectual shoulders of pioneers in regeneration research to provide sufficient stepping stones for the next generation of researchers in this fascinating field.

Bruce M. Carlson

Acknowledgments

This book was made possible through the sharing of information over many years with colleagues and students all around the world. I am particularly indebted to those who have generously provided me with reprints, not only of their own work, but hard-to-locate reports from often quite obscure sources. The present book grew out of an earlier one (Carlson, BM. 1986. Moscow: *Regeneratsiya,* Nauka), which was published only in Russian, and I thank Dr. Victor Mitashov for suggesting that I write that one.

Thanks to the University of Michigan's retirement furlough program, this is the first book that I have been able to write during the daytime hours. Special thanks also to my wife and colleague, Jean, for her continuous support of my book-writing projects.

The present book has benefited greatly from the many fine original illustrations that were prepared by Shayne Davidson, with whom I had previously worked on a number of smaller projects. It's always a pleasure to work with a true professional. Thanks are also due to Jasna Markovac and Tari Broderick at Elsevier for accepting the manuscript and supporting the actual writing. Fran Levy skillfully and cheerfully guided the transformation of the manuscript into a book. To her, many thanks are due.

List of Abbreviations

cAMP	Cyclic adenosine monophosphate
ANG	Angiopoietin
AP	Activator protein
ARIA	Acetylcholine receptor inducing activity
BDNF	Brain-derived neurotrophic factor
BHH	Banded hedgehog
BMP	Bone morphogenetic protein
CAM	Cell adhesion molecule
CNS	Central nervous system
CNTF	Ciliary neurotrophic factor
DHKA	Dictyostelium histidine kinase
DNA	Deoxyribonucleic acid
DOB	Devoid of blastema—a zebrafish mutant
ECM	Extracellular matrix
EGF	Epidermal growth factor
EGFR	Epidermal growth factor receptor
EPH	Ephrin
ERK	Extracellular signal-regulated kinase
ES	Embryonic stem cells
FGF	Fibroblast growth factor
FGFR	Fibroblast growth factor receptor
G	Gap, with respect to the mitotic cycle
GAP-43	Growth-associated protein
GDNF	Glial cell–derived neurotrophic factor
GFP	Green fluorescent protein
GRHL	Homolog of *Drosophila grainy head* gene
GTPase	Guanosine triphosphatase
HB-EGF	Heparin-binding epidermal growth factor
HGF	Hepatic growth factor, also known as scatter factor
HIF	Hypoxia-inducible factor
HOX	Homeobox-containing transcription factor
HSP	Heat shock protein
IGF	Insulin-like growth factor
IHH	Indian hedgehog

KGF	Keratinocyte growth factor
KROX	A gene that regulates hindbrain development
L1	A cell adhesion molecule (NgCAM)
LF	A transcription factor downstream of WNT
M	Mitosis—component of the cell cycle
MAG	Myelin-associated glycoprotein
MAP	Microtubule-associated protein
MDR	A surface antigen found on hematogenous stem cells
MEIS	Homeobox genes activated by retinoic acid
MMP	Matrix metalloproteinase
MPS	A mitotic checkpoint kinase
MRF	Myogenic regulatory factor
MRL	Strain of mouse with a high regenerative capacity
MSX	Ortholog of *Drosophila* Msh (muscle segment homeobox)
MYF	A myogenic regulatory factor
MYOD	A myogenic regulatory factor
NGF	Nerve growth factor
NGFR	Nerve growth factor receptor
NGR	Nogo receptor
NK	Natural killer cells
NKX	A transcription factor regulating heart development
NO	Nitric oxide
Nogo	An antigen from myelin degeneration products that inhibits axonal growth
NT	Neurotrophic factor
NVTBOX	*Notophthalmus viridescens* T-box gene
OCT	Octa box—a small transcription factor
OMGP	Oligodendrocyte-myelin glycoprotein
PAX	Paired homeobox-containing transcription factor
PDGF	Platelet-derived growth factor
PNS	Peripheral nervous system
PROD	An ortholog of CD59, part of the complement group
PRX	Called Mhox earlier; a member of group of Aristaless-like homeobox genes
RAF	A serine/threonine kinase
RAS	A small GTPase
Rb	Retinoblastoma
RGD	Arginine-glycine-asparagine sequence of importance in cell adhesion
RHO	A cytoplasmic regulatory protein
RNA	Ribonucleic acid
S	Synthesis (DNA) phase in the mitotic cycle
SCA	An antigen found on stem cells
shh	Sonic hedgehog
SIS	Small intestine submucosa—a natural tissue substrate

SOX	An HMG box-containing transcription factor
SPARC	Secreted protein rich in cysteine—an extracellular matrix protein
T	T cell—a lymphocyte that matured through the thymus, involved in cellular immunity
TBX	T box transcription factors
TGF	Transforming growth factor
THY	Thymidylate synthase complementing protein
TIMP	Tissue inhibitor of matrix metalloproteinase
TNF	Tumor necrosis factor
VEGF	Vascular endothelial growth factor

CHAPTER 1

An Introduction to Regeneration

If there were no regeneration, there could be no life. If everything regenerated, there would be no death.

—Richard J. Goss (1969)

BRIEF HISTORY

Regeneration is one of the oldest fields in biology (Dinsmore, 1991; Vorontsova and Liosner, 1960). The phenomenon was known to the ancients and was described in the writings of both Aristotle and Pliny, but the first scientific observations of regeneration were made and reported in 1712 by René-Antoine Ferchault de Réaumur, who made a detailed description of limb regeneration in crayfish. During that time, the preformation controversy was raging, and Réaumur hypothesized that the regenerating limbs arose from the expansion of tiny preformed limbs that resided at the base of the limb. The next half century saw a variety of seminal investigations on regeneration in a variety of different organisms—hydra by Abraham Trembley in 1740, annelids by Charles Bonnet in 1745, amphibians by Spallanzani (1769), and planarians by P. S. Pallas in the 1770s. The phenomenon of regeneration made such an impression on the philosophic discussions of the time that it spilled into the popular arena. Late in the eighteenth century, members of the French nobility took scissors into their gardens and amputated heads of land snails to watch them regenerate.

By the early nineteenth century, the early naturalists had described regeneration in many kinds of animals, but until the elaboration of the cell theory in 1838 by Matthias Schleiden and Theodor Schwann and the development of histologic techniques, research on regeneration was principally confined to gross observations. Charles Darwin conducted regeneration studies on planaria while on his famous voyage aboard the *Beagle*, and later regeneration was (and continues to be) discussed in terms of evolutionary theory. In the pre-Mendelian era of genetics, Weismann (1892) elaborated one of the first theories of morphogenesis in limb regeneration. At the end of the nineteenth century, Thomas Hunt Morgan (1901) led an active school of regeneration research before he left the field in favor of his famous studies on *Drosophila* genetics.

The twentieth century began with detailed histologic descriptions of a variety of regenerative processes in many species, but after a couple of decades, interest in mammalian regeneration waned because of the seeming lack of ability of many human tissues to regenerate. Studies on invertebrate regeneration were heavily influenced by the metabolic gradient theories of Child (1941), and the field of amphibian limb regeneration moved from pure description to experimental studies on the role of specific tissue components in regeneration, the role of dedifferentiation, and morphogenesis. The period around World War II saw a burst of attempts to stimulate limb regeneration in frogs and other nonregenerating forms. Interest in regeneration waned in the second half of the century, although seminal work was being done on regeneration in many systems, including mammalian tissues. Only toward the end of the century did new insights on morphogenesis stimulate general interest in the field. The molecular revolution was late to affect the field of regeneration, and even now our knowledge of molecular events in almost any system of regeneration is sketchy. Around the turn of this century, the sudden awareness of stem cells and their potential in promoting the regeneration of human tissues has given the field new life; currently, expectations are high.

WHAT IS REGENERATION?

The word *regeneration* means different things to different people. According to a terse dictionary definition from *Stedman's Medical Dictionary, regeneration* is the "reproduction or reconstitution of a lost or injured part" or "a form of asexual reproduction." Within such a definition, however, is encompassed a broad spectrum of natural phenomena that operate through dramatically different mechanisms. In recent years, there has been considerable interest in building missing or damaged tissues and organs in humans through the application of bioengineering principles or through the use of stem cell technology. This field is now being called *regenerative medicine,* and most major medical research centers have set up regenerative medicine units. Such applications of biomedic technology use approaches and techniques that can be quite discipline specific. Ideally, one would like to devise regenerative techniques that take maximal advantage of natural regenerative processes that occur in the body.

For years, the literature on regeneration has been replete with discussions of types of regenerative processes. Such classifications have their value when well-understood processes are delineated and compared, but unfortunately, many regenerative processes are so poorly understood that it is difficult to assign them to any particular category. What follows in this chapter is an attempt to outline a variety of types of named regenerative processes, without trying to fit all known varieties of regeneration into them. This will provide a basic vocabulary for the more in-depth treatment of specific subjects that will follow later in this book. Figure 1-1 provides a mini-taxonomy of many of the named varieties of regeneration that are generally recognized. Interestingly, a number of well-studied regenerative phenomena have not had generic names applied to them and would not, therefore, appear in such an artificial taxonomy. An example is regeneration of the lens in the eye of urodele amphibians, about which much more will be written in this book.

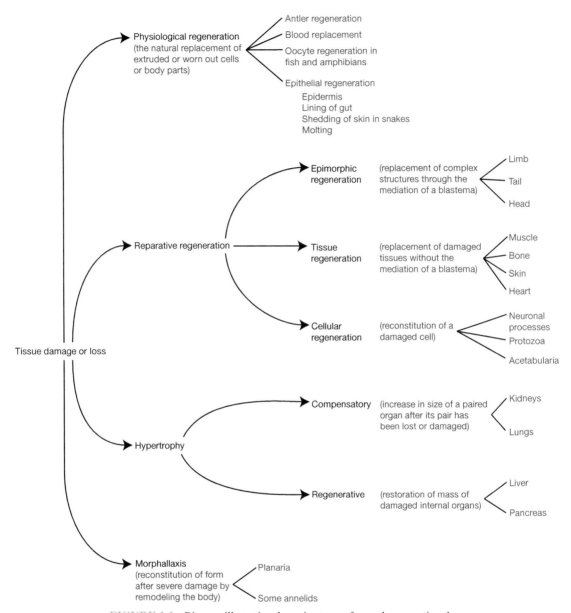

FIGURE 1-1 Diagram illustrating the major types of named regenerative phenomena.

Physiological Regeneration

Physiological regeneration, the natural replacement of extruded or worn-out body parts, is a process that occurs in many of our body systems and is often studied in the form of cellular turnover. Some of the best examples of mammalian physiological regeneration are the shedding cycles of the epidermis or the epithelial cells lining the gut, the renewal of the endometrium after a menstrual period, and the replacement of blood cells, which have well-defined, finite life spans. Even these examples embody quite different cellular processes and control mechanisms (Fig. 1-2), but their common feature is the maintenance of cellular or tissue equilibrium in the body.

By its nature, physiological regeneration must be highly attuned to bodily needs. The control of erythropoiesis through the erythropoietin pathway (Migliaccio et al., 1996) is one of the best understood regulatory mechanisms of physiological regeneration, but many other examples of physiological regeneration operate under quite different circumstances and respond to very different stimuli. One of the most striking examples is the annual regeneration cycle of antlers (see Fig. 5-4), a complex process that follows a well-defined cycle based on light (Goss, 1983). Other examples of physiological regeneration in nonmammalian species are the shedding cycles of crustacean exoskeletons and snake skin, the annual molting and replacement cycle of feathers in birds, and the regeneration of oocytes in fishes and other lower vertebrates that lay large numbers of eggs each year.

An important characteristic of many varieties of physiological regeneration is that their intensity can be adjusted to meet physiological needs. One of the most dramatic examples is the greatly increased output of red blood cells after hemorrhage or ascent into high altitudes. At another level, the formation of a callus in the epidermis of the hand or foot is a reflection of a local control mechanism that responds to mechanical conditions.

In dealing with the limits of labels, it is important to understand that the term *physiological regeneration* does not imply a specific type of mechanism. There is a tremendous difference between the relatively simple replacement cycle of an epidermal cell and that of a deer antler, which involves much of the complexity of the regeneration of an entire limb. *Physiological regeneration* should, therefore, be looked at as a term of convenience that embodies a variety of processes designed to maintain the normal equilibrium of the body tissues.

Reparative Regeneration

Reparative regeneration is the term that has been applied to most varieties of post-traumatic regeneration. Reparative regeneration can occur at levels from the single cell to major parts of the body. Some of the most dramatic examples of reparative regeneration are the regeneration of the amputated limb or tail of a salamander or newt, or the reconstitution of the entire body of a planarian from a fragment less than 1/200 of the original mass. As is the case in physiological regeneration, reparative regeneration can

A.

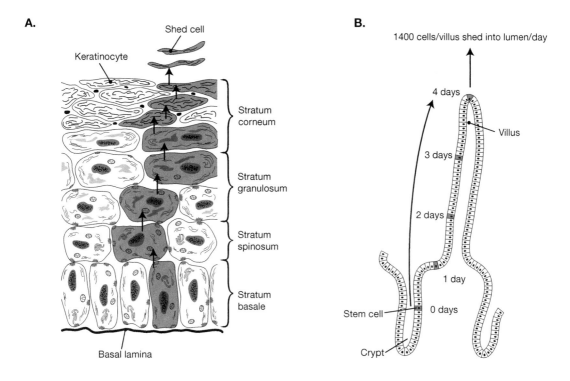

B.

C.

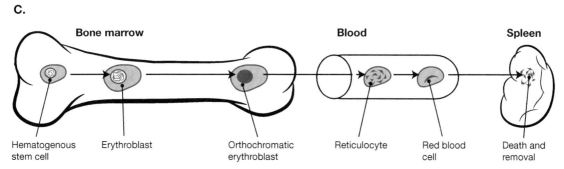

FIGURE 1-2   Three types of physiological regeneration in mammals. *(A)* The turnover cycle of an epidermal cell. *(B)* The shedding cycle of an epithelial cell on a villus in the small intestine. *(C)* The replacement cycle of a red blood cell.

occur by a wide variety of means. Thus, the common element in reparative regeneration is the replacement of a lost or damaged part of the body. Several varieties of reparative regeneration are presented below, not only to illustrate the categories, but to provide a basic descriptive vocabulary that will be useful for understanding material presented in subsequent chapters.

Epimorphic Regeneration

The term *epimorphic regeneration* was coined by Thomas Hunt Morgan (1901, p. 23) to refer to "cases of regeneration in which a proliferation of material precedes the development of the new part." Morgan contrasted epimorphosis with *morphallaxis* (see later), "in which a part is transformed directly into a new organism, or part of an organism without proliferation at the cut surfaces." Over time, the working understanding of both of these terms has changed considerably from Morgan's original definitions, which served their purpose at a time when the collective knowledge of regeneration was much more limited.

In practice, epimorphic regeneration is now applied to phenomena that are characterized by the formation of a regeneration blastema[1] that arises through epithelial mesenchymal interactions and that contains and expresses intrinsic morphogenetic information. The classic example of epimorphosis is the regenerating amphibian limb (see the following section).

Regenerating Amphibian Limb

One of the most fascinating processes in nature is the way in which the stump of an amputated amphibian limb deals with its loss by mobilizing cells at the amputation surface to form a regeneration blastema that then goes on to produce an almost exact replica of the amputated limb (Fig. 1-3). Iten and Bryant (1973), Schmidt (1968), Tank et al. (1976), Tsonis (1996), and Wallace (1981) provide detailed descriptions of limb regeneration, including staging systems.

Phase of Wound Healing. Immediately after amputation, the soft tissues retract, largely because of contractions of the muscle stumps within the limb; seconds later, bleeding from major vessels stops through contractions of the vascular walls (Fig. 1-4, A). Within hours, the epidermal cells at the margins of the amputation wound become mobilized and begin to migrate across the amputation surface, which is largely covered by a substrate of products from serum and extracellular fluids (Repesh and Oberpriller, 1978, 1980). The sheet of migrating epidermis completely covers the amputation surface within hours or a day or two, depending on the size of the amputated limb. At a purely observational basis, there appears to be little, if any, difference between the epithelialization of an amputation surface and the epithelial healing of a nonregenerating skin wound in the same individual.

Phagocytosis and Demolition. For a brief period after the amputation wound has healed, the histologic appearance of the distal end of the limb stump is quite unremarkable (see Fig. 1-4, B). The most distal ends of the muscles, nerves, bones, and skin

---

[1]Unfortunately, the term *blastema* is also applied to structures or stages that appear in other reparative or developmental circumstances. For example, the blastema that is referred to in the literature of fracture healing of bone is not at all the equivalent of a limb regeneration blastema, because it does not arise through epithelial mesenchymal interactions and it does not appear to contain intrinsic morphogenetic information.

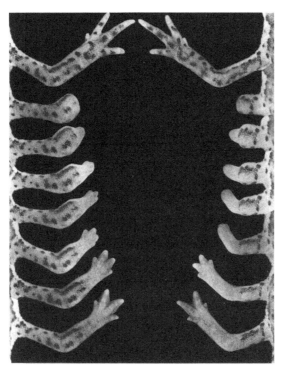

FIGURE 1-3 Successive stages in the regeneration of newt arms amputated at upper (right) and lower (left) arm levels. Starting below the normal arms at the top, the intervals of regeneration are 7, 21, 25, 28, 32, 42, and 70 days after amputation. (Reprinted from Goss R.J. 1969. *Principles of regeneration.* New York: Academic Press, by permission.)

show obvious evidence of recent trauma, and the soft tissues are lightly infiltrated by inflammatory cells, but to a far lesser extent than would be a similarly damaged structure in a mammal. The wound epidermis, which is not underlain by a basement membrane, begins to thicken, initially as a response to the tissue debris beneath it (Singer and Salpeter, 1961). Despite the unremarkable appearance of the limb stump at this early stage, many of the events that convert a wound healing response to a regenerative response are gearing up. Prominent among them are the reorganization of the distal extracellular matrix through the actions of matrix metalloproteinases (Kato et al., 2003) and later the influence of serine proteases, such as thrombin, in stimulating formerly quiescent cells to reenter the cell cycle (Brockes and Kumar, 2002). Although this is the period during which the decision whether to regenerate is made in the amputated limb, less is known about this phase in the regenerative process than almost any other phase.

Dedifferentiation. The phase of dedifferentiation (see Fig. 1-4, C), historically one of the most controversial in the field of regeneration, occurs when the limb stump

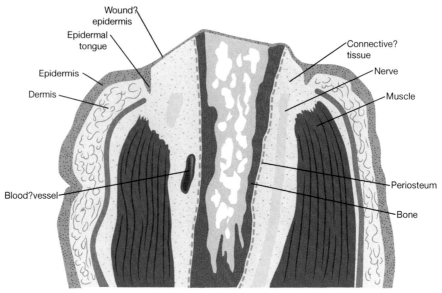

A.

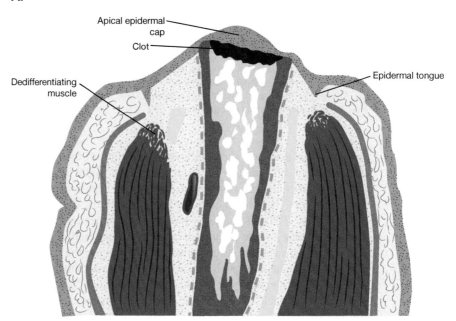

B.

FIGURE 1-4   Semischematic drawings of stages of regeneration of the amputated upper forelimb in the mature axolotl (adapted from photomicrographs). *(A)* Early wound healing. A wound epidermis has covered the amputation surface. Epidermal tongues mark the boundaries between original and wound epidermis. *(B)* Phase of demolition. The wound epidermis has thickened, and a few inflammatory cells are present in the distal region of the limb stump. The apical epidermis is beginning to thicken into an apical epidermal cap.

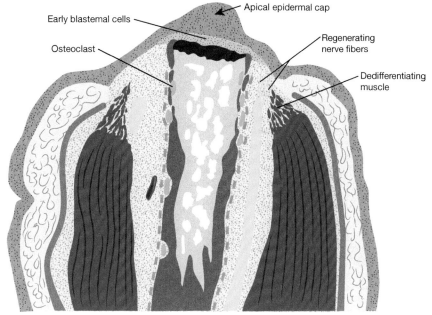

C.

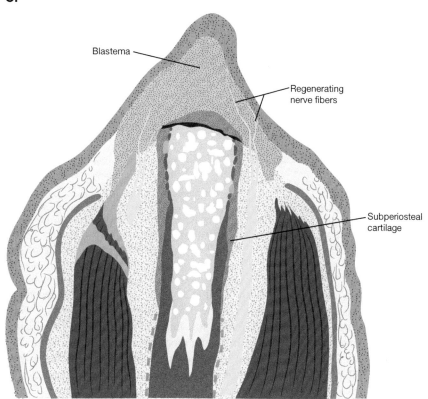

D.

FIGURE 1-4 *(Continued)* *(C)* Phase of dedifferentiation. Osteoclasts (blue ovals) have begun to chew away on the bone, and the distal tissues of the limb are being swept into the dedifferentiative response. The wound epidermis has thickened into an apical epidermal cap, and nerve fibers are regenerating into the most distal stump regions. *(D)* Blastema formation. A mass of homogeneous-appearing blastema cells has begun to protrude past the amputation surface (see Fig. 3-7 for photomicrograph), and a cuff of subperiosteal cartilage has formed around the shaft of the humerus.

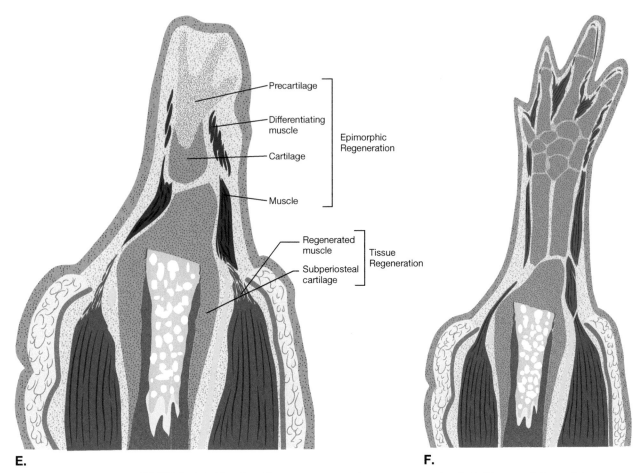

FIGURE 1-4 *(Continued)* *(E)* Morphogenesis of the regenerating limb. The overall form of the regenerating limb has been set, and individual tissue elements (muscle and skeleton) are differentiating according to a proximodistal gradient. Areas where regeneration is occurring by the epimorphic and tissue modes are indicated. *(F)* A mature regenerate. Morphogenesis of skeleton and muscles is complete. The regenerated skeleton is still cartilaginous. The regenerate has still not grown to its final complete size.

has made a clear commitment to engage in an epimorphic regenerative process. In the broadest descriptive sense, the dedifferentiation phase occurs when the tissues at the distal end of the stump have lost many of their mature histologic characteristics and have begun to produce cells of a more embryonic morphologic character (future blastema cells). The essence of the historical controversy has been whether at a single cellular level a differentiated cell can lose its specialized characteristics and revert to a more generalized form that could perhaps redifferentiate into another cellular phenotype. This issue is discussed in greater depth in Chapter 2.

At the structural level, dedifferentiation is characterized by the osteoclastic erosion of bone, the dissolution of cartilage matrix in larval forms, or both. The wound epidermis has become noticeably thickened in the form of an apical epidermal cap and has been penetrated by regenerating nerve fibers. Muscle fibers at the amputation surface have lost their striations, and the areas vacated by the formerly differentiated muscle fibers are becoming infiltrated with immature-looking cells of the type that will ultimately constitute the blastema. The thickened dermis of the skin remains clearly recognizable, but similar cellular changes are occurring in the dermal fibroblasts.

Blastema Formation.   As the immature-looking cells accumulate during late dedifferentiation, they become concentrated distally beneath the thickened apical epidermal cap. This aggregation of cells, which produces a budlike outgrowth at the tip of the limb stump, is called the *regeneration blastema* (see Fig. 1-4, D). In both structure and properties, the blastema is reminiscent of the embryonic limb bud. Like the mesenchymal cells of the limb bud, blastema cells are structurally almost indistinguishable from one another, despite the fact that distinct subpopulations are known to exist. Initially, the regeneration blastema is relatively avascular, but as it matures, a network of capillaries forms within it (Peadon and Singer, 1966; Rageh et al., 2002). The basal layer of cells within the apical epidermal cap often takes on a regular columnar morphology. This is reminiscent of other types of epithelia, such as that of the cornea, which secrete substances or material into the underlying tissue matrix.

Of greatest importance is the unseen pattern-forming activity that occurs within the blastema. Through mechanisms that remain little understood, an approximate recapitulation of the events that occur during embryogenesis of the limb is beginning to occur. (Models of pattern formation within the blastema are discussed later.)

Morphogenesis.   Morphogenesis represents the morphologic fruition of the pattern-forming activity that has taken place within the regeneration blastema (see Fig. 1-4, E). It is reflected by rapid outgrowth of the blastema and the condensation of some of the blastema cells into precartilaginous models of the major skeletal elements of the regenerating limb. Slightly lagging skeletal morphogenesis, a complete musculature of the regenerating limb is laid down in a pattern that is virtually identical to that seen during embryogenesis (Grim and Carlson, 1974). Pronounced proximodistal and anteroposterior gradients of differentiation at both the tissue and cellular levels are apparent during this period. At the end of the period of morphogenesis, the regenerating limb is structurally almost identical to the limb that was removed.

Growth.   In regenerating limbs of larger individuals, once morphogenesis is completed, the regenerated limb is perfectly formed, but much smaller than the original limb. Over succeeding weeks, growth of the regenerate continues until ultimately the regenerate has attained the same dimensions as a normal limb. Limbs of amphibian larvae are often so small that by the time morphogenesis is complete, there is little need for a subsequent growth phase (see Fig. 1-4, F).

Other Varieties of Epimorphic Regeneration

Epimorphic regeneration is not just confined to the amputated amphibian limb. Regeneration of a major body part mediated by a regeneration blastema is found in species representing a variety of major taxonomic divisions within the animal kingdom (Fig. 1-5). Although there are differences in details, epimorphic regeneration in any of these species is guided by certain common elements.

The first element is the covering of the amputation surface by a wound epithelium. In addition to simply sealing off the wound from outside influences, which is also a function of nonregenerative healing, the wound epithelium appears to provide a morphogenetic boundary or reference point that is important in organizing the developmental information required to form a blastema (Maden, 1977). In a number of systems that have been investigated, the apposition of epithelia of different qualities (e.g., dorsal and ventral in the case of planaria [Agata et al., 2003] and from opposite sides of the limb in the case of amphibians [French, Bryant, and Bryant, 1976]) is a critical first step in initiating an epimorphic regenerative process.

The source of cells that will make up the blastema appears to differ among species. Although dedifferentiation at the tissue and probably the cellular level is the generator of blastemal cells in species such as amphibians and sabellid worms, neoblasts or reserve cells represent the primary cellular source of the blastema in planaria. Of critical importance is what these cells bring with them as they aggregate into a blastema. Do they carry with them morphogenetic information or specific information about their tissue of origin, or are they wiped clean of their former developmental history?

Regardless of the origin and former history of the blastemal cells, the cells of the blastema proliferate under the influence of the overlying epithelium and take on many of the properties of the embryonic precursors of the structure that was amputated. A main difference is that the blastema is tied to a mature body and must ultimately become integrated with it.

Most characteristic of epimorphic regeneration is that a morphogenetic pattern is set up within the blastema, and it is expressed as the blastema grows. The unfolding of the regenerate typically follows a morphologic pattern that closely recapitulates the ontologic one.

Tissue Regeneration

The term *tissue regeneration* has at some time been applied to the repair of most damaged tissues within the body, with the greatest emphasis on mammalian tissues because of the medical implications (McMinn, 1969). The regeneration of some tissues (e.g., skeletal muscle or bone) has received a great deal of attention, but other tissues have been largely neglected. This section focuses on elements common to the regeneration of a number of tissues but will use skeletal muscle and bone as specific examples.

Tissue regeneration can be initiated by a wide variety of traumatic means. Mechanical trauma is the most common, but extremes of heat or cold or a wide variety of chemical insults, including toxins, are also causes of tissue damage. Free

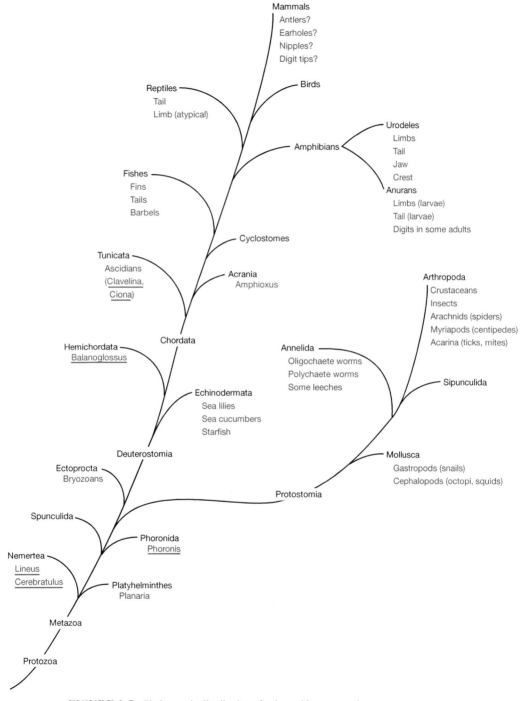

FIGURE 1-5   Phylogenetic distribution of epimorphic regeneration.

(nonvascularized) transplantation of a variety of tissues stimulates tissue regeneration because the devascularized tissues of the graft typically undergo partial or complete necrosis caused by the interruption of the blood supply. In this particular situation, the source of cells for regeneration may be either surviving elements within the graft itself or cells that immigrate into the graft, in which case, the graft essentially serves as a natural substrate rather than a regenerating system itself.

Regardless of the nature of the initial traumatic stimulus, the reactions of the traumatized tissues in general follow a similar course (Table 1-1). Bleeding may or may not be present within the area of tissue damage, but regardless, one of the early reactions is edema and swelling, followed by an inflammatory response. The nature of the initial inflammatory response bears a critical relation to the success of tissue regeneration. A massive focus of acute inflammation, characterized by large accumulations of neutrophils, is inimical to a healthy tissue regenerative response, probably because of the highly destructive environment that surrounds such a focus of inflammatory cells. The macrophage-based phase of inflammation that follows the initial acute phase is important. In addition to removing cell and tissue debris, macrophages secrete large numbers of growth factors that appear to activate precursor cells for tissue regeneration.

The cellular precursors for tissue regeneration are specific for a particular tissue, but in most cases, tissue regeneration involves additional cells beyond the dominant cell type. For example, in skeletal muscle, the principal tissue-specific precursor cell is the satellite cell (Mauro, 1961), but the tissue regeneration of muscle typically includes a number of other cell types, such as fibroblasts, endothelial cells, smooth muscle cells, and Schwann cells. Tissue regeneration almost always involves regeneration of the local microvasculature, reconstruction or remodeling of the connective tissue stroma, and in the case of skeletal muscle, the regeneration of its nerve supply. Whereas in skeletal muscle, the satellite cells are scattered throughout the tissue (beneath the basal laminae of the individual muscle fibers), in bone, the cellular precursors are concentrated in the periosteum and endosteum. Cardiac muscle regeneration in amphibians involves mitosis of remaining functional cardiomyocytes to produce additional cells for the regenerative process (Oberpriller and Oberpriller, 1991). The

TABLE 1-1

Common Features in Tissue Regenerative Processes

 1. Trauma (e.g., mechanical, chemical, thermal)
 2. Localized posttraumatic ischemia and edema
 3. Local inflammation and the removal of damaged tissues by phagocytosis
 4. Activation of the cellular precursors of regeneration
 5. Revascularization of the traumatized region
 6. The extracellular matrix as a substrate for regeneration
 7. Increase in the number of regenerating cells by proliferation
 8. Differentiation of the regenerating tissue
 9. Morphogenesis of the regenerating tissue
10. Functional restoration

actual or potential contribution of various forms of stem cells to natural tissue regenerative processes remains highly controversial.

Early regeneration for many tissues requires an environment that both stimulates
precursor cells to divide and allows the progeny of the precursor cells to be situated in
such a manner that they collectively become integrated into a functional tissue once
regeneration is completed. In most regenerative processes, there is a broad transition
from an environment in which tissue destruction and removal are dominant to one in
which cellular proliferation and reconstruction predominate. One of the most important
elements in this transition is the return of a functional microvasculature. In areas of
ischemia, damaged tissues typically undergo a certain degree of intrinsic destruction,
but then things come to a halt until a blood supply is restored. As blood vessels begin
to grow into the ischemic area, they bring with them macrophage precursors that leave
the circulation. Slightly in advance of the tips of the ingrowing vascular sprouts, macrophages move into the damaged tissue and begin to phagocytize cellular and tissue
debris. Almost simultaneously, the presence of macrophages is associated with the
activation of precursor cells for regeneration. In the case of fractured bone, removal of
necrotic osseus matrix depends on the presence of monocyte-derived osteoclasts and
a fluid environment that allows the removal of dissolved bone salts, both of which are
a function of the returning microvasculature.

Of critical importance to a successful tissue regenerative response is an appropriate
relation between the regenerating cells and the tissue matrix in which they are proliferating and differentiating. Here, considerable tissue-specific variation exists. In the
case of skeletal muscle, the most important scaffold consists of the persisting basal
laminae of the original muscle fibers. These basal laminae serve a variety of functions,
principally by acting as semipermeable cellular filters. By definition, the satellite cells
are located beneath the basal lamina of the muscle fiber. After degeneration of a
damaged muscle fiber and the phagocytosis of its remains by macrophages, the satellite
cells within the basal lamina become activated and begin to divide, probably in
response to stimulatory factors released by the macrophages, although the release from
an inhibitory influence by the muscle fiber cannot be excluded (Fig. 1-6). For the most
part, in a regenerating system, the persisting basal laminae appear to prevent the entry
of fibroblasts from the outside, although macrophages appear to be able to penetrate it
with little difficulty. Only when the satellite cells have been inactivated do collagen
fibers form within the empty basal laminae (Carlson and Carlson, 1991). Another
important function for the persistent basal laminae in regenerating muscle is their role
as a scaffold for the orientation of the regenerating muscle fibers (Vracko, 1974). Initially, the orientation of newly regenerating muscle fibers (myotubes) corresponds to
that of the original basal laminae. Ultimately, regenerating muscle fibers secrete their
own basal laminae, but during the early days of regeneration, persisting basal laminae
are the dominant players. Later, mechanical forces may reorient the regenerating
muscle fibers (Carlson, 1972b).

In the regeneration of skeletal tissues, much different relations are seen between the
regenerating tissue-specific cells and the extracellular environment, because for the
most part, the cells secrete an abundant matrix as the tissue is forming or regenerating.

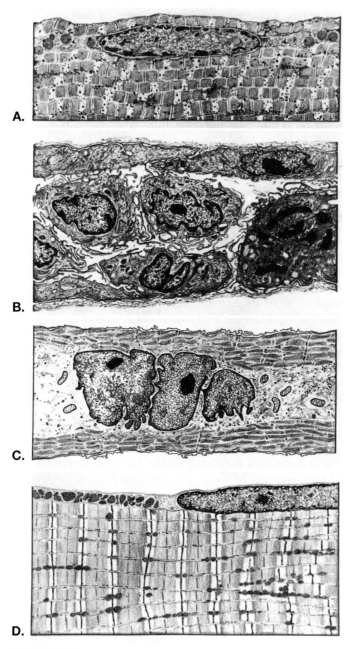

FIGURE 1-6 Series of drawings showing stages in the degeneration and regeneration of a mammalian skeletal muscle fiber. *(A)* Initial intrinsic degeneration of an ischemic muscle fiber. The nucleus is becoming pyknotic, and individual sarcomeres are separating because of the action of proteases. *(B)* Phagocytosis of the degenerating muscle fiber by numerous invading macrophages. Beneath the persisting basal lamina (top) can be seen two spindle-shaped, activated satellite cells. *(C)* Regenerating myotube with central nuclei and newly synthesized bundles of contractile filaments, all beneath the remains of the original basal lamina. *(D)* Regenerated muscle fiber.

Particularly in the case of a regenerating long bone, the series of tissue replacements progressing from a clot, to a cartilaginous callus, to the regenerated bone itself involves different dynamics between cells and matrix at each iteration (see Figs. 8-2 to 8-4). The cells are initially less constrained by the architecture of the original bone; instead, they are highly responsive to the overall mechanical environment surrounding them early in the regenerative process (Murray, 1936).

Cellular proliferation is critical to the success of any tissue regenerative response. Important elements in understanding the role of proliferation in tissue regeneration are: (1) what kinds of cells are proliferating, (2) where proliferation occurs, and (3) what stimulates proliferation. In many cases, these questions have been remarkably difficult to resolve. The nature of the proliferating cells that will ultimately form the regenerated tissue has proved to be controversial for almost all tissues investigated. For years, the field of skeletal muscle regeneration was consumed with the issue of dedifferentiation of muscle fibers versus satellite cells as the cellular source of new muscle (see Mauro, 1979; Mauro et al., 1970). When it was finally demonstrated that satellite cells do contribute to regenerating mammalian muscle (Snow, 1977), the discussion died down without ever resolving the issue of whether other types of cells could also serve as precursor cells. Similar histories could be told about the search for the progenitor cell in other types of tissue regeneration. With increasingly richer stores of cell-specific antibodies and other types of intracellular markers, it is becoming easier to determine both the nature and location of proliferating precursor cells in tissue regeneration.

In vivo studies on the cause of proliferation during tissue regeneration have been hampered because of the complexity of the tissue environment. As a result, many mechanistic studies have been performed on cultured cells. Typically, various individual and combinations of growth factors have been added to the culture medium, and their effect on the cells in question noted. As might be expected, many patterns of growth factor–induced proliferation *in vitro* exist. In the case of skeletal muscle, which has been carefully investigated, the growth factors that stimulate proliferation are quite different from those that promote differentiation (Hawke and Garry, 2001; Johnson and Allen, 1990).

One major difference between tissue regeneration and epimorphic regeneration is that overall morphogenesis of a structure regenerating by the tissue mode is under the control of local, often mechanical factors. In no tissue is this more readily apparent than in regenerating bone, where both the external form and the orientation of the osteons within the regenerated segment conform to the mechanical environment that is applied to the regenerating bone (see Fig. 8-1). Although commonly only part of a structure is injured and is subsequently replaced by tissue regeneration, there are circumstances when the integrity of an entire structure is essentially destroyed and that structure regenerates by the tissue mode. One example is the regeneration of mammalian ribs. In surgical practice when bone tissue is needed, it is common to remove a rib bone, leaving only the periosteum, which in time will regenerate a new bony rib within it. Another example is the regeneration of entire muscles in rodents after mincing or destruction of all of the muscle fibers by toxins or local anesthetics (Carlson, 1968; Carlson and Faulkner, 1996).

Ideally, tissue regeneration is followed by full functional return. This, however, is not always the case, and there is considerable tissue variability in the degree of functional return. In the case of bone, the regenerated segment may ultimately recover its preinjury level of mechanical strength, but it is common for regenerated skeletal muscle to generate far less force than the preinjured muscle. In muscle, in particular, the level of functional return depends on many factors, including the degree of scarring at the site of injury and the extent of reinnervation of severely damaged muscle fibers. Nevertheless, in certain experimental models, it is possible to recover full strength in entire muscles regenerated by the tissue mode (Carlson and Faulkner, 1996).

In urodele amphibians, the same tissue may be regenerated by either the tissue or epimorphic mode. Table 1-2 outlines the major characteristics of tissue versus epimorphic regeneration for limb muscles in the axolotl.

TABLE 1-2

Comparison of Tissue and Epimorphic Regeneration of Muscle

Characteristic	Tissue regeneration	Epimorphic regeneration
Initial stimulus	Muscle fiber damage	Anything that stimulates the regeneration of a limb
Cellular source of myoblasts	Satellite cells; marrow-derived stem cells; other?	Dedifferentiation; stem cells; other?
Removal of damaged cytoplasm	Phagocytosis plays a prominent role	Phagocytosis much less prominent
Regeneration blastema	Absent	Present
Relation of regenerating muscle cells to basal lamina	Most regeneration occurs within the confines of old basal laminae	Most regeneration occurs in the absence of old basal laminae
Time course	Fast	Slow
Relation to nerves	Early differentiation and morphogenesis independent of nerves. Final differentiation requires motor nerves	Nerves (any type) required for blastema formation; morphogenesis is independent of nerves
Relation between amount of damaged and regenerating muscle	Fairly direct between minimum and maximum thresholds	Amount of muscle in regenerates is independent of amount of damaged muscle in stump
Gradients	Related to patterns of blood supply, often centripetal	Pronounced proximodistal gradient of decreasing maturity. A lesser preaxial to postaxial gradient
Development of function	Development of contractile properties recapitulates the ontogenetic pattern	Unknown
Morphology of regenerate	Usually imperfect	Perfect
Amount of connective tissue	Above normal	Normal
Morphology of mature muscle fibers	Central nuclei commonly persist	Normal at the histologic level
Morphology of development	Unlike that in embryo above the cellular level	Close recapitulation of ontogenetic development
Morphogenetic control	Gross morphogenesis and internal architecture can be accounted for by physical factors	Morphogenetic controls similar to those operating in the embryo
Role of function in morphogenesis	Functional environment improves the quality of the regenerate	Function not needed for normal morphogenesis
Positional information	Present, but not expressed in amphibians	Present and expressed
Interactions between regenerative processes	Suppressed by epimorphic regeneration	Dominant over tissue regeneration

Cellular Regeneration

Cellular regeneration refers to the reconstruction of a single cell that has been trauma-tized. Classic examples in the regeneration literature are the reconstitution of protozoa after resection or natural fission (in this case, a form of asexual reproduction) and the regeneration of transected or otherwise damaged axons of peripheral nerves. However, some cells severely damaged by toxins or an unfavorable local environment may also retain the capacity to repair themselves.

The Protozoa as a group are certainly the champion cellular regenerators. One of the most spectacular examples of this prowess is seen in *Stentor,* a goblet-shaped ciliate protozoan with a remarkably complex structure (Fig. 1-7). On the opposite end from an attenuated holdfast is a mouth region surrounded by a rim of adoral cilia. The "body" is characterized by a pattern of alternating dark and light stripes, and a macronuclear chain is strung out along the length of the body. *Stentor* reproduces asexually by fission, so that each half of the subdivided cell must be able to reproduce the missing half. Although a single cell, *Stentor* is sufficiently large that it can be subjected to experi-mental surgery, including grafting procedures (Tartar, 1961). What is remarkable about *Stentor* is its ability to restore and regulate its complex pattern of stripes and the adoral ciliary rim. This ability of pattern regulation within a single cell must be taken into account by theories concerning morphogenetic control mechanisms, which are usually applied to multicellular systems (Frankel, 1974).

Another noteworthy case of regeneration in a single-celled organism is seen in *Acetabularia,* a unicellular alga that looks like an inverted umbrella with a nucleus located in a rootlike holdfast region at its base (Fig. 1-8). In the premolecular era, *Acetabularia* was the object of a remarkable series of regeneration experiments (Hämmerling, 1963). If the cap of an *Acetabularia* is removed, a new one will regener-ate from the stalk (see Fig. 1-8). If the stalk is separated from both the cap and the basal holdfast region containing the nucleus, regeneration fails to occur. However, if the cap is removed and then several days later the basal region containing the nucleus is also removed, regeneration of a cap takes place. During the 1950s, experiments of

A.   B.   C.   D.   E.

FIGURE 1-7 Posttraumatic regeneration in *Stentor coerulus.* (A) Removal of the oral end. (B–E) Suc-cessive stages in re-formation of the adoral region and cytostome.

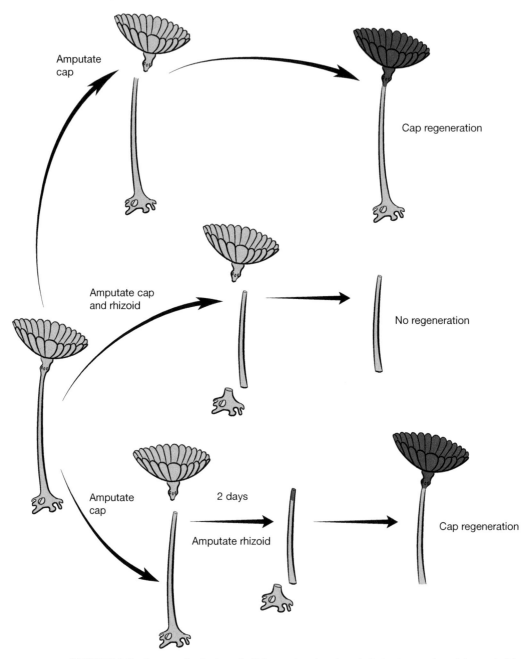

FIGURE 1-8 Regeneration in the unicellular marine alga, *Acetabularia*. Amputation of the cap is followed by regeneration (red). After simultaneous amputation of the cap and the nucleus (yellow)-containing rhizoid, the stalk does not regenerate. If the rhizoid is amputated 2 days after amputation of the cap, cap regeneration occurs. (Based on Hämmerling's [1963] experiments.)

this type provided valuable insight into the role of nuclear products in guiding the course of complex cellular processes, such as regeneration.

Among mammals, axonal regeneration is certainly the most spectacular example of cellular regeneration. Transection of an axon (or dendrite) is followed by a short period of Wallerian degeneration, which gives way to a prolonged phase of axonal extension averaging 1 mm/day (Sunderland, 1978). Axonal regeneration is a complex interactive process that, in addition to the production of new cytoplasmic material, involves extensive interactions between the axonal process (growth cone) and its immediate environment. In a typical peripheral nerve, the Schwann cells lining the axonal sheath respond actively to nerve transection, and they change their own synthetic profile to produce a variety of growth factors that stimulate outgrowth of the axon (Gillen et al., 1997; Grothe and Nikkhah, 2001; Ide, 1996). In addition to interactions with surrounding cells, the regenerating axon is also highly sensitive to its physical substrate, a property of great importance to surgeons involved in peripheral nerve repair. Once the regenerating axon arrives at the end organ—a muscle fiber in the case of a motor axon—other interactions with the local extracellular matrix and molecular signals from the end organ determine the pattern of settling down and functional reconnection with the end organ.

One of the few other clear-cut examples of cellular regeneration is repair of the damaged skeletal muscle fiber, a multinucleated structure. Damage to one portion of a muscle fiber, which may be several millimeters long, is commonly repaired by a combination of myonuclear migration and the fusion of activated satellite cells to the damaged end of the fiber. The repair of mononuclear cells other than neurons has rarely been considered in the mammalian regeneration literature.

Regeneration by Induction

The concept of regeneration by induction traces its roots back to the embryologic experiments of Spemann (1938) and his school. The essence of this largely experimental approach is that tissue-specific regeneration can be stimulated by the application of tissues or materials with specific inductive properties. A variety of early experiments in this area were summarized by Levander (1964), who himself conducted a series of experiments involving the implantation of tissues soaked for 24 to 48 hours in trypan blue into sites of tissue damage. Levander believed that such treatment stimulated the inductive properties of the soaked tissues.

An instructive example of regeneration by induction involves regeneration of mammalian skull bone (Polezhaev, 1972a, pp. 68–84). In young mammals, defects in the skull may heal by the formation of new bone, but in the adults of larger mammals, such as dogs or humans, defects in the skull are partially filled in by a connective tissue scar without the formation of new bone. In a method developed by Polezhaev (1972b, English summary) and Matvccva (1958, 1959b), a large defect is created in the skull bones of dogs, exposing the dura mater. Then other bone is ground up to the consistency of sawdust (done with a kitchen hamburger grinder in Polezhaev's laboratory), mixed with a solution of penicillin, and placed into the defect that was created in the skull.

The ground-up bone can be either skull bone or long bone material, obtained from the same or other species, and can be lyophilized, but not boiled. Within 7 to 10 days after implantation, new trabeculae of bone form in the defect, and ultimately, the defect heals with well-formed bone. According to Polezhaev, the bone sawdust serves as a source of inductive material; the reacting system is cells of immature connective tissue (local or bloodborne) that find their way into the defect; and the presence of the dura mater is a necessary condition for the induction of new bone (Fig. 1-9). In the absence of the dura, new bone does not form.

This method was used on human patients, and in the mid-1960s in Moscow, I saw a presentation of a patient who had a large traumatic defect (perhaps 8 × 12 cm) of the

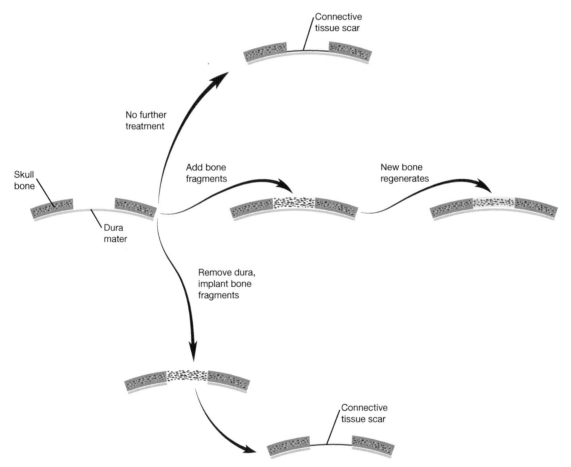

FIGURE 1-9   Regeneration by induction of a defect in the adult mammalian skull. Simple removal of bone results in the formation of a connective tissue scar. If the defect is filled in with a mass of finely ground bone, new bone regenerates within a few weeks. If the dura mater is removed, regeneration fails to occur despite the addition of ground bone into the defect. (Based on Polezhaev's [1972b] experiments.)

parietal bone that had been successfully treated in this manner (Strebkov, 1966). It has been used on other series of patients in Russia as well, where cadaver material served as the source of the ground bone (Polezhaev, 1982).

Later, this method of bone healing was independently rediscovered several times, most notably by Urist (1965, 1997), whose continued experiments ultimately led to the identification and characterization of bone morphogenetic protein as the active agent in the inductive process (Urist et al., 1979, 1984). It should not be forgotten, however, that the experimental induction of bone in soft tissue had been accomplished decades earlier. Perhaps the most striking example was Huggins's demonstration of the induction of bone by urinary epithelium (Huggins, 1931).

Aside from studies on bone and possibly dentine, regeneration by induction has had a controversial history. Claims of regeneration of cardiac and skeletal muscle (Levander, 1964; Polezhaev, 1977b) were disputed because of problems in specifically identifying the tissues that were supposedly regenerated by the inductive process. One of the problems in repeating this work is that these earlier workers described their best results as obtained by treating the inducing tissues with a specific batch of trypan blue (#2691 by Gurr), which may no longer exist unless lots have been preserved in old collections of histologic stains. In this age of stem cell biology, it may be unwise to dismiss these old experiments completely, because it is possible that these treatments had some influence on a population of potential progenitor cells that was not even being considered at the time.

HYPERTROPHY

Many internal organs have the capacity to increase their mass after damage or partial removal, or if one member of a pair (e.g., kidneys) is removed. Similarly, in the absence of overt injury or tissue removal, many of the same organs increase their mass in response to increased functional demand. The field of hypertrophy is exceedingly complex and difficult to encapsulate, but nevertheless one can make certain generalizations that distinguish hypertrophy from other forms of regeneration.

A key element to the hypertrophy of most internal organs is that an increase in functional mass, rather than a restoration of external form, is typically the most important outcome. A classic example of this is seen in Higgins and Anderson's surgical model of liver regeneration in the rat (1931). In this model, two of the major lobes of the liver (70% of mass) are removed. Within 1 to 2 weeks, the remaining lobes of the liver have enlarged to compensate for the overall loss in mass, but the lobes that have been removed do not grow back. Forty days after removal of one kidney in the rat, the remaining kidney enlarges to about two-thirds the combined mass of the two original kidneys (Addis and Lew, 1940). One of the significant differences between hypertrophy and epimorphic regeneration is that in the latter, restorative activity occurs close to the amputation surface, whereas in hypertrophy, the restorative response occurs throughout the remainder of the organ.

How does a hypertrophying organ get larger? At both the level of the individual cell or the internal functional unit (e.g., nephrons in kidney, alveoli in lung, or lobules in liver) there can be an increase in number (hyperplasia), increase in size (hypertrophy), or both (Goss, 1966) (Fig. 1-10). At the level of a whole organ, the number of functional units can remain constant, but an increase in size of these units could be caused by cellular proliferation within that functional unit. In addition, an increase in fluid or the amount of extracellular matrix could contribute to an overall increase in mass.

Origins of the new cells in hypertrophic organs are discussed in greater detail in Chapter 2, but according to traditional thinking, the division of parenchymal cells accounts for the greatest proportion of the new cellular mass. However, for almost every organ in which hypertrophy occurs, the role of stem cells, whether resident or immigrating, in the restoration of these structures is currently being discussed. As a generalization, proliferation of parenchymal cells is commonly the primary mechanism, but if this response is inadequate, local or marrow-derived stem cells may then proliferate to make up the deficit.

Most controversial is the question of the mechanism(s) underlying the hypertrophy of internal organs. According to one point of view (Goss, 1964b), most hypertrophy, whether regenerative or merely adaptive, is based on increased functional demand. Such a basis still leaves room for a variety of specific biochemical and molecular control mechanisms. Hypotheses have centered around two main strategies: removal of an inhibitor of growth (Bullough, 1967; Weiss and Kavanau, 1957) or the action of a positive stimulatory factor, such as hormones or growth factors (Malt, 1983; Poole, 1966; Sidorova, 1978).

The partially resected rat liver has been by far the most intensively studied system of hypertrophy, and from these studies, it is apparent that hypertrophy involves the interactions of a multitude of stimulatory and inhibitory influences (Fausto et al., 2006; Michalopoulos and DeFrances, 1997; Taub, 2004). Although normal hepatocytes are mitotically inactive, within 12 hours and peaking at 24 hours, up to 95% of the cells begin to enter mitosis. Entry into mitosis is stimulated by a variety of factors extrinsic to the liver (Fig. 1-11). Within the liver, three principal cell types are primarily involved. First are the Kupffer cells, phagocytic cells that line the sinusoids. The earliest ("priming") phase begins with activation by gut-derived lipopolysaccharides, which leads to activation of the Kupffer cells by tumor necrosis factor-α. The early role of cytokines in the priming phase continues with the release of interleukin-6 (IL-6) by the Kupffer cells and its binding to IL-6 receptors on the surface of the hepatocytes. Certain fractions of the complement system (C3 and C5) appear to be important in the early phases of stimulation of liver regeneration as well. Next, the action of growth factors pushes the mitotically quiescent hepatocytes into the cell cycle. Primed by IL-6, the hepatocytes become competent to respond to hepatocyte growth factor and transforming growth factor-α (TGF-α), which are released from the periportal extracellular matrix or secreted by stellate cells, that is, pericyte-like cells that line the hepatic sinusoids (Balabaud et al., 2004). In response to these, as well as additional extrinsic factors, such as epidermal growth factor, supplied by the thyroid, duodenum, pancreas, and adrenal gland (see Fig. 1-11), the hepatocytes enter the mitotic cycle. Together with

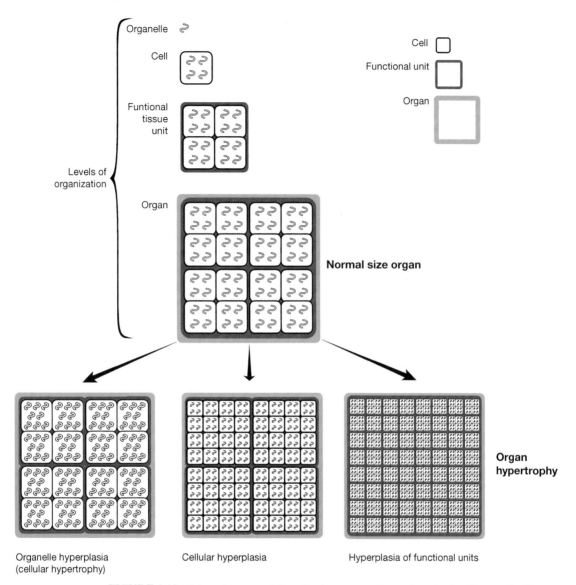

FIGURE 1-10 Schematic representation of various means of organ hypertrophy. The same final hypertrophic organ mass can be attained by (1) an increase in the number of organelles (cellular hypertrophy) without an increase in the number of cells or functional units, (2) by cellular hyperplasia without increasing the size of the cells or number of functional units, or (3) by increasing the number of functional units without increasing cell size or number of organelles per cell. (After Goss R.J. 1966. Hypertrophy vs. hyperplasia. *Science* 153:1615–1620, by permission.)

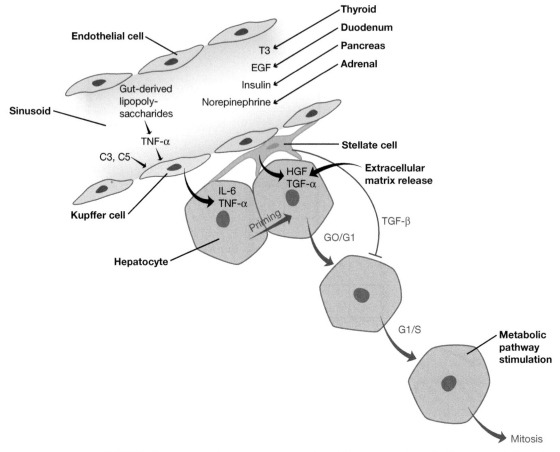

FIGURE 1-11 Factors involved in the initiation of liver regeneration. Bloodborne signals from the sinusoids act through Kupffer and stellate cells to transmit signals leading to the proliferation of hepatocytes. C—complement; EGF—epidermal growth factor; HGF—hepatic growth factor; IL-6—interleukin-6; TGF-β—transforming growth factor-β; TNF-α—tumor necrosis factor-α.

growth factors, the overall body metabolic needs and pathways also influence the extent of hepatocyte proliferation. This may be the basis for the functional demand theory of hypertrophy (Goss, 1964b). Proliferation of nonhepatocytes in the liver typically lags that of the hepatocytes by a couple of days, with the result that small clusters of newly produced hepatocytes take form before they interact with other elements of the liver to reconstitute the normal internal hepatic architecture. Less is known about what brings cellular regenerative activity in the liver to a halt, but TGF-β, produced by the stellate cells, is known to inhibit the reentry of the hepatocytes into the mitotic cycle. In the regenerating human liver, function is recovered before mass, and the process is more prolonged than liver regeneration in the rat. In some situations of chronic toxic damage to the liver (e.g., alcoholic cirrhosis, repeated exposure to carbon tetrachloride), hepatocyte regeneration is not sufficient to prevent fibroblastic cells within the liver from producing connective tissue scarring.

MORPHALLAXIS

Ever since Morgan coined the term *morphallaxis,* its application as a descriptive term has proved to be a challenge to regenerationists. In his original definition, Morgan attributed two properties to the phenomenon. One was that the "part is transformed directly into a new organism, or part of an organism," and the other was that this occurs "without proliferation at the cut surfaces" (Morgan, 1901, p. 23). With more than a century of scattered research since the term was originally defined, it is now clear that the most relevant part of this definition is that concerning the reorganization of a part of the body during regeneration.

Classical morphallaxis is a phenomenon that is confined mostly to invertebrate regenerating systems, but it is a striking phenomenon that is poorly understood at the mechanistic level. An excellent example of morphallaxis is seen in *Sabella,* a marine fan worm. *Sabella* is characterized by a head from which numerous fronds radiate to make a fanlike structure; thoracic segments in which bristles, or setae, protrude in one direction; and abdominal segments, in which the setae clearly project in a different direction from the thoracic ones. Sabellids regenerate well. If as few as 4 to 10 abdominal segments (of a total of about 200 segments) are isolated from the worm, a new head and anterior end regenerate from the anterior cut surface, and a new posterior end regenerates from the posterior cut surface (Berrill, 1931). From more recent research (Hill, 1970), we now know that anterior regeneration occurs through the formation of a blastema in classic epimorphic fashion. However, during the regenerative process, another striking phenomenon occurs. The remaining abdominal segments undergo a complete internal and external reorganization so that by the time regeneration is completed, the original abdominal segments have become completely transformed into thoracic segments, as indicated by the orientation of the setae (Fig. 1-12). This phenomenon represents the morphallactic transformation of a body region. Commonly,

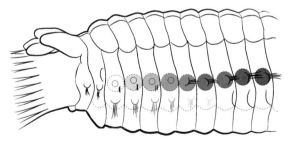

FIGURE 1-12   Morphallactic transformation of abdominal into thoracic segments during head regeneration in *Sabella.* Blue (from right to left): successive stages of dissolution of abdominal setae and setigerous bulbs. Red (from right to left): successive stages in the formation of thoracic setae. (Based on Berrill's (1931, 1978) experiments.)

morphallactic phenomena occur in the transitional zone between an epimorphically regenerating structure and the remainder of the body. In vertebrates, there are less striking transitions in the zone between the domain of an epimorphically regenerating structure and the transected tissues of the stump, but whether they would qualify as morphallaxis is questionable.

In certain other invertebrate systems that are able to regenerate an entire body from a small part, the entire piece appears to undergo morphallactic reorganization as it is regenerating. A classic example is the nemertean worm *Lineus,* in which small sections shrink and undergo a gross form of dedifferentiation, only to reemerge at the end of the process as small, but anatomically complete, worms (see Fig. 15-1, B). This process of morphallactic reorganization is so efficient that when small regenerated worms are themselves sectioned, miniature regenerated worms of less than 1/200,000 of the original volume of the original worm can be obtained (Coe, 1929).

Unfortunately, research funding patterns are such that, currently, little contemporary research is being conducted on morphallactic systems. Elucidation of the mechanisms underlying radical transformations of body regions could, nevertheless, provide important clues to some of the fundamental issues in morphogenesis.

ASEXUAL REPRODUCTION

Historically, asexual reproduction has commonly been treated together with regeneration, because there are often striking similarities between the two processes (Ivanova-Kazac, 1977; Vorontsova and Liosner, 1960). Among many invertebrates, asexual reproduction represents a means of increasing the number of individuals without the involvement of gametes. The two principal modes of asexual reproduction are fission and budding, with fragmentation being a third (Fig. 1-13).

Fission, as the name implies, represents the natural subdivision of a single body into one or more parts and the reorganization of each of the parts into a complete individual. For many protozoa, this represents the standard mode of reproduction, which requires some form of distribution of the genetic material and major cytoplasmic elements into

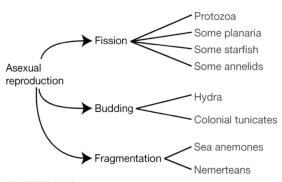

FIGURE 1-13   Major modes of asexual reproduction among invertebrates.

each of the potential future fragments before cellular fission occurs. In platyhelminths and annelids, fission or fragmentation are the most common modes of asexual reproduction. Such fission is usually transverse, and the location of the plane of fission varies from species to species. In most planaria, there is no indication of the future location of the plane of fission; but in others, the plane of fission is preceded by the formation of a groove, about which vital organelles accumulate. Fission, especially among sexually immature planaria, is often stimulated by an unfavorable environment. Once the planarian has undergone fission, the fragments then regenerate entire bodies, using combinations of epimorphosis and morphallaxis. Even animals as large as certain starfish are able to reproduce by fission. In *Linckia multiflora*, adults will discard individual rays and then proceed to regenerate a central body disk together with the remaining rays (Vorontsova and Liosner, 1960, p. 54).

Studies on an annelid worm *(Pristina leidyi)* that undergoes fission midway along the length of its body (called *paratomic fission*) have shown that, at the plane of fission, cellular and molecular events reminiscent of regeneration occur (Bely and Wray, 2001). *Engrailed,* a gene thought to be phylogenetically involved in segmentation, is expressed on both sides of the plane of fission, whereas *orthodenticle,* a gene that defines anterior structures during early embryogenesis, is expressed early in the fission process on the side of the fission plane where a new head will ultimately form at the anterior end of the posterior half of the worm.

Asexual reproduction by budding is common in many types of coelenterates, bryozoans, and tunicates, whether single or colonial forms. In hydroids, in particular, an axial patterning mechanism that underlies the formation of buds is closely linked with that which guides regeneration after transection (Bode, 2003).

SUMMARY

Regeneration is a fundamental property of life, but there are many varieties of regeneration and many different means by which regeneration is accomplished. The loss of cells through normal attrition is replaced on a continuous basis by processes designated as physiological regeneration. Posttraumatic loss is dealt with by a variety of mechanisms. Some species, scattered throughout the animal kingdom, have developed or retained the ability to respond to the loss of complex parts by forming a regeneration blastema, which often reexpresses developmental pathways to faithfully recreate the missing structure. This is called *epimorphic regeneration.* Many other tissues respond to damage by activating internal cellular resources to restore, with varying degrees of success, the disrupted regions. Damaged or functionally insufficient internal organs recover functional mass through the proliferation or enlargement of cells or functional units within the remaining part of the organ in a process called *hypertrophy.* Some invertebrates are able to completely reorganize the remaining portions of badly damaged or shrinking bodies through a poorly understood phenomenon called *morphallaxis.* For many invertebrates, multiplication through asexual reproduction shares many similarities with regeneration.

CHAPTER 2

Origins of Cells in Regenerating Systems

Omnis cellula e cellula.

—Rudolph Virchow (1855)

Where do the cells that participate in regeneration come from? Surprisingly, the answer to this seemingly straightforward question has proved to be one of the most elusive in the entire field of regeneration. Approaches to this question have varied, often in relation to the historical period in which the question was asked. Similarly, the results of experimental approaches have often been interpreted in the context of currents of thought prevailing at the time, and often what now would appear to be obvious options were not even considered.

One way of looking at cellular origins could be termed a *topographic approach*—that is, where the cells that participate in regeneration come from. A second, more specific approach is to ask what the cellular precursors of particular regenerating tissues or their components are. It is becoming apparent that in complex biologic phenomena, such as regeneration, there may not be only a single answer to such questions. A common property of many regenerating systems is that if one mechanism of replacement is eliminated, another that accomplishes the same purpose may take its place.

WHERE DO REGENERATING CELLS COME FROM?

A Topographic Approach

One of the earliest-asked questions in the study of regeneration was: "Where do the cells that participate in regeneration come from?" In this case, "where" referred to locality and not types of precursor cells. Do the cells that account for reparative regeneration arise from the local area of injury or do they migrate to the site of injury from distant sites? Experiments on two epimorphically regenerating systems, the amphibian limb and planaria, illustrate the experimental approach to this question.

In one of the earliest experiments involving the use of x-radiation in the study of regeneration, Butler (1935) irradiated and then amputated limbs of larval *Ambystoma*. The irradiated limbs failed to regenerate (Fig. 2-1, A). He then transplanted normal

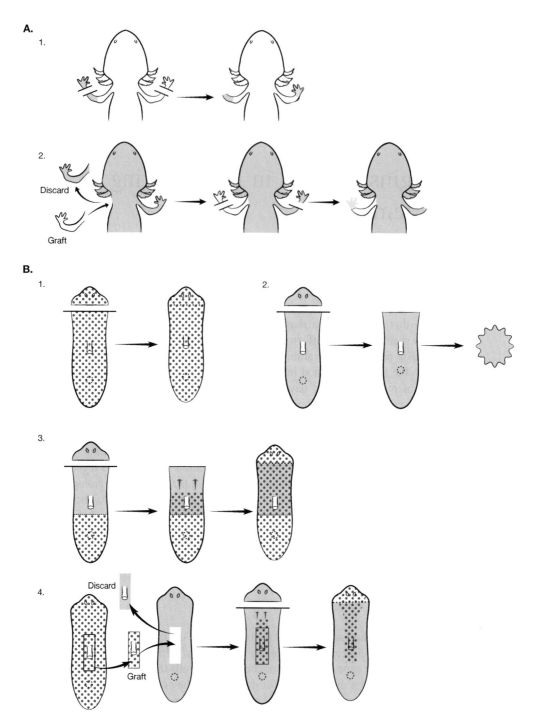

FIGURE 2-1 Irradiation experiments designed to demonstrate the topographic origins of regeneration cells. *(A)* If amputation is performed through a distally irradiated limb (blue) of an *Ambystoma* larva, regeneration fails to occur (1). If a normal limb is grafted to a totally irradiated host (blue), regeneration occurs after amputation through the grafted limb. (Based on Butler's [1935] experiments.) *(B)* Amputation through the head of a normal planarian (1) results in complete regeneration, whereas after heavy irradiation (5000 r), not only does head regeneration fail to occur, but the animal ultimately disintegrates (2). If the anterior half of a planarian is irradiated (blue) and the head is subsequently amputated, regeneration occurs after a period of several weeks (3). If a rectangular piece from a healthy planarian is grafted into an irradiated host, an amputated head regenerates (4). The red circles in *B* represent neoblasts, which migrate from healthy into irradiated regions of the worm and ultimately constitute the regenerate. (After Wolff and Lender [1962].)

limbs onto completely irradiated larval hosts and found that the nonirradiated transplanted limbs were able to regenerate after amputation. From these and similar experiments, Butler concluded that the cells responsible for amphibian limb regeneration have a local origin in the vicinity of the plane of amputation. These results contradict suggestions by Hellmich (1930) and earlier authors, who propose that in addition to local mesenchyme, some of the cells participating in limb regeneration come from leukocytes, specifically lymphocytes, in the blood. Butler's experiment definitely shows that local cells near the amputation surface are sufficient to regenerate a limb and that cells from outside that region are not. What it could not show is whether cells from outside the narrow terminal zone of regeneration in the limb stump could participate in the regenerative process, even in small amounts.

Irradiation experiments on planaria led investigators to exactly the opposite conclusion. When planaria are totally irradiated with a dose of 5000 roentgen (r), amputated parts do not regenerate, and eventually, the entire animal disintegrates. However, if the anterior half of a planarian is irradiated and the head is then amputated, a blastema appears after a delay of several weeks, and a new head is formed (Wolff and Lender, 1962). Similarly, if a rectangular piece of a nonirradiated animal is grafted to a totally irradiated host with an amputated head, cells from the graft migrate through the irradiated tissues to form a blastema that regenerates a new head (see Fig. 2-1, B). The migrating cells in planaria have been traditionally termed *neoblasts,* and because of their size and staining characteristics, they are relatively easy to follow. These irradiation experiments on planaria clearly indicate that distant cells can participate in blastema formation and regeneration, but they do not rule out the participation of locally derived cells. In fact, experiments that involve chromosomal cellular markers (Saló and Baguñà, 1989) have shown that the division and subsequent migration into the blastema of cells located within 200 to 300 μm of the amputation surface is sufficient to account for the number of cells located in the planarian blastema. Yet this type of experiment does not rule out the possibility of participation of more distantly located cells.

Examination of other regenerating systems shows considerable variation in the sites of origin of the cells that participate in regeneration. In *Hydra,* for instance, normal body form is maintained through processes of cell streaming toward the head region and toward the peduncle on the other end (Bode, 2003). After amputation, similar cellular streaming provides the cells that reconstitute the missing part.

In internal organs that regenerate by hypertrophy, such as the liver, the mitotically active cells are not located at the wound surface, but rather are scattered throughout the remainder of the organ (Sidorova, 1978). In the amputated cardiac ventricle of the newt, a blastema does not form, but mitotically dividing cardiomyocytes are concentrated in the vicinity of the traumatized tissue (Oberpriller and Oberpriller, 1974). Skeletal muscle contains myogenic precursor cells (satellite cells) scattered throughout the muscle. When muscle fibers are damaged, satellite cells in the area of injury become activated, but satellite cells also possess a certain ability to migrate from more distant sites toward sites of injury, particularly along the length of a muscle fiber.

One of the most controversial questions concerning the topographic origin of regeneration cells is whether cells from the bone marrow or circulating blood, or both, can participate in a regenerative process and, if so, the extent to which this normally occurs. Although earlier suggestions that bloodborne cells may contribute to the regeneration of tissues, such as skeletal muscle (Hellmich, 1930; Bateson et al., 1967), were treated with considerable skepticism, other, more recent research has provided evidence that this may, indeed, occur. In various experiments or clinical situations, there have been a number of reports that bone marrow-derived cells can contribute to the regeneration of tissues as diverse as skeletal (Ferrari et al., 1998) and cardiac muscle (Laflamme et al., 2002; Orlic et al., 2002; Quaini et al., 2002), liver (Ferry and Hadchouel, 2002; Kleeberger et al., 2002), bone (Friedenstein and Lalikina, 1973), and brain (Eglitis and Mezey, 1997; Kopen et al., 1999). However, the quantitative importance of such contributions to the normal regeneration of these tissues remains the subject of considerable discussion. Nevertheless, that cells originating some distance from a tissue can reach that tissue and participate in some aspect of its regeneration seems to be firmly established. Issues to be discussed later are the nature of the bone marrow-derived stem cells and circumstances that lead to greater participation of such cells in regenerative processes.

Cellular Origins

No question in regeneration has engendered more controversy over the years than that of the origins of the cells that constitute a regenerating structure. The details may differ for different structures and organs, but the basic questions remain the same. For cell types, the major categories of options are: (1) dedifferentiation of mature cells in the remaining tissue, (2) proliferation of remaining parenchymal cells without dedifferentiation, (3) proliferation of stem (progenitor) cells resident in the injured tissue, and (4) influx of stem cells originating outside the damaged tissue.

One important consideration when reviewing experimental data is differentiating between what normally happens after injury and what can happen when the system is stressed. There is increasing evidence for the existence of cellular backup systems that can come into play when the normal cellular sources of regenerating material are deficient. An example of this comes from the early limb regeneration literature. Several investigators (Bischler, 1926; Fritsch, 1911; Weiss, 1925) removed the skeleton from the stumps of amputated limbs of salamanders. Despite that, the regenerated limbs contained a perfectly normal skeleton, demonstrating that in this system, at least, there does not have to be a tissue-by-tissue correspondence in regeneration.

DEDIFFERENTIATION

One of the most controversial concepts in regeneration is that of dedifferentiation. The essence of the idea is that cells or tissues lose their differentiated characteristics and produce immature-looking cells that then participate in the regenerative process. Much of the controversy has arisen from the implication that the dedifferentiated cells can then go on to form cells of a phenotype different from that of the original cells or tissue (i.e., transdifferentiation or metaplasia). Although in recent decades skeletal muscle, which traditionally has been considered to be a terminally differentiated tissue, has been the object of most discussion and research, dedifferentiation was studied and documented much earlier in other tissues.

Amphibian Lens and Retina

The classic example of dedifferentiation is commonly called *Wolffian regeneration* of the urodele lens from the dorsal iris. First described by Colucci in 1891, the model is quite straightforward. The lens can be easily extirpated through a slit in the cornea. For the next 4 weeks, a regular sequence of changes in the epithelial cells along the rim of the dorsal iris results in their transformation from heavily pigmented epithelial cells to clear elongated lens cells containing the crystalline proteins that characterize the normal lens (Fig. 2-2, Table 2-1). In one of the pioneering descriptions of lens

TABLE 2-1

Major Cellular Events in Regeneration of the Newt Lens

Sato stage[a]	Days after lentectomy	Events in iris epithelial cells (IECs) and their progenies *in situ*
0	0	Fully differentiated as melanocytes; nondividing nucleolar activation; transition from G0 to G1 phase
I	3–6	Nucleolar activation; transition from G0 to G1 phase; start of proliferation; start of macrophage infiltration
II	6–10	Active proliferation; reduction of melanosome numbers caused by phagocytic activity of macrophages
III	8–10	Appearance of depigmented cells at the middorsal region of the iris epithelium; continued depigmentation and proliferation
IV	9–15	Formation of lens vesicle by depigmented progenies of IECs at the middorsal margin of iris; proliferation and depigmentation in other parts of iris
V	12–15	Thickening of internal layer of growing lens vesicle; withdrawal of cells in this area from cell cycle; start of synthesis of γ- and β-crystallins in these cells; addition of depigmented IECs to lens vesicle; continues proliferation of other IEC cells
VI	12–16	Growth of lens vesicle; start of lens fiber differentiation, with accumulation of γ- and β-crystallins; further addition of depigmented IECs to the lens vesicle
VII	15–18	Formation of the lens fiber complex in the internal layer; high mitotic activity in the external layer; absence of mitotic activity in the lens fiber complex; reduced proliferation and depigmentation in iris epithelium
VIII	15–19	Appearance of β-crystallins in external layer of lens; start of appearance of nondividing secondary lens fibers that accumulate both γ- and β-crystallins at the margin of the external layer
IX	18–20	Further accumulation of γ- and β-, as well as α-crystallins in lens fibers; presence of β- and α-crystallins in the external layer (lens epithelium); cessation of proliferative activity in iris epithelium
X	18–25	Definitive lens tissue; continued production of secondary lens fibers; persisting mitotic activity only in lens epithelium

[a]Yamada's modification of Sato stages for *Notophthalmus viridescens*.
Modified from Yamada, T. 1977. *Control mechanisms in cell-type conversion in newt lens regeneration.* Basel, Switzerland: Karger, by permission.

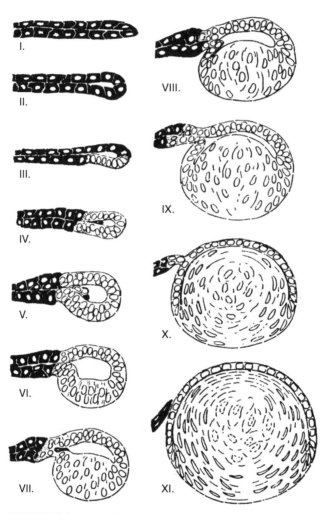

FIGURE 2-2   Stages of lens regeneration from the dorsal iris in the newt. The Roman numerals indicate Sato's regeneration stages. For orientation, each figure shows a section through the dorsal iris, with the corneal side above and the side facing the retina below. The rim of the iris that faces the normal lens is to the right. Pigmented iris cells are indicated in black, and depigmented iris cells and regenerating lens cells are indicated in white. Incompletely depigmented iris cells are dotted. (Reprinted from Yamada, T. 1967. Cellular and subcellular events in Wolffian lens regeneration. *Curr Top Dev Biol* 2:247–283, by permission.)

regeneration, Wolff (1895) used the term *Rückdifferenzierung,* which he described as a return to an earlier developmental stage. The fundamental biology of urodele lens regeneration has been well described by Reyer (1954, 1977) and Yamada (1967, 1977). Interestingly, in some species of anurans *(Hynobius, Xenopus),* the lens regenerates from the cornea rather than the dorsal iris (rev. by Reyer, 1977).

Fibroblast growth factor (FGF) exerts a powerful effect on lens regeneration. Inhibition of fibroblast growth factor receptor 1 (FGFR-1) in the eye blocks lens regeneration from the dorsal iris (del Rio-Tsonis et al., 1998a), whereas the addition of continuous FGF-1 to the eye results in excessive dedifferentiation of dorsal and ventral iris, as well as retina (Yang, Wang, and Tassava, 2005). If administered to the intact eye, FGF-2 provokes the development of a second lens from the dorsal iris in the absence of removal of the normal lens (Hayashi et al., 2004).

Dedifferentiation in the amphibian eye is not confined to tissues creating the lens. The urodele retina regenerates very well, and classic experimentation has shown that the basis for retinal regeneration is depigmentation of cells in the pigment layer of the retina and their later differentiation into cells characteristic of the neural retina (Reyer, 1977; Stroeva and Mitashov, 1983). More recently, Ikegami et al. (2002) have developed an *in vitro* culture model of the newt pigmented retinal epithelium and have directly observed the proliferation and depigmentation of these cells and their eventual transformation into cells with the immunocytochemical properties of retinal neurons. This dedifferentiative phenomenon is not confined to the eye of the adult newt. In embryonic birds (Park and Hollenberg, 1989; Pittack et al., 1991), fetal rats (Zhao et al., 1995), and larval *Xenopus* (Sakaguchi et al., 1997), exogenously applied FGF-2 *in vitro* or *in vivo* stimulates a similar cellular transformation.

Skeletal Muscle

Without question, one of the thorniest issues in the field of vertebrate regeneration has been whether during any kind of regenerative process a multinucleated skeletal muscle fiber can truly dedifferentiate—that is, break up into mononucleated fragments, which can then participate in the regeneration of new tissue. After the discovery of the satellite cell by Mauro (1961), the discussion became polarized around the issue of muscle fiber dedifferentiation versus the satellite cell as the origin of new muscle fibers (Mauro et al., 1970). As is so often the case in long-standing disagreements in science, both sides proved to be partially correct, but as Carlson (1979a) points out, other possibilities for the cellular origin of regenerating muscle were ignored. Now, at least stem cells must be added to the equation.

Although by the late 1800s a number of authors had already described many of the reactions of damaged muscle in a variety of regenerating systems (rev. Fraisse, 1885), Thornton's (1938) careful histologic observations led him to the conclusion that, in the regenerating limb, mononucleated fragments break off from the ends of the damaged muscle fibers and join the other cells of the regeneration blastema. From her ultrastructural studies, Hay (1959) supports Thornton's conclusions. Confirmation of this conclusion, which was based on descriptive observations, had to await experimental studies.

In 1968, Steen examined the stability of differentiation of both muscle and cartilage during limb regeneration. Using both triploid and H^3-thymidine labels individually or in combination (the best markers available at the time), he transplanted the labeled tissues into freshly amputated limbs. Labeled cells from both cartilage and muscle grafts were found in the blastema, indicating that these tissues had dedifferentiated. Whereas labeled chondrocytes differentiated almost exclusively into cartilage in the regenerates, cells derived from muscle grafts were seen in regenerated cartilage as well. Steen points out, however, that the muscle grafts contained numerous cell types and that one could not infer exact cellular origins from that experiment. In a later report, Steen (1970) illustrates triploid myotube nuclei in limbs containing grafts of triploid fin connective tissue. More precise studies on dedifferentiation at the cellular level could not be done until new cell culture and marking techniques were developed and applied 25 years later.

Effective dedifferentiation of a muscle fiber involves three separable functional components: reentry of myonuclei into the cell cycle; cellularization (i.e., fragmentation of the multinucleated fiber into mononuclear cellular units); and the loss of muscle-specific differentiation markers, such as the myosins (Brockes and Kumar, 2002; Echeverri and Tanaka, 2002b). These events are followed by the mitotic division of the mononuclear cells formed from the fragmenting muscle fiber (Fig. 2-3).

Reentry into the cell cycle was demonstrated by exposing cultured newt myotubes to serum (Tanaka et al., 1997). This activates a pathway that leads to inactivation by phosphorylation of retinoblastoma protein, a cell cycle repressor in muscle. Brockes et al. (2001) hypothesize that thrombin in serum is a major player in the extrinsic sequence of events that initiates the dedifferentiative process. Fragmentation of the muscle fiber was demonstrated by implanting myotubes labeled with either rhodamine dextran or a retroviral marker into the regenerating limb of a newt and observing labeled mononucleated cells in the blastema (Brockes et al., 2001). In amputated tails of larval axolotls, Echeverri et al. (2001) injected fluorescent dyes into individual muscle fibers and followed their fragmentation and the contribution of the resulting mononuclear cells into the regeneration blastema. They estimated that almost 30% of the blastemal cells in the regenerating tail arise from dedifferentiated muscle.

Recent experimentation has shown that, under some conditions, mammalian myotubes can also dedifferentiate *in vitro* (Odelberg, 2002). By inducing ectopic expression of the transcription factor, Msx-1, in cultured mouse myotubes, Odelberg et al. (2000) note both fragmentation and the differentiation of the resulting mononucleated cells into not only muscle, but cells expressing chondrogenic, adipogenic, and osteogenic markers. Later, the same group (McGann et al., 2001) obtained similar dedifferentiation of mammalian myotubes that had been exposed to an extract derived from regenerating newt limbs.

The exposure of cultured myotubes to small organic molecules shows promise of identifying means of dissecting the process of muscle fiber dedifferentiation (Ding and Schultz, 2004). Three molecules, myoseverin (Rosania et al., 2000), a triazine 2 compound, and also nocodazole (Duckmanton et al., 2005), can induce the fragmentation of cultured myotubes into mononucleated cells through their depolymerizing action on microtubules. According to Duckmanton and colleagues (2005), the nucleated fragments derived through the action of myoseverin retain the expression of myosin heavy chain, a differentiation marker of the myotubes, but fail to reenter the cell cycle.

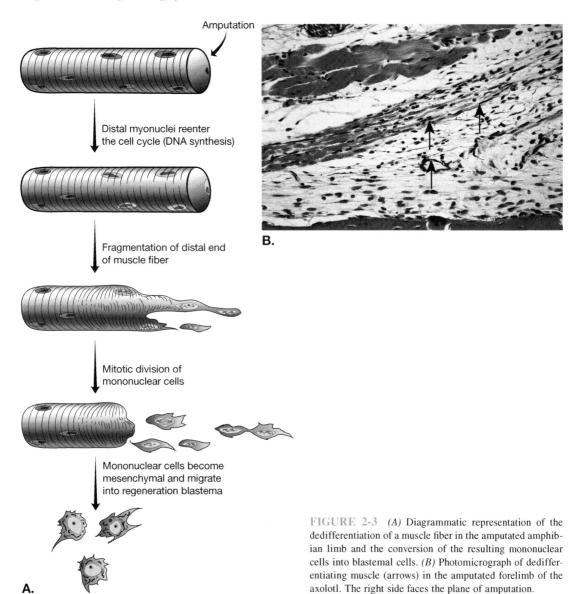

FIGURE 2-3 *(A)* Diagrammatic representation of the dedifferentiation of a muscle fiber in the amputated amphibian limb and the conversion of the resulting mononuclear cells into blastemal cells. *(B)* Photomicrograph of dedifferentiating muscle (arrows) in the amputated forelimb of the axolotl. The right side faces the plane of amputation.

Kumar and coworkers (2004) have investigated the relation between the fragmentation or cellularization of urodele muscle fibers isolated *in vitro* and the expression of Msx-1. They found a high correlation between Msx-1 expression and the muscle fibers that undergo fragmentation in culture. In addition, when they prevented the depolymerization of microtubules by Taxol, fragmentation was greatly reduced. Kumar and coworkers (2004) suggest that microtubules might be a downstream target of Msx-1. This study also showed that reentry into the cell cycle is not a necessary consequence of the initial fragmentation response, in which smaller, multinucleated fragments of muscle break off from the main muscle fiber. Surprisingly, however, neither overexpressing nor down-regulating Msx-1 *in vivo* appears to influence the course of muscle fiber dedifferentiation in the regenerating axolotl tail (Schnapp and Tanaka, 2005). The reason for the discrepancy between the results of the *in vitro* and *in vivo* experiments currently is not understood.

Two other molecules, ciliary neurotrophic factor (Chen et al., 2005) and a small molecule named *reversine* (Ding and Schultz, 2004), have been shown to act on committed mammalian mononuclear myoblasts in a manner that causes their dedifferentiation into stem cell-like progenitors. These dedifferentiated progenitor cells can then go on to redifferentiate *in vitro* into adipocytes, osteoblasts, glial cells, and neuronal cells.

We do not currently have sufficient information to map out the pathway(s) leading to the fragmentation and dedifferentiation of either mammalian or amphibian muscle fibers. However, that fragmentation of mammalian muscle fibers can be brought about experimentally through the manipulation of single variables suggests that even though not fully functional under normal circumstances, much of the pathway by which dedifferentiation can be accomplished has been preserved through mammalian evolution. Other elements of the dedifferentiation process that must be included in an overall scheme are reentry into the cell cycle and the loss of molecules characteristic of the differentiated state. From experiments that involve denervation after amputation and simple wound healing in *Xenopus,* Suzuki and colleagues (2005) have suggested that expressed *Tbx5* and *Prx1* might be good markers for the dedifferentiated state.

A much more clear-cut model of muscle dedifferentiation is seen in the mononucleated striated muscle fibers found in the umbrella of jellyfish medusae (Schmid, 1992; Schmid and Reber-Müller, 1995). Portions of the umbrella containing muscle cells apposed to their normal extracellular matrix (the mesoglea) remain stable in culture for prolonged periods. However, if the matrix is digested away by enzymes, such as collagenase or pronase, the muscle tissue becomes destabilized and the cells lose most or all of their myofibrils. They then undergo mitosis and thereafter typically transdifferentiate into two new cell types: smooth muscle and a sensory nerve cell. Under other conditions of culture, the muscle cells can regenerate an entire tentacle or manubrium, with as many as seven to eight cell types.

Msx is involved in the dedifferentiation of jellyfish striated muscle, but the circumstances of its expression differ considerably from those in vertebrate muscle. Msx is strongly expressed in quiescent striated muscle, but when the muscle-extracellular matrix relation is disrupted by incubation in pronase, Msx transcripts are completely degraded within an hour (Galle et al., 2005). Silencing of the *Msx* gene occurs at the same time when the stem cell gene *Cniwi* is rapidly up-regulated. Reactivation of *Msx* occurs when the regenerating cells begin to reestablish relations with the extracellular matrix. This pattern of expression is seemingly the reciprocal of that which occurs in dedifferentiating vertebrate skeletal muscle and in vertebrate tissues, in general, where *Msx* expression appears to prevent mesenchymal cells from exiting the cell cycle and undergoing terminal differentiation.

Structural dedifferentiation is not confined to skeletal muscle. Differentiated cardiac muscle cells in both amphibians and mammals are capable of undergoing mitosis (Oberpriller et al., 1991; Rumyantsev, 1991). However, before undergoing cytokinesis, the cardiomyocytes undergo a partial dedifferentiation by breaking down a substantial portion of their contractile apparatus. Well-ordered arrays of myofilaments that aggregate into myofibrils re-form after cell division is complete. Even among the protozoa, local dedifferentiation of the motor apparatus must occur before reconstruction of the damaged cell can occur (Vorontsova and Liosner, 1960, p. 146). An apparent exception to the rule that a cell must undergo at least partial dedifferentiation before dividing is seen in hair cells within the organ of Corti in mammals. Mammalian hair cells do not normally divide, but if retinoblastoma protein is inactivated, hair cells divide without losing their differentiated characteristics (Matsui et al., 2005).

Schwann Cells

Schwann cells are highly specialized cells that ensheathe the axons of peripheral nerves. In addition to their production of the myelin wrappings, they secrete large numbers of growth factors and other active molecules that maintain the integrity and function of the axons that they enclose (Scherer and Salzer, 2001). After a nerve has been transected or damaged, the portions of the axons distal to the lesion degenerate, leaving a sheath composed principally of Schwann cells and extracellular matrix (see p. 82). Without the trophic support of the axons, the Schwann cells shed their baggage of myelin and undergo structural and functional dedifferentiation within 48 hours. This is followed by a massive mitotic response at 3 days after injury.

Activation of the Ras/Raf/ERK signaling pathway stimulates the dedifferentiative response, even in the presence of axons (Harrisingh et al., 2004). Dedifferentiation of Schwann cells is accompanied by the down-regulation of myelin-related mRNA and the up-regulation of molecules such as nerve growth factor receptor and GAP-43, which are associated with axonal growth and maintenance (Scherer and Salzer, 2001). *In vivo,* neuregulins, which are important mediators of axon–Schwann cell interactions, do not seem necessary to induce dedifferentiation, but they may be involved in the subsequent proliferative response. After the regenerating axons begin to grow into the regions of the nerve stump distal to the lesion (see p. 83), the Schwann cells begin to redifferentiate. Under the influence of cyclic adenosine monophosphate, they up-regulate Oct-6 and Krox-20 and begin the process of remyelinating the regenerated axons.

Slime Molds

Slime molds *(Dictyostelium)* have a complex life cycle during which individual ameboid cells undergo aggregation and then, within about 24 hours, develop over several stages into a mature fruiting body that releases spores, which germinate to form individual cells that repeat the life cycle. Under certain conditions, such as dissociation, the cells of a slime mold undergo massive dedifferentiation. Because of their simplicity and the relative uniformity of the dedifferentiative reaction, aspects of cellular dedifferentiation in slime molds are much easier to analyze than they would be in most animal systems.

Katoh and coworkers (2004) used microarrays to study patterns of gene expression in slime molds dedifferentiating from disaggregated cells taken from different early stages of development of the fruiting body. They identified three major stages of dedifferentiation. The first was an early stage, lasting only about an hour, in which developmental genes continued to be expressed. This phase quickly transitioned into a second phase, characterized by the down-regulation of the developmental genes and degradation of their mRNA. The duration of this phase was proportional to the time that the cells had developed before they were disaggregated. The third phase began when the cells started to synthesize growth genes as they entered DNA synthesis and underwent proliferation. These investigators found 122 genes that were coordinately regulated during dedifferentiation but not during development; they also found that at least one gene, *dhkA,* was essential for dedifferentiation to occur.

Compared with vertebrates, slime molds are quite poorly integrated organisms. It remains to be seen whether vertebrates possess an active generic dedifferentiation program that crosses tissue boundaries or whether individual tissues, such as the dorsal iris or skeletal muscle, make use of tissue-specific programs to elicit the dedifferentiative response.

PROLIFERATION OF PARENCHYMAL CELLS

Many internal organs regenerate through the proliferation of remaining parenchymal (usually epithelial) cells. Of these organs, the liver provides the classic example (Bucher and Malt, 1971; Sidorova, 1978). Simple histologic examination shows that most cells in the liver consist of the epithelial parenchyma. Dividing cells are only rarely encountered in the stable adult liver, but after partial hepatectomy in the rat, the parenchyma becomes quickly galvanized into a mitotic response (Grisham, 1962). After a short lag period, a burst of DNA synthesis occurs in 30% of the hepatocytes by 20 hours. This is followed by a peak of mitosis at 26 to 28 hours (Fig. 2-4). Other paired organs, such as kidney and lung, show similar early peaks of mitosis in their parenchymal cells after the removal of one member of the pair (Sidorova, 1978).

A parenchymal origin of new cells also occurs in a number of forms of physiological regeneration. In several cases, such as the epidermis, the intestinal epithelium, and the uterine endometrium, the replacement cells are, for the most part, if not all, derived from existing epithelium, but in the case of the epidermis and intestinal epithelium, the cells that create new epithelial cells have a stemlike quality.

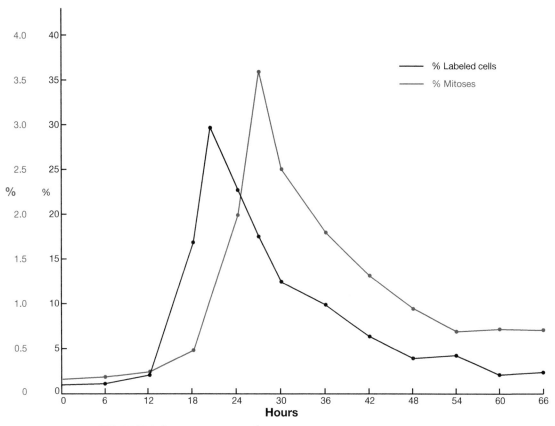

FIGURE 2-4   DNA synthesis (H^3-thymidine–labeled cells) and proliferation of parenchymal cells in the regenerating rat liver. (Based on Bucher and Malt's [1971] data.)

PROLIFERATION OF RESIDENT PROGENITOR CELLS

One of the common cellular strategies for both the growth and repair of tissues consists of the presence of a population of progenitor cells that can readily differentiate into the cell type that provides the functional basis for that tissue. Both the nature and location of the progenitor cells vary from tissue to tissue. In some tissues, the progenitor cells are located in discrete, morphologically identifiable layers, such as the perichondrium or periosteum in cartilage and bone or the endosteum in bone. At the other extreme are progenitor cells that are scattered throughout the tissue. A classic example of this type is the satellite cells of mammalian muscle, which are dispersed beneath the basal lamina of each muscle fiber. Another example is the oval cell in the liver, which appears to serve as a backup progenitor cell if the proliferation of parenchymal cells proves to be insufficient. Intermediate between these two arrangements are those in which progenitor cells are localized to certain regions within a tissue, but do not form an anatomically discrete layer. Included in this category are the zone of epithelial progenitor cells within intestinal crypts (see Fig. 1-2, B) and the cells of the ciliary margin of the retina of vertebrates whose eyes are characterized by continuous growth. Examples of many forms of resident progenitor cells are given in Table 2-2.

Beyond simply identifying progenitor cells and mapping their locations in tissues, one of the major issues in progenitor cell biology is determining what conditions keep them in an equilibrium status and what stimulates them to enter the cell cycle and divide (Cai et al., 2004; Dhawan and Rando, 2005). Their pattern of proliferation is also important. Mechanisms for both producing differentiated progeny and replenishing the supply of progenitor cells must exist to maintain a viable population of progenitor cells (Fig. 2-5). Indirect evidence from a number of systems suggests that within an overall population of progenitor cells there may be cells with different properties. Certainly within the bone marrow there are precursor cells with different degrees of "stemness." Some cells can produce the entire spectrum of mesenchymal cell types, whereas others have undergone a certain degree of restriction (see Fig. 12-3).

TABLE 2-2

Patterns of Tissue-Specific Progenitor Cells

Located in discrete layers
Perichondrium of cartilage
Periosteum of bone
Endosteum of bone
Pigment epithelium of retina

Scattered throughout the tissue
Satellite cells of skeletal muscle
Hematopoietic stem cells of bone marrow
Oogonia in ovaries of fish and amphibia
Hematopoietic centers in liver and spleen

Localized in specific regions of tissues
Stem cells of intestinal crypts
Ciliary margin of retina
Spermatogonia in basal layer of seminiferous tubules in testis
Oval cells of liver
Basal layer of epidermis
Basal layer of esophageal epithelium
Sheath (bulge region) of hair follicle
Subependymal/subventricular cells in central nervous system
Neural stem cells in hippocampus and amygdala
Equatorial epithelium of lens
Intercalated duct cells of pancreas and salivary glands
Mammary ducts in breast
Bronchiolar/alveolar stem cells of lung

A. Replacment

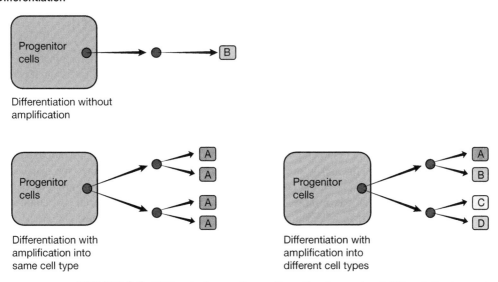

B. Differentiation

FIGURE 2-5   Different schemes of progenitor cell replacement and differentiation in generic organs. *(A)* Progenitor cell replacement models. *(B)* Differentiation models. The progenitor cells are indicated in red.

Transplantation studies of progenitor cells of skeletal muscle have shown that most of these cells die within hours of transplantation, whereas a small number survive and populate the muscle into which they have been transplanted (Skuk and Tremblay, 2003). These surviving cells may represent a population of cells that have broader developmental capabilities than do the others.

Starting with the experiments of Snow (1977), little doubt exists that satellite cells represent the main cellular precursors of regenerating mammalian muscle. In amphibians, the situation is murkier. Although satellite cells were first described in frog muscle, it was not clear that salamanders had satellite cells until Popiela's (1976) and Cherkasova's reports (1982), which documented the existence of a form of satellite cell completely surrounded by a basal lamina (which they called *postsatellite cells*) in newts. Meanwhile, Carlson and Rogers (1976) had described typical satellite cells in limb muscle of the axolotl.

Morrison and colleagues (2006) have shown that, in the newt, postsatellite cells are activated during the dedifferentiation process after limb amputation. This leads to the question of whether both satellite cells and products of muscle fiber dedifferentiation contribute to the regeneration blastema and, if so, in what proportions. Pure muscle regeneration in both the axolotl and newt (Carlson, 1970a; Griffin et al., 1987) does not involve a blastema. Although the definitive experiments have not been done, it would be surprising if this form of regeneration is not satellite cell based.

INFLUX OF PROGENITOR (STEM) CELLS ORIGINATING OUTSIDE THE DAMAGED TISSUE

Much attention has been given to the possibility that certain cells originating outside a given tissue or organ can differentiate into cell types specific to that tissue, if provided with a proper environment. That environment could be the tissue or organ itself, an ectopic *in vivo* site, or a culture dish containing an appropriate mix of media and growth factors. These cells can potentially migrate into a damaged or even undamaged tissue, or they can be collected, expanded *in vitro,* and introduced into specific target areas in the body. The technology and biology of stem cell introduction into damaged tissues is a basis for the newly developing field of regenerative medicine (see discussion in Chapter 12). This section concentrates on the evidence that cells originating outside a tissue or organ can migrate into it and provide the cellular material that could potentially take part in its regeneration. There are two possible scenarios. One is the migration of extrinsic cells into the tissue before it is injured. After injury, these immigrant cells could then participate in the regeneration of that tissue. The other scenario is the migration of such cells into the tissue only after it has been injured. Where in the tissue these cells reside is another important question.

For many years, research on cellular migration or translocation within the body was hindered by the lack of specific permanent cell markers. More recently, markers have been developed and are beginning to be used in such studies. These include Y-chromosomal markers, green fluorescent protein, genetic markers with tissue-specific regulatory elements, and a variety of antigenic stem cell markers. To date, the most compelling experimental evidence for the translocation of progenitor cells into an organ has come from transplantation studies. In both experimental animals and humans, analysis of individuals who have had bone marrow transplants has shown the presence of marrow-derived cells in a number of organs. Conversely, host-derived cells have been noted in transplanted organs, such as liver and heart, when there is a disparity in the sex between donor and host (Körbling et al., 2002; Murry et al., 2002; Theise et al., 2000).

To summarize an increasingly large body of literature, there is evidence of the immigration of cells from the bone marrow and possibly other parts of the body in a

TABLE 2-3
**Tissues or Organs in Which Bone-Marrow–
Derived Cells Have Been Demonstrated**

Adipose tissue	Liver
Blood vessels	Lung
Brain	Pancreas
Digestive tract	Skeletal muscle
Heart	Skeleton
Kidney	Skin

remarkably wide range of tissues and organs (Table 2-3), but what these cells are and how much they contribute to the regeneration of these structures is, in most cases, still an open question. In most reported cases, such externally derived cells constitute only a small percentage of the cells in the organ, typically in the range of 1% or less.

SUMMARY

The origins of the cells that take part in regeneration are almost as diverse as the varieties of regeneration. From a purely topographic standpoint, the cells participating in regeneration can arise locally at the site of damage (e.g., the regenerating amphibian limb); throughout the remainder of the organ (e.g., the regenerating liver or heart); or at some distance from the site of damage, in the case of marrow-derived, stem cell–mediated regeneration or the migration of neoblasts in planaria. Cellular sources of regeneration material are as varied as proliferating parenchyma in the regenerating liver, tissue-based progenitor cells (satellite cells) in skeletal muscle, or distant stem cells (bone marrow in mammals, scattered neoblasts in planaria). There is now little doubt that in some forms of regeneration true dedifferentiation can also occur, for example, in lens and retina regeneration or the epimorphic regeneration of skeletal muscle. In some systems (e.g., the regenerating liver), there are hierarchies of cellular precursors, so that if one level fails, a second backup system is activated.

CHAPTER 3

Epithelialization

Yet even in this simple structure [epidermis], repair of injury involves a combination of processes and maneuvers almost as complex as a military campaign.

—Paul Weiss, 1959 Harvey Lecture

Many types of trauma that ultimately lead to regeneration involve the disruption of an epithelially lined tissue surface, whether on the inside or the outside of the body. The immediate biologic response to such a wound is the initiation of a sequence of events leading to the restoration of epithelial continuity. For wounds of the skin, epithelialization is important in the prevention of infection. Equally important, however, is the minimization of water loss in terrestrial animals or providing a barrier against osmotic imbalance in aquatic animals.

Epithelial wound healing occurs regardless of whether a more organized regenerative process will ultimately ensue. Little detectable difference exists between epithelial wound healing in amputated limbs that can regenerate and those that cannot. A major challenge in the study of limb regeneration is to determine at what point there is a divergence between the epimorphic regenerative response to amputation and the tissue healing that occurs in nonregenerating limbs. It is important to understand whether the wound epithelium is a direct participant in the decision whether to regenerate a limb or it is merely a tissue that facilitates or responds to that decision.

GENERAL FEATURES OF EPITHELIALIZATION

Throughout the animal kingdom there is a strong tendency for defects in an epithelial surface to be restored. In fact, it is often difficult to prevent the epithelial healing response. Of all the questions on epithelial wound healing, one of the least controversial concerns the origin of the epithelial cells that cover the wound. Virtually all investigators agree that the wound epithelium arises from other epithelial cells of the same type. A major source of wound epithelium in the skin is the epidermis at the margin of the wound, but in shallow skin wounds of mammals, epithelial stem cells located in remaining portions (bulge region) of hair follicles and sebaceous glands also constitute an important source (Alonso and Fuchs, 2003; Krawczyk, 1971; Winter, 1972). The process of epithelial wound healing consists of the migration of epithelial cells over an appropriate substrate until the continuity of the epithelial surface is restored

(Table 3-1). Broad biologic questions are: (1) What stimulates the migratory behavior of the epithelial cells? (2) What controls their path of migration? and (3) What causes the migratory activity to cease? A critical permissive factor in epithelialization is the nature of substrate over which the epithelial cells migrate. The control of mitotic activity required for complete epithelial healing is important from both practical and theoretic standpoints.

Over the years, views of mechanisms underlying epithelialization have changed considerably. In the premolecular era, epithelial wound healing was framed by cellular behavior. A prominent conceptual framework was that of "coaptation" (Weiss, 1950). According to this concept, cells of a simple epithelium are stable—that is, in a state of good coaptation—if these conditions are present: (1) their basal surfaces reside on a substrate (basal lamina), (2) their apical surfaces are open to the air or a fluid medium, and (3) their lateral surfaces are apposed to the lateral surfaces of cells of the same type. As applied to epithelial healing, the principle of coaptation permits one to describe the behavior of epithelial cells under normal circumstances and after the creation of a wound. In an undisturbed epithelial sheet, the cells are said to be in a state of perfect coaptation, with their lateral surfaces in contact with other epithelial cells of the same type. The creation of a wound disturbs the normal coaptive relationship. The epithelial cells at the margin of the wound sense that they are no longer in contact with like cells and become mobilized. This mobilization results in an outward migration, which continues until the cells make contact with epithelial cells of the same or a compatible type, at which time the state of coaptation is said to be restored. Conversely, if the epithelial cells come into contact with nonliving material (Lash, 1956) or with noncompatible types of epithelia (e.g., epidermis meeting intestinal epithelium; Chiakulas [1952]), instability remains and the epithelium tends to pile up around or to overgrow the cells or foreign material (Fig. 3-1).

Also prominent during this era were theories of growth control that involved the actions of inhibitors, sometimes called *chalones* (Bullough, 1967). The chalone concept was elaborated principally to explain the control of mitotic activity in epidermal cells. It originated from studies on reactions to epidermal wounding in mice. It had been

TABLE 3-1

Stages in the Epithelialization of a Simple Skin Wound

1. Retraction of the edges of the wound.
2. Rapid covering of the wound by an exudate or a blood clot.
3. Detachment of marginal epidermal cells from their underlying basal lamina.
4. Mobilization of epithelial cells closest to the margins of the wound.
5. Migration of epidermal cells over a simple exudate or under a blood clot.
6. Cessation of migration when migrating epithelial cells from opposite sides of the wound meet.
7. Initiation of mitotic activity in epidermal cells along the edge of the wound.
8. Thickening of wound epithelium by further migration or mitotic activity of the wound epithelium itself.
9. Re-formation of basal lamina and attachment of basal cells to it.
10. Final structural and functional remodeling.

A.

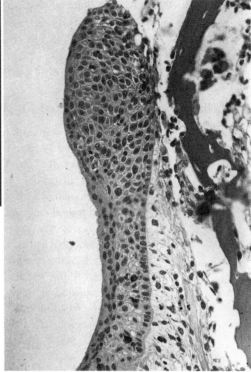

FIGURE 3-1   Example of epithelial piling up around the base of a protruding bone in the amputated limb of an axolotl. *(A)* Gross photograph. (left) Early blastema formation after normal amputation. (right) Cuff of thickened epidermis at the base of a protruding bone. *(B)* Photomicrograph showing the thickened free edge of an epidermal cuff along the edge of a protruding bone. Distal is on top.

B.

known for some time that an increased number of epidermal cells undergo mitosis along the margins of a wound. Bullough and Lawrence (1960) attributed the mitotic activity to the loss of an inhibitor of cell division that was produced by the removed epidermal cells themselves. The inhibitor (chalone), thought to be a low-molecular-weight glycoprotein, was considered to be tissue-specific, but not species- or even class-specific. According to the chalone concept, the absence of epithelial cells from a wound would result in a reduction or the absence of chalones from the wound margin, thus allowing the epithelial cells at the wound margin to undergo mitosis until the cells of the healed wound produced chalones in sufficient concentrations again to produce equilibrium conditions within the healed epithelium. The inhibitory concept replaced an earlier idea of unspecified "wound hormones" that were supposed to stimulate healing and regeneration.

The 1970s and 1980s was the era of the extracellular matrix, and the information obtained from matrix research considerably clarified the relationship between the migrating wound epithelium and the underlying substrate. This period was followed by an explosion of new information on growth factors and cytokines. As a result, wound healing has most recently been viewed as a cellular phenomenon occurring in a veritable soup of growth factors and their receptors. Much mechanistic research, however, needs to be done beyond descriptive localization of these factors in components of a wound.

MODEL SYSTEMS OF EPITHELIAL HEALING

Epithelialization of the Amputated Amphibian Extremity

Immediately after amputation, the soft tissues of the limb retract, leaving the protruding end of the bone as an indication of the real level of amputation. Bleeding from the severed blood vessels subsides within a minute. During the first hour or two after amputation, the distal cut ends of the stump tissues become covered with a fibrinous exudate that, together with a few extravasated blood cells, provides the substrate on which the epidermal cells will migrate to seal off the amputation surface.

The sequence of events in postamputational epithelial healing was carefully described by Repesh and Oberpriller (1978, 1980). As early as an hour after amputation, the epidermal cells at the wound margin show signs of mobilization for migration, such as widened intercellular spaces, changes in cell shape, the disappearance of hemidesmosomes (Norman and Schmidt, 1967), and the detachment of basal epidermal cells from the underlying basal lamina. Within 2 hours, evidence of early epidermal migration is seen. Extending out from the margin of the cut epidermis, a thin sheet of flattened epidermal cells has begun to stretch out across the fibrin substrate (Fig. 3-2). The advancing edge of the wound epidermis may be several cell layers thick, and at the leading edge of the basal cells is a broad lamella that is in close contact with the underlying substratum (Fig. 3-3). The spaces between migrating epidermal cells

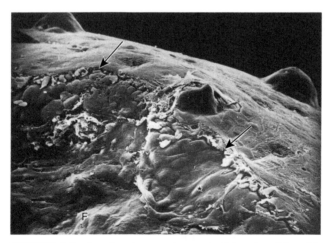

FIGURE 3-2   Scanning electron micrograph of the edge of the epidermis (arrows) 2 hours after limb amputation in the newt. Epidermal cells (asterisk) are beginning to migrate over the fibrin-covered amputation surface (F). (Reprinted from Repesh, L.A., and J.C. Oberpriller. 1980. Ultrastructural studies on migrating epidermal cells during the wound healing stage of regeneration in the adult newt, *Notophthalmus viridescens. Am J Anat* 159:187–208, by permission.)

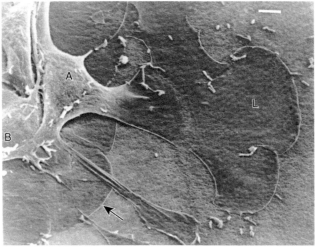

A.

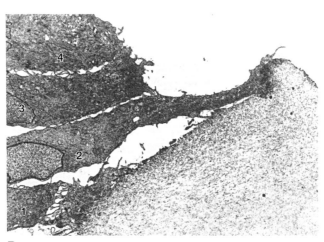

B.

FIGURE 3-3 *(A)* Scanning electron micrograph of epidermal cells of the newt migrating *in vitro* over a Nuclepore filter in Holtfreter solution. L indicates a large lamellipodium from cell A. Beneath it (arrow) are lamellipodia, probably extending from cell B. *(B)* Transmission electron micrograph (TEM) showing the advancing edge of the migrating wound epidermis after limb amputation in the newt. Cell 2 overlies cell 1 and is extending a process over the fibrinous substrate that covers the wound surface. (A: Reprinted from Donaldson, D.J., and M.K. Dunlap. 1981. Epidermal cell migration during attempted closure of skin wounds in the adult newt: Observations based on cytochalasin treatment and scanning electron microscopy. *J Exp Zool* 217:33–43, by permission; B: Reprinted from Repesh, L.A. and J.C. Oberpriller JC. 1980. Ultrastructural studies on migrating epidermal cells during the wound healing stage of regeneration in the adult newt, *Notophthalmus viridescens. Am J Anat,* 159:187–208, by permission.)

are greatly reduced, and many of the normal desmosomal connections appear to be broken. This pattern of migratory activity persists over the next few hours, as the epidermis covers more of the wound (Fig. 3-4). A consistent feature during this period of healing is the close contact between the lamellar edge of the leading epidermal cells and the underlying fibrin sheet.

Depending on temperature and the cross-sectional area of the limb, the amputation surface is sealed off within 9 to 13 hours after amputation. When they meet, the epidermal cells lose their streaming profiles and become rounded as they temporarily pile up at the site of their meeting (Fig. 3-5). During the migratory phase, the epidermal cells that cover the amputation surface neither synthesize DNA nor undergo mitosis (Chalkley, 1954; Hay and Fischman, 1961).

One feature of postamputational healing that has often attracted notice is the presence of epidermal "tongues" that extend downward from the wound surface just at the base of the distal edge of the original skin (Fig. 3-6). This has been considered to be an important indication of the ability of an amputated structure to undergo epimorphic regeneration, even in mammals. In reality, such tongues represent the early path of migration of the wound epidermis over a depression created by soft-tissue retraction; the migrating epidermis has to fill in the trench between the dermis and the subjacent soft tissues so that migration can proceed across the remainder of the amputation surface. Thus, the tongues are a morphologic reflection of the substrate requirements of migratory epithelial cells. Epidermal tongues have been seen and similarly interpreted in skin wounds of mammals by Hartwell (1955). Epithelial projections into deeper tissues can also be seen in areas of great tissue damage or around bits of foreign material, but these projections are related to epithelial phagocytosis (see later).

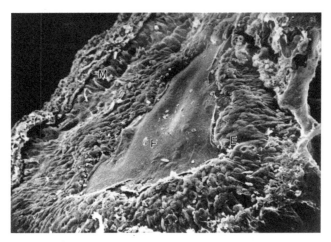

FIGURE 3-4 Scanning electron micrograph of advancing wound epithelium (E) covering the fibrin (F)-covered wound 4 hours after limb amputation in the newt. M indicates wound margin. (Reprinted from Repesh, L.A. and J.C. Oberpriller. 1980. Ultrastructural studies on migrating epidermal cells during the wound healing stage of regeneration in the adult newt, *Notophthalmus viridescens. Am J Anat* 159:187–208, by permission.)

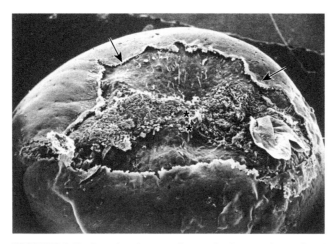

FIGURE 3-5 Scanning electron micrograph of amputation surface 9 hours after limb amputation in the newt. Migrating epidermis has covered the wound and is piling up in the center (asterisk). Arrows indicate the margin of the wound. (Reprinted from Repesh, L.A. and J.C. Oberpriller. 1980. Ultrastructural studies on migrating epidermal cells during the wound healing stage of regeneration in the adult newt, *Notophthalmus viridescens. Am J Anat* 159:187–208, by permission.)

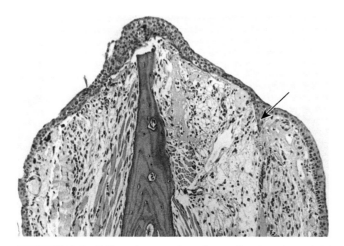

FIGURE 3-6 Epidermal tongue (arrow) extending into the underlying tissues at the margin of the amputation surface in the newt. Such tongues represent epidermal cells that have filled in a depression created by retraction of soft tissues in the hours after limb amputation.

Shortly after epidermis has covered the amputation surface, DNA synthesis and mitosis are seen in the cells of the wound epithelium. This results in a considerable thickening of the epithelium into an apical epithelial cap, which still has not reconstituted a basement membrane beneath it. Thornton (1957) inhibited limb regeneration in *Ambystoma* larvae by surgically removing the apical epidermal cap, and a number of subsequent experimental models have shown that the apical cap is necessary for the continuation of regeneration. Students of regeneration have equated the apical epidermal cap in regenerating urodele limbs with the apical ectodermal ridge, which controls limb outgrowth in the embryos of higher vertebrates. The basal layer of epidermal cells in the apical cap has a strikingly regular columnar morphology (Fig. 3-7), with the Golgi apparatus interposed between the nucleus and the basal surface, suggesting that these cells are adapted for secretion of molecules directed toward the underlying tissues. Christensen and Tassava (2000) have postulated that the basal columnar cells represent the regenerating urodele limb's equivalent of the apical ectodermal ridge.

Epithelialization of Mammalian Skin Wounds

The epithelialization of mammalian skin wounds has been the subject of many laboratory studies (rev. Martin, 1997; Singer and Clark, 1999). Although there are slight

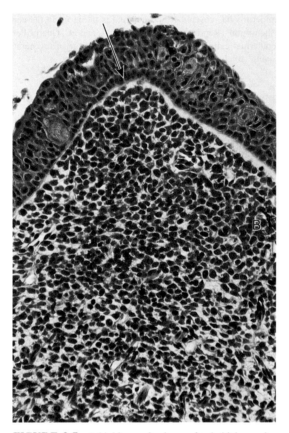

FIGURE 3-7   A highly regular layer of cuboidal or columnar cells (arrow) in the basal layer of the apical epidermal cap that covers the blastema (B) in an amputated axolotl limb. These cells may be the source of secreted growth factors.

variations in the process, depending on the model of wounding that is used, there emerges a general pattern of cellular activity that is remarkably similar to that which was described previously for healing of the amputated amphibian extremity.

Mammalian epidermal healing occurs in two main variants. One is the case in which the surface of the wound is allowed to dry, resulting in the formation of a scab. In the other variant, the surface of the wound remains moist. The latter (moist) condition occurs when a dressing is placed over the wound or in a healing model made by creating suction blisters on the skin (Krawczyk, 1971).

Regardless of the model, the initial reactions of the wound epidermis are quite similar. Within minutes after wounding, the defect is filled by a clot consisting of platelets and other blood cells embedded in a matrix of fibrin, with lesser amounts of fibronectin, thrombospondin, and vitronectin. The clot serves as the source of a complex mix of growth factors and cytokines, which initiate many aspects of the wound-healing response within the epidermis and the underlying dermis (see discussion later in this chapter). One of the first reactions of the epidermis at the margins of the wound is detachment of the basal hemidesmosomes from the underlying basal lamina. Within a few hours of wounding, the epithelium at the margins of the wound thickens because of an increase in volume of the cells and a widening of the intercellular spaces; the latter is due, in part, to the loosening of the desmosomes that bind the epithelial cells (keratinocytes) to one another. Depending on the species, the initial outgrowth of ruffled cytoplasmic lamellae from the cells at the wound margin may not extend into the wound until 24 hours after its creation. Although this is delayed in comparison with the initiation of migration in the amphibian wound epidermis, the morphology of epidermal outgrowth is remarkably similar. According to Garlick and Taichman (1994), it is the suprabasal keratinocytes that migrate to cover the wound, whereas the basal cells remain at the wound margin and undergo mitosis.

The substrate for epidermal migration is a fibrin clot interposed among elements of the damaged dermis. In a noncovered incision wound, the epidermis does not migrate across the dried surface of the wound; rather, it selects a plane between the necrotic tissue that lines the edges of the wound and the healthy tissue that surrounds it. In the absence of major necrosis, the wound epithelium migrates just beneath or through the blood clot within the wound (Fig. 3-8). The wound itself is filled with a hard scab, which becomes detached later in the healing process.

As the wound epidermis spreads out beneath the hard crust of dead tissue and debris, the cells are loosely connected to one another by desmosomes (fewer than the normal number). The lamellae of the leading edge of the cells are in close contact with the fibrinous substratum. As the migratory phase proceeds, the basal lamina is reconstituted from the edges of the wound toward the center, and hemidesmosomes appear on the basal surfaces of the epidermal cells. In contrast, no specialized contacts are seen between the outer surface of the migrating wound epithelial cells and the scab. No mitotic figures are seen in migrating epithelial calls, but there is abundant evidence of phagocytic activity by these cells. When epidermal cells from opposite sides of the wound meet, there is a temporary piling up of cells at the line of contact, but soon the epidermis returns to normal thickness. One of the last phases in the wound healing

A.

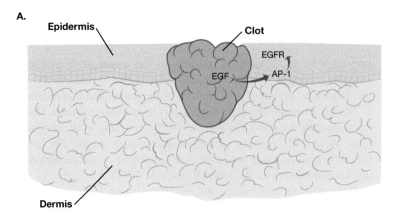

B.

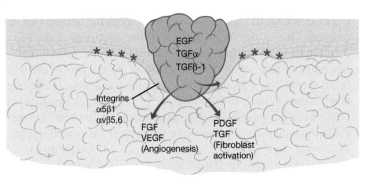

C.

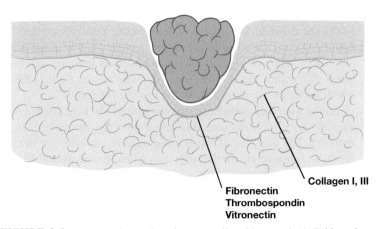

FIGURE 3-8   Stages in the healing of a mammalian skin wound. *(A)* Epidermal growth factor (EGF) from the wound is one of the factors that initiates mobilization of the wound epidermis. AP-1, activator protein-1; EGFR, epidermal growth factor receptor. *(B)* Growth factors from the clot stimulate not only epidermal migration, but also angiogenesis and fibroblast activation. Newly expressed integrins facilitate adhesion of the migrating epidermal cells to the substrate. Mitosis (asterisks) occurs in basal keratinocytes near the wound margin. FGF, fibroblast growth factor; PDGF, platelet-derived growth factor; TGF, transforming growth factor; VEGF, vascular endothelial growth factor. *(C)* Having digested their way through the clot, sheets of wound epidermal cells migrate over extracellular matrix molecules and meet near the center of the wound. The remains of the clot are then shed.

process is the layering and then keratinization of the wound epidermis. This process leads to sloughing of the overlying scab.

In a covered (moist) skin wound, the overall process of healing is similar, but epidermal mobilization and migration are more rapid. With the absence of a hard crust, the epidermal cells are more likely to migrate over the surface of the wound.

ANALYSIS OF EVENTS IN THE EPITHELIALIZATION OF WOUNDS

Mobilization

In most cases of epithelial healing, regardless of species or location, the morphologic features of mobilization along the edge of the wound are remarkably similar. The cells become more rounded, they become detached from the basal lamina, and they become separated from one another by wide intercellular spaces. The separation is facilitated by dissolution of intercellular desmosomes. Internally, keratin filaments become reorganized, and bundles of actin filaments become arranged about the periphery of the cells to facilitate future movement. The keratinocytes at the leading edge also express new integrins that act as receptors for substrate molecules. These allow the migrating cells to adhere to their new substrate within the open wound (Martin, 1997).

Surprisingly, few hypotheses have been advanced to account for premigratory behavior, and there remains little concrete information on mechanisms underlying the mobilization phase of wound healing. One of the big issues is how the presence of a "free edge" is translated into the mobilization phenomenon and ultimately into migration. Weiss's (1950) concept of coaptation is really a framework for describing and predicting cellular events. It does not imply specific control mechanisms. A critical first step is to determine whether the cells at the free edge of the wound are released from some sort of inhibition or whether they respond to positive stimuli from the wound site or elsewhere, or both.

On the one hand, the epithelial cells at the edge of the wound can be viewed as being freed from the constraints of contact inhibition, which would release a natural tendency of the cells to spread out. On the other hand, the same cells could be responding to some of the multitude of growth factors and cytokines present within the wound. What is reasonably certain is that stimuli from the blood clot are not necessary to bring about the initial epithelial reaction, because it also occurs in wounds that do not involve bleeding and that leave the epidermal basal lamina intact.

Some of the dynamics of the edge of the wound epithelium have been clarified by research initially directed at understanding mechanisms of closure of the eyelids in mouse embryos. Mutations of a number of molecules (e.g., epidermal growth factor receptor [EGFR], transforming growth factor-α [TGF-α], fibroblast growth factor receptor [FGFR], and others) result in the "open eye" phenotype. This is caused by the lack of migration of the epithelium at the margins of the developing eyelids across the cornea. Li and colleagues (2003a) have shown that loss of c-Jun in the skin of mice results in both open eyes and defects in epidermal migration during wound healing. c-Jun appears to be a critical component of an autocrine/paracrine signaling pathway whereby epidermal growth factor (EGF) in the wound influences the expression of activator protein-1 (AP-1), a transcription factor that controls the cellular response to injury, including epidermal migration. AP-1, in turn, influences the expression of EGFR and heparin-binding EGF (HB-EGF), one of its ligands (see Fig. 3-8, A). Full function of EGFRs appears to be a key to the initiation of epidermal cell migration, because with reduced or absent function, epidermal cell migration is defective.

Migration

Within a few hours after wounding, epithelial cells begin to migrate centripetally to cover the wound. Although the circumstances of wounding can vary considerably, a common element in the healing process is the migration of the epithelial cells over a meshwork that consists principally of noncellular matrix material, mainly various proportions of fibrin, fibronectin, and vitronectin. In the case of mammalian skin wounds, the migrating epithelial cells must also digest their way through the scab that fills the wound.

At even a cellular level, the mechanism of epithelialization is still not fully understood. Discussion has centered around two competing hypotheses. According to one hypothesis (Lash, 1955, 1956; Odland and Ross, 1968; Weiss, 1961), the wound epithelium migrates as a sheet, with the cells advancing in rank order, like soldiers on parade, with cells retaining their original positions relative to one another. Proponents of this idea generally believe that each cell is actively involved in the migration, rather than being pulled along by the efforts of cells in the front row or being pushed by cells behind. Mahan and Donaldson (1986) demonstrated by direct observation that cells on the leading edge of early skin wounds in the adult newt remained at the edge, even after considerable migration had occurred. Such a mechanism appears to operate in amphibians and possibly in embryonic wound healing, which has been described as occurring in a purse string–like fashion.

The second hypothesis attributes epithelial migration to a leap-frog mechanism, in which migrating epidermal cells advance by passing over those that have already made connections with the substratum (Gibbins, 1978; Krawczyk, 1971; Winter, 1972). According to this proposed mechanism, after a particular epithelial cell (cell A) has made a firm connection with the substratum, its leading lamella retracts, and then it begins to elaborate hemidesmosomes and basal lamina material. While this is occurring, the next cell back from the epithelial margin (cell B) extends a lamella over cell A and makes contact with the substratum just ahead of it (Fig. 3-9). Then that cell (cell B) becomes the leading cell until it becomes bypassed by cell C, thus repeating the cycle. The best evidence in support of this model comes from mammalian systems.

Of critical importance to epithelial migration is the substrate (Donaldson and Mahan, 1988). To a certain extent, the specific nature of the substrate depends on the type of wound and the time in the healing cycle. In the simplest type of epithelial

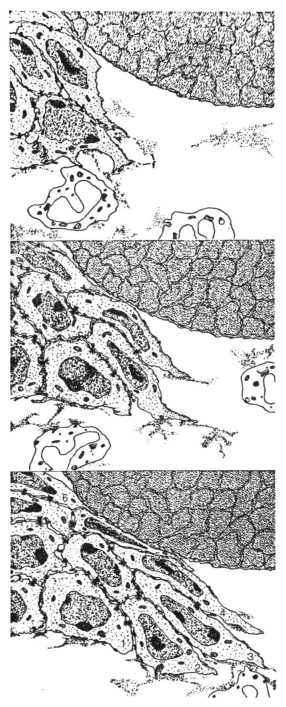

FIGURE 3-9   Representation of epidermal healing movements beneath the coagulum (top right) in suction-induced blisters in 2-day-old mice. Numbers 1 to 6 represent the sequence of progression of epidermal cells as they migrate over one another to cover the wound surface. (Reprinted from Krawczyk, W.S. 1971. A pattern of epidermal cell migration during wound healing. *J Cell Biol* 49:247–263, by permission.)

wound, the substrate consists of basal lamina, which is rich in molecules such as type IV collagen and laminin. More commonly, the substrate consists of a mixture of blood products—fibronectin, vitronectin, and thrombospondin—embedded in a meshwork of fibrin, as well as elements of the normal dermal connective tissue, such as types I and III collagen (see Fig. 3-8, C). After epithelial migration has begun, the epithelial cells themselves become capable of secreting fibronectin, a better substrate for cell migration than laminin.

Epithelial cell migration requires not only an appropriate substrate, but the ability of the epithelial cells to adhere to that substrate. Leading-edge epithelial cells extend long lamellipodia, which adhere to the substrate through the expression of new integrins, $\alpha_5\beta_1$ and $\alpha_v\beta_6$ fibronectin/tenascin receptors and the $\alpha_v\beta_5$ vitronectin receptor (Mehendale and Martin, 2001). The epithelial cells are pulled forward by the contraction of bundles of actomyosin filaments that connect to adhesion points in the lamellipodia.

Among the multitude of growth factors and cytokines liberated within the wound site, several appear to facilitate epithelial migration (Werner and Grose, 2003). Members of the EGF family (e.g., EGF, TGF-α, and HB-EGF) activate a small GTPase, Rac, which serves as a molecular switch that controls lamellipodia extension and the assembly of its focal adhesion complexes with the underlying substrate. Other factors that appear to stimulate epithelial migration are keratinocyte growth factor (KGF) and TGF-β. Knockout studies that involve individual growth factors (e.g., KGF or TGF-α) have shown relatively little effect on epithelialization, but like so many important biologic functions, redundancy may largely compensate for the absence of a single factor.

In the healing of a mammalian skin wound, the progression of epithelial migration requires the leading edge of the wound epithelium to penetrate the hard material of the scab. To a great extent, this is accomplished by the production and secretion of lytic enzymes by the epithelial cells. These enzymes are directed at both constituents of the clot and the associated connective tissue matrix. Migrating keratinocytes up-regulate both tissue- and urokinase-type plasminogen activators, which stimulate the production of the fibrinolytic enzyme, plasmin, from its precursor, plasminogen (see Fig. 4–1). This is a major mechanism for digesting a pathway for epidermis through the clotted blood in a scab (Li et al., 2003a). The cells that lead the migrating epithelial sheet also produce a variety of matrix metalloproteinase enzymes (e.g., collagenases, stromelysin), which attack specific components of the dermal extracellular matrix that is in the pathway of the advancing wound epithelium.

Recent studies (Henry et al., 2003; Bandyopadhyay et al., 2006) have emphasized the importance of the plasma-to-serum transition in a fresh wound in stimulating the migration of human keratinocytes. Under normal circumstances, plasma constitutes the normal fluid environment underlying the epidermis. The environment of an acute wound is plasma based, but as the plasma becomes rapidly converted to serum through clotting, the keratinocytes become activated and begin to migrate to cover the wound. This serum-based migratory activity is correlated with a high level of TGF-β_3 in serum, which conversely inhibits the migration of dermal fibroblasts and microvascular endothelial cells in the tissues underlying the wound. After epidermal cell migration is completed, the serum-rich fluid environment returns to one that is plasma based and low in TGF-β_3. This stabilizes the formerly motile keratinocytes, but stimulates the migration of dermal fibroblasts and endothelial cells to fill in the deep spaces created by the wound.

A couple of natural systems characterized by the nonmigration of epidermis over wounds merit mention. Weiss and Matoltsy (1959) created skin wounds in chick embryos and found no epithelial migration until the 10th day of embryonic life. The free margin of the epidermis failed to adhere to the underlying dermis, and cysts or polyps of epidermis formed along the margin of the wound. At the 10th day of development, epidermal migration began, whether the wound was fresh or had been made as early as day 4. The initiation of epidermal migration in old wounds showed that a fresh wound is not necessary for migration. Yet, despite its inability to migrate *in vivo,* the same epidermis was able to migrate *in vitro.* More recently, Chernoff and Robertson (1990) found that the period at which epidermal migration becomes possible coincides with an increase in EGF content of the skin.

Another instance of nonmigration of epidermis over a defect occurs when the antlers of deer lose their covering of skin (shed their velvet) during the mating season. Not until the antler has been shed and the new cycle of regrowth is ready to begin in the spring does the marginal epidermis resume its ability to migrate over the wound created by the shed antler (Goss, 1978). A natural situation worthy of further investigation is the lack of migration of oral epithelium over the teeth.

Cessation of Migration

Epithelial migration over a wound ceases when the leading edges from opposite sides of the wound meet. The signal to stop migration does not appear to be transmitted immediately, however, because a number of investigators have made the incidental observation that epithelial cells sometimes pile up at the point of junction (see Fig. 3-5). Soon, however, the wound epithelium becomes smooth, and no trace of the point of meeting of the epithelial sheets remains. If the migrating epidermis meets a nonepidermal structure, it continues to pile up around it (see Fig. 3-1).

In a test of Weiss's (1950) coaptation model, Chiakulas (1952) opposed cuffs of different types of epithelia and observed their behavior when the two epithelial edges met. He found an affinity between epithelial cells of the same type or between types of epithelia that normally contact one another in the body (e.g., between epidermal and oral epithelial cells). However, many other combinations were incompatible, showing poor coaptation and stability along the junction of the epithelial edges.

Completion of the epidermal phase of wound healing involves reconstruction of the normal layering of the healed epidermis. Recently, *Grhl-3,* a homolog of the *grainy head* gene in *Drosophila,* has been found to control the formation (and re-formation) of the barrier function of the epidermis by promoting the expression of transglutaminase 1 (TGase 1), which acts as a cross linker in the outer stratum corneum of the epidermis (Moussian and Uv, 2005; Ting et al., 2005). Mice with a mutation of *Grhl-3* not only showed reduced expression of TGase 1, but also exhibited impaired migration of epidermal cells over skin wounds. This suggests that, like AP-1, *Grhl-3* might play a much broader role in the overall epidermal healing reaction.

Mitotic Activity in Healing Wounds

Regardless of the species, a characteristic pattern of mitotic activity is seen in healing epithelial wounds. It is almost universally accepted that neither DNA synthesis nor mitosis occurs in epithelial cells actively migrating to cover a wound (Christophers, 1972; Hay and Fischman, 1961). The first burst of cell proliferation in the healing of an open epithelial wound occurs in the cells just around the margin of the wound. In mammals, the burst begins toward the second day after wounding (Fig. 3-10). If samples are examined at frequent enough intervals, it can be seen that mitotic activity is not constant, but rather proceeds in a series of short waves (Epstein and Sullivan, 1964; Giacometti, 1967). During peaks of mitotic activity, the duration of the S-phase and mitosis is reduced in guinea pig and primate epidermis (Christophers, 1972; Giacometti, 1967); but in healing of the lens epithelium in mice, the cell cycle does not differ from normal (Rafferty and Smith, 1976). In the wound epithelium itself, mitosis is not resumed until a given cell has ceased migrating and has become stabilized (see Fig. 3-10).

Considerable evidence now is available that epithelial mitosis in a healing wound is largely stimulated by some of the many growth factors produced by tissues within the healing wound rather than by the removal of chalones or other inhibitors, although some participation of inhibitory substances has not been rigorously excluded. The most

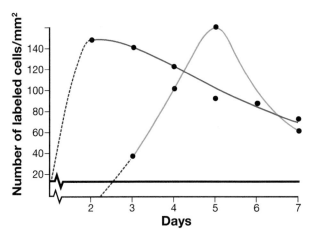

FIGURE 3-10 Proliferative activity of two populations of epidermal cells during the healing of a mammalian skin wound. Red indicates incorporation of H^3-thymidine in epidermal cells at the margin of the wound; green indicates incorporation of H^3-thymidine in cells of the actual wound epidermis. (From Christophers's [1972] data.)

important growth factors that affect epithelial mitosis are EGF, HB-EGF, KGF fibroblast growth factor-7 (FGF-7), and TGF-α (Singer and Clark, 1999). It is of interest that different components of the wound produce different factors. Whereas EGF is given off by the platelets in the clot, KGF is produced by fibroblasts underlying the epidermis. TGF-α is not only secreted by macrophages present in the area of tissue damage, but later it is also produced by the marginal epidermal cells themselves. Because cells must be exposed to these mitogens for a number of hours before proliferation is stimulated, this may account, in part, for the delay in the proliferative response of the marginal epithelial cells. It is also highly likely that neither migrating epithelial cells nor the unstable cells at the wound margin produce the requisite receptors for the growth factors until the cells settle down after the initial migratory phase.

Several studies have shown that EGF, in particular, can accelerate the healing of epithelial wounds not only in skin, but in the cornea and gastric epithelium (Schultz et al., 1991). Conversely, the inhibition of KGF receptor function delays reepithelialization and reduces the mitotic response of the epithelial cells at the edge of the wound (Werner et al., 1994).

Re-formation of the Basal Lamina and Epithelial Attachment over the Wound Surface

Shortly after the time of meeting, the cells of the leading epithelial edges must change their patterns of synthesis to cease producing molecules that facilitate migration. In a normal skin wound, the formerly leading epithelial cells begin producing basal lamina material and also elaborate the hemidesmosome connections that cause strong adherence of the epidermis to the underlying tissues. In mammalian skin healing, especially in elderly adults, the junction between wound epidermis and underlying dermis is often relatively smooth, without the dermal pegs that characterize the normal epidermal–dermal junction. Interestingly, the completed wound epidermis of the amputated amphibian limb is said not to form a basal lamina, suggesting that there may be a fundamental difference in this aspect of the wound healing process that precedes epimorphic regeneration.

ELECTRIC FIELDS AND EPITHELIALIZATION

Transepithelial electric potentials are common to all types of epithelia. When an epithelium is wounded and there is a break in its continuity, current flows out of the wound, and the endogenous electric field typically has strength in the range of 40 to 100 mV/mm (Nuccitelli, 2003; Vanable, 1989). Many experiments have shown that epithelial cells are highly sensitive to electric fields and that *in vitro* they migrate toward the cathode (negative pole). For more than a century, various devices have been used to stimulate the epithelialization of human skin wounds by the application of a wide variety of electric currents and potentials to the region of the wound (Gardner et al.,

1999; Ojingwa and Isseroff, 2002), but only in recent years has there been evidence of specific cellular effects. Earlier studies on a variety of mainly amphibian systems showed that interference with wound currents by ion channel-blocking agents, such as benzamil or amiloride, impaired epithelial wound healing (Vanable, 1989). Vanable's group has shown that by changing field strength above or below normal levels, the rate of epithelial healing can be significantly altered (Sta Iglesia et al., 1996; Sta Iglesia and Vanable, 1998).

An *in vivo* study on corneal wounds in the rat showed some remarkable cellular effects of electric fields on epithelial behavior and healing (Song et al., 2002). By the use of various pharmacologic agents that altered electric fields through different mechanisms, this group found that the axes of epithelial cell divisions were highly oriented at the edge of a wound and that the degree of orientation (the cleavage planes were perpendicular to the field vector) of the cells was proportional to the strength of the field. The same study showed that both the rate of epithelial cell migration and the frequency of mitotic cells are proportional to field strength. A real challenge for the future is to determine how the striking effects on epithelialization by both growth factors and electric fields interrelate.

EPITHELIAL PHAGOCYTOSIS AND HISTOLYSIS

During a typical wound healing process, the epithelial cells may engage in activities other than those directly connected with covering the wound. Prominent among these activities is phagocytosis. Two varieties of phagocytic activity have been noted.

One variety, seen in most species of vertebrates, consists of the incorporation of small masses of damaged cells or pieces of fibrinous substrate into the epithelial cells that are migrating over the wound (Odland and Ross, 1968; Singer and Salpeter, 1961). In addition to intracellular inclusions, small masses of debris are commonly found in the intercellular spaces.

The other variety of epithelial phagocytosis occurs on a much larger scale, and it is best represented in the amphibian epidermis (rev. Singer and Salpeter, 1961). When there has been massive tissue damage or when foreign objects are placed beneath the epidermis, large tongues of epidermis grow down into the deeper tissues and surround the foreign object or damaged tissue (Carlson, 1967; Scheuing and Singer, 1957). Once surrounded, the material is extruded from the body as a large mass (Fig. 3-11). Because of this property of amphibian epidermis, it is difficult to use skin tags for the purpose of marking animals or to maintain subcutaneous implants (e.g., of carcinogenic substances) in place for prolonged periods.

A major but poorly investigated question concerns the role of the wound epidermis in an amputated amphibian limb in promoting histolysis of the underlying tissues. A number of early Russian investigations (Adova and Feldt, 1939; Orechowitsch and Bromley, 1934; Polezhaev, 1936) produced evidence that the wound epidermis possesses histolytic properties, and the investigators believed that this played a role in the initiation of the overall process of demolition and dedifferentiation. Later, Taban (1955)

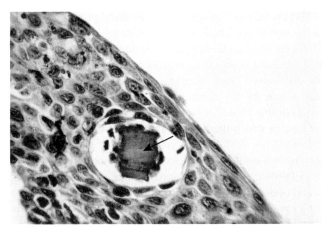

FIGURE 3-11   Excessively thickened wound epidermis in the newt shown engulfing a small fragment of bone (arrow). The animal had earlier been given an injection of actinomycin D.

reported that the amphibian wound epidermis has the ability to liquefy a blood clot. Collagenolytic activity has been found in the mammalian wound epidermis (Grillo and Gross, 1967) and also in salamander limb stumps (Grillo et al., 1968). Although collagenolytic activity was not found during the immediate postamputation period, Grillo and coworkers (1968) speculated that the wound epithelium might be involved later in its production. Stocum (1995) has reviewed the literature relating to the remodeling of the distal limb stump tissues before and during dedifferentiation. Despite evidence of the presence of a variety of enzymes that can destroy extracellular matrix or break down tissue, in general there is surprisingly little direct evidence connecting the wound epidermis either to the direct production and secretion of such lytic enzymes or to the production of specific growth factors that would stimulate the production of such enzymes by the underlying distal cells of the limb stump.

REGENERATION OF EPIDERMAL APPENDAGES

One of the distinguishing characteristics of true regeneration of skin as opposed to simple healing is the presence of epidermal appendages, such as hair, feathers, sebaceous glands, or sweat glands. In the embryo, such structures arise from inductive interactions between ectoderm and underlying mesenchyme, whether mesodermal or neural crest in origin (Carlson, 2004). Interpreting the results of wounding models is often difficult, because if the wound is shallow enough, epidermal appendages can arise from stem cells originating in deep remnants left behind at the time of wounding. Despite this ambiguity, there are several models in which true regeneration of epidermal appendages appears likely, if not certain.

Markelova (1953, 1960) and Aspiz (1954) first discovered that hairs and sebaceous glands are formed in regenerates from full-thickness skin wounds in rabbit ears.

Breedis (1954) created skin wounds in rabbits and prevented closure of the wound so that migrating wound epidermis was in contact with the underlying granulation tissue. New hair follicles were widely scattered throughout the wound epidermis. Later, Goss (1981) noted the occasional appearance of hairs and sebaceous glands in the skin that regenerates over punch holes in rabbit ears. The MRL strain of mouse also fills in punch holes in ears by regeneration, and Heber-Katz (1999) has suggested that hair growth occurs in this model as well.

The one case in which there is no doubt about the *de novo* formation of hair is the antlers of deer (Goss, 1978). Every year, when the antlers regrow, the skin overlying the bone forms thousands of tiny hairs, thus accounting for the designation "velvet" to the covering of growing antlers. It is apparent that, in this case, the interactions between epidermis and underlying tissue required to induce hair growth are operating. Li and Suttie (2001) have stressed the embryonic characteristics of the growing antler. This is a model that could benefit by further research.

RELATION BETWEEN EPITHELIALIZATION AND EPIMORPHIC REGENERATION

One of the most perplexing issues in the field of regeneration is the relation between a wound epidermis and the initiation of an epimorphic regenerative process. It has been known for more than 100 years that the presence of a wound epidermis is an absolute requirement for the initiation of limb or tail regeneration in salamanders. This was demonstrated by covering the amputation surface with full-thickness skin (Efimov, 1931; Godlewski, 1928; Polezhaev, 1936; Tornier, 1906) or by daily removal of the thickened apical epidermis (Thornton, 1957). Goss (1956a) amputated forelimbs of newts and inserted the distal epidermis-free ends into the body cavity. These limbs did not regenerate. The inhibition of regeneration by full-thickness skin is not confined solely to amphibians. The natural regeneration of amputated human fingertips is completely inhibited if the injury is repaired surgically by sewing a flap of skin over the amputation surface (Doletsky et al., 1976; Douglas, 1972; Illingworth, 1974) (see Fig. 5-3).

A potential early relation between the wound epidermis and histolysis of the underlying tissues is noted earlier in this chapter. Although the wound epidermis does not appear to be required for the initiation of dedifferentiation, Tassava and Mescher (1975) and Mescher (1976) propose that the wound epidermis keeps the dedifferentiated cells in the cell cycle for several rounds of cell division, which would allow dedifferentiation at the tissue level to continue and progress into blastema formation. Borgens et al. (1977b) have measured a major outflow of electric current through the wound epidermis of amputated amphibian limbs and suggest that this is an important mechanism in the initiation of regeneration. Alitzer and coworkers (2002) note the correlation between inhibition of limb regeneration by skin flaps and the suppression of the peak of current outflow after amputation. However, the relation between electric current and subsequent regenerative events remains to incompletely defined.

Another potential role of the wound epidermis in epimorphic regeneration is suggested by some of the models of morphogenetic control in regeneration, especially those that integrate morphogenetic phenomena with early events in the regenerative process. A particularly intriguing approach has been the class of "averaging models," which have been devised to explain events that occur along the proximodistal axis of the limb (Maden, 1977; Stocum, 1978b). These models assume that the tissues of the limb possess a gradient of positional values, and that the wound epidermis serves as a reference point for the underlying tissues. If there is a large discrepancy between positional values of the distalmost deep tissues and the fixed value of the wound epidermis, the distal deep tissues begin to dedifferentiate and divide. The cellular progeny acquire positional values that are intermediate between those of the mature stump tissues and the wound epidermis (see Fig. 7-14). This averaging process continues until the missing positional values are filled in through the formation of a blastema and subsequent outgrowth.

These averaging models, which were highly influential in their time, now require reexamination in view of the discovery of prod 1 (CD59), which appears to both mark and control proximodistal aspects of morphogenesis (see p. 149) (Morais da Silva et al., 2002). Later research (Echeverri and Tanaka, 2005) showed that even in the early blastema, cells with the most distal signatures are found at the end of the blastema (see Fig. 7-15). This suggests that averaging may occur among cells of the blastema, rather than between the wound epidermis and distalmost cells of the stump. It is still possible, however, that interaction with the wound epidermis may help to set the distalmost prod 1 values of the cells of the early blastema.

Other models of morphogenesis relate regeneration to properties along the transverse plane of the amputated structure. In its most simple form, the interaction of morphogenetic opposites is required for the initiation of regeneration. For example, Chandebois (1957, 1979) has reported that in planaria, regeneration of the head will not occur if the dorsal and ventral epidermis are not allowed to meet during the healing of an amputation wound. A modification of the polar coordinate model of morphogenesis (see p. 154) (Bryant et al., 1981) makes specific assumptions about wound healing patterns that are important for the initiation of regeneration in both insects and amphibians, although in this case, the morphogenetically active agents may be mesenchymal cells underlying the wound epidermis.

Underlying many of the experiments and hypotheses outlined in the preceding paragraphs is the following question: Does the wound epidermis have any specific properties that are needed for the initiation of epimorphic regeneration, or will any type of epithelial covering suffice? The answer remains unclear.

Several strategies have been used. One consists of replacing the skin of the limb with skin from some other region of the body and allowing the foreign epidermis to migrate over the amputation surface. Although a wound epidermis from most areas of the urodele body permits regeneration, skin from the head and sometimes the back does not support regeneration (Polezhaev, 1936; Thornton, 1962). Carlson (1982) tested the ability of epidermal tissue from newly metamorphosed *Rana pipiens* (stage 24) to support and participate in limb regeneration. He grafted forearm skin onto homologous locations in small axolotls. Although the donor frogs were no longer able to regenerate limbs, the axolotl limbs covered by frog skin went through the early stages of regeneration normally, and a few even formed well-defined blastemas and early digital condensations before the frog epidermis was displaced by epidermis migrating in from the host.

One of the best understood characteristics of a wound epidermis capable of supporting regeneration is that it produces FGF (Christen and Slack, 1997). Many studies on limb development (Martin, 1998) have shown the importance of FGF (mainly FGF-2, -4, and -8) in stimulating limb outgrowth, and subsequent research on urodele limb regeneration has also demonstrated the need for production of FGF by the wound epidermis. The finding of the expression of FGF-8 by the wound epidermis of nonregenerating frog limbs (Endo and Ide, personal communication, cited in Muller et al., 1999, p. 411) provides a possible explanation for the earlier results obtained by Carlson (see preceding paragraph), but they also suggest that the inability of frog limbs to regenerate is more complex than simply the lack of a supportive wound epidermis. A recently described FGF-20a null mutant in zebrafish, *dob* (devoid of blastema), is characterized by the absence of fin regeneration (Whitehead et al., 2005). This mutant is characterized by defective epidermal migration after fin amputation, leading to a greatly thickened wound epidermis without the characteristic regular basal layer of cuboidal epidermal cells. The basal layer of the *dob* wound epidermis did not express *Lef1,* a transcription factor downstream of Wnt, or *sparc,* a matrix protein. Interestingly, however, the FGF-20a that is missing in this mutant is normally expressed in mesenchyme at the epidermal–mesenchymal junction. This study suggests that the normal basal layer of the wound epidermis produces and secretes something that is necessary for the initiation of regeneration, but it is possible that this property of the wound epidermis is stimulated by signals coming from underlying mesenchymal cells.

We are still far from understanding the role of the wound epidermis in the initiation of limb regeneration. The wound epidermis is certainly necessary, but further research is required to determine whether some intrinsic property of the epidermis represents the initial trigger for epimorphic regeneration or whether some signal from cells underlying the wound epidermis conditions it to fulfill its critical role in supporting dedifferentiation and blastema formation.

SUMMARY

Almost any time a wound disrupts the continuity of an epithelium, the epithelial cells at the margins of the wound mobilize and migrate to cover the wound. Although the initial stimulus for epithelial migration is not completely defined, the migrating epithelium is responsive to a variety of growth factors emanating from the wound and to the nature of the substrate on which it is migrating. On completion of migration of the epidermis in a skin wound, the cells resume mitotic activity and produce growth factors, which act on the underlying tissues. In amputated amphibian limbs, epidermal healing looks the same regardless of whether limb regeneration will ensue. How the wound epidermis is involved in stimulating or allowing limb regeneration is still little understood, although it is known to produce both histolytic agents and growth factors that may help to set up the subsequent phase of dedifferentiation. What is clear is that in the absence of a wound epidermis, epimorphic regeneration will not occur.

CHAPTER 4

Role of the Substrate in Regeneration

The extracellular matrix represents nature's scaffold for tissue development and repair.

—Stephen Badylak (2002)

Regenerative processes are heavily dependent on interactions between the cells that participate in regeneration and the substrate on which they function. Interestingly, a major distinction appears to exist in substrate relations between tissue and epimorphic regeneration. In tissue regenerative processes, specific substrate requirements must often be met for regeneration to proceed. This is true for both natural regeneration and induced regeneration, the latter of which often involves purely artificial or processed tissue substrates. In contrast, it is becoming apparent that a precondition for successful epimorphic regeneration is the removal of many existing substrate constraints in the area where dedifferentiation or blastema formation will occur. Nevertheless, the most important generalization is: For any regenerative process the relationship between the regenerating cells and their substrate is critical. That relationship can mean the difference between true regeneration and a scarring response. This chapter examines the role of natural substrates in several regenerating systems. The use of artificial substrates in controlled regenerative processes is covered in more detail in Chapter 13.

PRINCIPLES OF CELL–SUBSTRATE INTERACTIONS

The relation between a cell and its substrate is important for both the maintenance of normal tissue functions and the migrations or changes in state that are part of healing and regenerative processes. The word *substrate* itself merits some discussion. As normally used, it has definite biomechanical connotations. One thinks of basal laminae, fibrin clots, or fibrillar components of the extracellular matrix (ECM). In certain systems, however, the substrate can be other cells or even tissue fluids. It is becoming increasingly apparent that a major function of many natural substrates is the binding of growth factors or other biologically active molecules.

One of the principal ways in which cells interact with their substrates is through cell-surface adhesion molecules. These membrane-bound glycoprotein molecules mediate attachments between cells and other cells or components of the ECM. There are four major classes of cell adhesion molecules (CAMs) (Table 4-1), but there may be many members within each class. For example, the integrins consist of heterodimers,

TABLE 4-1

Classes of Cell-Surface Adhesion Molecules

Class	Interaction type	Ca^{++} or Mg^{++} involvement	Ligand
Cell adhesion molecules (CAMs)	Homophilic/ heterophilic	No	Other CAMs
Cadherins	Homophilic	Yes	Identical cadherins
Selectins	Heterophilic	Yes	Sialyl structures, in blood vessels only
Integrins	Heterophilic	No	Various: collagen, fibronectin, vitronectin, laminin, thrombospondin, complement, fibrinogen, von Willebrand factor

with various combinations of α and β subunits. With at least 15 forms of α subunits and 8 forms of β subunits, the number of possible variants of integrin dimers is immense.

Some classes of cell-surface adhesion molecules, for example, the cadherins and CAMs, mediate cell–cell interactions, whereas the integrins are principally involved in adhesive interactions between cells and components of the ECM. In some systems, for example, regenerating axons, the specific nature of the substrate and the attachment of the cells to it can make an enormous difference in the success of the regenerative process. Other components of the ECM, such as hyaluronic acid and other water-binding molecules, can have a more general influence on cellular behavior in regenerating systems. In both embryonic and regenerating systems, the accumulation of hyaluronic acid is typically a prelude to events that involve cellular migration.

The importance of nanoscale interactions between cells and their substrates is becoming increasingly apparent (Stevens and George, 2005). Cells are bound to their substrates by focal adhesions between receptors and specific elements of the substrate, for example, the RGD (Arg-Gly-Asp) sequence on fibronectin and other matrix molecules. Factors as subtle as the nanometer spacing between different receptors can significantly influence the activity of cells such as osteoblasts (Benoit and Anseth, 2005). Similarly, exquisite mechanisms of sensing the stiffness of the substrate through interactions between focal adhesion points and substrate-bound ligands can strongly affect the differentiation of regenerating cells (Discher et al., 2005). Even mechanical deformation of the substrate can expose or hide substrate-bound growth factors or other biologically active molecules.

The substrate, however, is not simply a mechanical matrix. A variety of growth factors are bound to matrix molecules until they are released through the enzymatic activities of cells in the vicinity (Taipale and Keski-Oja, 1997) (Table 4-2). Others remain bound and require local interactions with cells to exert their effects. Because of the powerful effects of many of the growth factors, their binding to the substrate is

TABLE 4-2

Examples of Growth Factors and Cytokines That Are Bound to Components of the Extracellular Matrix

Bound to heparin/heparan sulfate proteoglycans	Bound to chondroitin sulfate chains
EGF	Platelet factor 4
FGF-1 to FGF-9	
GM-CSF	**Bound to matrix proteins**
HB-EGF	Collagen type IV: TGF-β, BMP-2, BMP-7
HGF/SF	Decorin: TGF-β
IL-2, -3, -4, -6, -8	Fibrin: TGF-β
IP-10	Fibronectin: TGF-β, TNF-α
KGF	IGF-binding proteins: IGF-1
MCP-1	Proteoglycan core proteins: TGF-β
MK	SPARC: PDGF-AB, PDGF-BB
NT-6	Thrombospondin: TGF-β
PDGF-A, -B	160-kDa protein: LIF
Platelet factor-4	
Purpurin	
Schwann cell growth factor	
TGF-β	
VEGF	

EGF, epidermal growth factor; FGF, fibroblast growth factor; GM-CSF, granulocyte-macrophage colony-stimulating factor; HB-EGF, heparin-binding epidermal growth factor; HGF/SF, hepatic growth factor, also known as scatter factor; IL, interleukin; IP-10, interferon-inducible protein-10; KGF, keratinocyte growth factor; LIF, leukemia inhibitory factor; MCP, monocyte chemotactic protein; MK, megakaryocyte; NT, neurotrophic factor; PDGF, platelet-derived growth factor; TGF, transforming growth factor; VEGF, vascular endothelial growth factor; TNF, tumor necrosis factor; BMP, bone morphogenetic protein; IGF, insulin-like growth factor; SPARC, secreted protein rich in cysteine, an extracellular matrix protein. Adapted from Adams, J.C., and F.M. Watt. 1993. Regulation of development and differentiation by the extracellular matrix. *Development* 117:1183–1198; Gailit, J., and R.A.F. Clark. 1994. Wound repair in the context of extracellular matrix. *Curr. Opin. Cell Biol.* 6:717–725; and Taipale, J., and J. Keski-Oja. 1997. Growth factors in the extracellular matrix. *FASEB J* 11:51–59, by permission.

a means of ensuring their availability in specific regions when called on, but also protects the nearby cells from their indiscriminate activities. The specificity can be such that fibroblast growth factor receptor 1 (FGFR-1), for example, is concentrated in focal adhesion points on cells (Plopper et al., 1995). The substrate serves as a medium for the transmission of local chemical signals, whether bound growth factors or paracrine signals secreted from nearby cells. Under other circumstances, for example, axonal elongation, signaling molecules bound to the substrate serve as determinants of the pathway of the outgrowing axon.

The physical nature of the substrate plays a significant role in regeneration. In the 1930s, Weiss (1933, 1934) demonstrated the importance of orientation of elements of the substrate, whether created by lines of mechanical tension on a plasma clot or scratches in the bottom of a tissue culture dish, on the direction of outgrowth of neurites *in vitro.* Research conducted both *in vivo* and *in vitro* has shown the importance of the reaction of the substrate to tension in both the differentiation and morphogenesis of regenerating and developing muscle (Carlson, 1972b; Vandenburgh, 1987). At a much lower level of organization, differences in nanotopography result in changes in the number and distribution of cellular microsensors, such as filopodia and microspikes, which protrude from the surfaces of cells (Curtis et al., 2004). Different cell types have different reactions to the physical nature of their substrate. These may be reflected in the varying orientation of fibrils in the matrix. For example, the overall orientation of fibrils in normal skin is meshlike. This is in striking contrast to the parallel fibers in a ligament or tendon, the orthogonal lattice in the cornea, and the concentric woven cylinders in bone.

An important consideration, especially for regeneration in mammals, is that the nature of the substrate, or microenvironment, is never constant. It changes with the individual's age, hormonal status, health and nutritional status, or macroenvironment, as well as many other factors. Tissue and cellular regenerative responses are particularly sensitive to the microenvironment. Epimorphic regeneration requires extensive removal of the original mechanical substrate near the amputation surface, but little is known about further specific substrate requirements, other than that an environment suitable for epidermal–mesodermal interactions and cell migration must be present.

THE SUBSTRATE IN EPIDERMAL WOUND HEALING

A simple type of epidermal wounding is the removal of epidermal cells over an essentially intact basal lamina (see Chapter 3). In this case, the substrate consists of the normal components of basal laminae, such as type IV collagen and laminin. In more severe and more typical skin wounds, the region of damage extends into the dermis, and the substrate for epidermal healing consists of injured dermis (later granulation tissue) covered by a scab consisting principally of clotted blood. Because the integrins on the migrating epidermal cells bind to matrix components of the dermis, rather than to fibrinogen and other components of the clot, the sheet of wound epidermis grows between the viable elements of the dermis and the overlying clot. Each element of this complex substrate interacts with the migrating epidermis in a different way.

Cellular components of the clot, principally platelets and macrophages, produce the growth factors that affect the migrating epidermis (epidermal growth factor [EGF], heparin binding EGF, transforming growth factor-α [TGF-α], and TGF-β), as well as the underlying vasculature (vascular endothelial growth factor [VEGF], fibroblast growth factor [FGF]) and fibroblasts (platelet-derived growth factor, TGF-β). In turn, the migrating wound epidermis, as well as the myofibroblasts and microvasculature in the damaged dermis and subsequent granulation tissue, all produce a variety of matrix metalloproteinases (MMPs) and plasminogen activators that act to dissolve the elements of the clot (Fig. 4-1). Plasminogen activators, both tissue-type and urokinase-type, catalyze the conversion of plasminogen in the clot to plasmin, a serine protease that degrades fibrinogen or fibrin into soluble breakdown products (Li et al., 2003b).

In the injured dermis, which is the viable substrate for epidermal wound healing, the initial substrate consists of a serum-rich, inflamed, connective tissue, which within days becomes converted to a highly vascularized granulation tissue. The serum-rich environment of the wound during the early phase of epidermal cell migration is inhibitory to the migration of dermal fibroblasts and endothelial cells, but when the fluid environment returns to one that is plasma based (with low levels of TGF-β3), then the fibroblasts and endothelial cells move in and rapidly begin to form granulation tissue (Bandyopadhyay et al., 2006). In addition to the secreted growth factors that act on the migrating epidermis (see Chapter 3 for details), components of the injured dermis secrete a variety of angiogenic factors (mainly VEGF and FGF) that stimulate

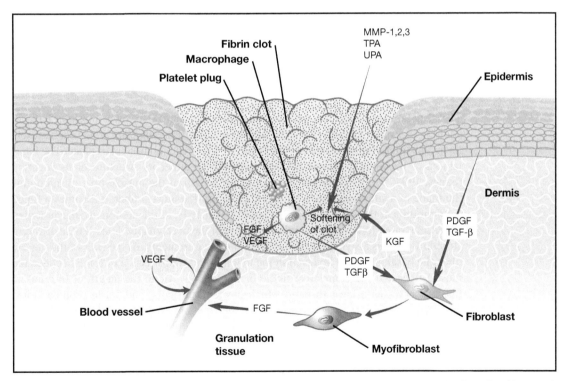

FIGURE 4-1   Expression of some major signaling molecules during the healing of a skin wound. Elements of the clot produce factors that stimulate angiogenesis, fibroblast activation, and epidermal cell activation, as well as soften the clot to permit epidermal migration through it. FGF, fibroblast growth factor; KGF, keratinocyte growth factor; MMP, matrix metalloproteinase; PDGF, platelet-derived growth factor; TGF, transforming growth factor; TPA, tissue plasminogen activator; UPA, urokinase plasminogen activator; VEGF, vascular endothelial growth factor.

the development of a rich microvascular network, as well as factors that convert fibroblasts into the myofibroblasts that are important in later contraction of the wound. Both the scab and the epidermis produce growth factors that contribute to these processes. Bundles of type I collagen also constitute an element of the tissue substrate for epidermal healing, and MMPs (some of which are collagenases) secreted by cells in both the epidermis and dermis facilitate the local remodeling of the fibrous component of the ECM.

The keratinocytes must change the expression of their integrins and attachment specializations to correspond to the environment in which they find themselves to effect epidermal healing. Initially, they have to divest themselves of the hemidesmosomes that bind the basal cells to the underlying basal lamina and the desmosomes that connect keratinocytes to their neighbors. They then need to reconstitute their expressed integrins to correspond to the raw dermal connective tissue over which they migrate. Finally, when epidermal healing is complete, the epidermal cells synthesize their own substrate (the underlying basal lamina) and reform the hemidesmosomes and integrins that firmly attach the epidermis to its immediate substrate.

THE SUBSTRATE IN THE REGENERATION OF A MUSCLE FIBER

The skeletal muscle fiber is encased in a basal lamina of its own making. This, in addition to a small amount of endomysial connective tissue that surrounds each muscle fiber, constitutes the substrate on which muscle fiber regeneration takes place. The basal lamina of the muscle fiber is not homogeneous in either its structure or its function (Sanes, 1994, 2003). Throughout much of the length of the muscle fiber, it maintains the shape of a simple cylinder, but at either end of the muscle fiber, it assumes a fluted configuration at the myotendinous junctions. Similarly, at the neuromuscular junction, the basal lamina follows the complex contours of the postsynaptic folds. These different structural regions also have different functional requirements. At the ends of the muscle fiber, the basal lamina is involved in the longitudinal transmission of force between the muscle fiber and the connective tissue of the tendon. Along the length of the muscle fiber, it serves as a layer through which oxygen, nutrients, and waste products must pass, and it also serves as a medium for the transmission of lateral forces within a contracting muscle. Finally, at the neuromuscular junction, the basal lamina is a barrier through which the neurotransmitter (acetylcholine) must pass. The enzyme acetylcholinesterase is bound to the basal lamina at the neuromuscular junction. In addition, the junctional basal lamina contains information that guides the differentiation of both the muscular and the neural sides of the neuromuscular junction (Burden et al., 1979; Sanes et al., 1978).

The muscle fiber basal lamina, which is a component of the overall basement membrane complex, contains the typical components of basal laminae, such as type IV collagen, laminin, proteoglycans, and assorted glycoconjugates. Many isoforms of laminin are differentially distributed within the basal lamina regions of the neuromuscular junction, the myotendinous junction, and the remaining extrajunctional areas (Patton, 2000). The components of the junctional basal lamina have been characterized extensively. Together with the usual components, it contains molecules such as acetylcholinesterase, which breaks down the transmitter, acetylcholine; ARIA (acetylcholine-receptor-inducing activity), which induces the synthesis of acetylcholine receptor subunits; agrin, which induces the clustering of acetylcholine receptor subunits; and laminin β_2 (s-laminin), which is a determinant of presynaptic differentiation of the motor nerve terminal (Ruegg, 1996).

After damage, the persisting basal lamina of the original fiber provides a microenvironment in which the regeneration of a new muscle fiber(s) is facilitated. Despite its persistence, the surviving basal lamina loses a significant number of its constituents after the muscle fiber is damaged. Gulati and coworkers (1983) have charted the disappearance of major components of the basement membrane after muscle damage. During the degenerative phase, fibronectin disappears first from the pericellular region. This is followed by the removal of laminin, type IV collagen, heparin sulfate proteoglycan, and then type V collagen from the basal lamina itself. Gulati (1985) speculates that enzymatic activity from either macrophages or myoblasts may be responsible for the removal of these components of the basal lamina. As muscle regeneration proceeds to the myotube stage, the myotubes secrete a new basal lamina around themselves, but this activity typically occurs within the confines of the persisting original basal lamina. The new basal lamina appears to contain all the normal components (Gulati, 1985; Kamińska and Fidziańska, 1990).

During the early days after muscle fiber damage, the persisting basal lamina serves as a semiselective cellular filter, which, for the most part, keeps satellite cells within it and keeps fibroblasts on the outside, but allows the passage of a massive number of macrophages, which remove the debris of the degenerating muscle fibers. Cellular activity, as well as properties of the basal lamina itself, may play a role in the selective cellular filtering of the basal lamina. During normal muscle fiber regeneration, collagen is never deposited inside the original basal laminae, but if regeneration of muscle fibers is prevented, large numbers of collagen fibers are deposited in some empty basal lamina tubes within 10 days (Carlson and Carlson, 1991). The early stages of muscle fiber regeneration are heavily dependent on the actions of growth factors, such as FGFs and TGFs, both to stimulate mitosis of myogenic cells and to promote fusion of myoblasts. Although these are known to be bound to extracellular matrices, there is currently little information on the degree to which important growth factors are bound to the muscle fiber basal laminae versus other elements of the endomysial connective tissue.

The information content of the basal lamina, especially at the neuromuscular junction, also plays a significant role in muscle regeneration. In the absence of motor innervation, regenerating muscle fibers still form the postsynaptic apparatus (junctional folds, aggregated acetylcholine receptors) beneath the persisting junctional basal lamina (Burden et al., 1979). Conversely, if muscle fibers are prevented from regenerating after damage, regenerating motor nerve terminals still selectively seek out the region of the junctional basal lamina to settle down (Sanes et al., 1978). In these instances, the informational molecules at the neuromuscular junction obviously survive within the basal lamina despite the loss of many other components.

One of the most important functions of the muscle fiber basal lamina is that of a mechanical scaffold (Vracko, 1974). The most impressive example of this is seen in minced muscle regeneration (Carlson, 1972b). When a muscle is minced into 1-mm^3 fragments, the organization of the fragments is totally random. This can be seen in the orientation of the regenerating myotubes several days after mincing (Fig. 4-2, A). Yet, within a couple of weeks, the normal mechanical factors operating within a minced muscle regenerate cause the regenerating muscle fibers to become oriented parallel to lines of tension within the regenerating muscle (see Fig. 4-2, B). Despite the early influence of the basal laminae, their presence is apparently not necessary for muscle fiber regeneration. Caldwell and colleagues (1990) injured muscle and removed basal laminae by trypsin injections and found that, despite a degree of disorganization during the early phases of regeneration, there was, in the long term, relatively little morpho-

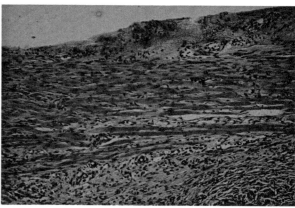

B.

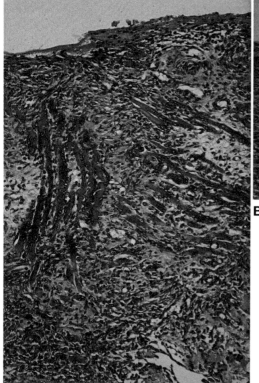

A.

FIGURE 4-2   The reorientation of regenerating muscle fibers after mincing the gastrocnemius muscle of a rat. In both *A* and *B*, the long axis of the muscle extends from left to right. *(A)* Five days after mincing, the regenerating myotubes are randomly oriented according to the original disposition of the minced pieces of muscle. *(B)* Nine days after mincing, the regenerating muscle fibers at the periphery of the mince are becoming oriented parallel to the long axis of the regenerate.

logic difference between muscles that regenerated in the presence or absence of basal laminae.

THE SUBSTRATE IN AXONAL REGENERATION

Axonal regeneration is a unique phenomenon because the process involves the extension of a highly elongated part of a single cell along a long length of substrate. To complicate matters further, axons can become severed or otherwise interrupted in two main components of the nervous system—the central (CNS) or peripheral nervous system (PNS)—and the reactions of the regenerating axons differ dramatically depending on the part of the nervous system in which the damage occurred. Under normal circumstances, axons regenerate well in the PNS but poorly in the CNS. Research since the mid-1980s has uncovered a great amount of information that provides at least a partial explanation for this difference. Because this knowledge is of great practical importance, it is now being translated into techniques designed to improve the efficacy of axonal regeneration in both the PNS and CNS.

With the dichotomy of regenerative responses to axonal damage in the CNS and PNS, a fundamental question concerned whether this difference was due to an intrinsic property or properties of the central and peripheral neurons or whether it could be accounted for by the environment in which axonal regeneration was taking place. The most clear-cut answer to this question came from the experiments of Aguayo (David and Aguayo, 1981), who made lesions in the CNS but provided the axotomized neurons the environment of a peripheral nerve in which to regenerate. The new substrate, which consists of segments of peripheral nerve, supported a remarkable degree of central axonal regeneration.

Regeneration of A Peripheral Nerve

In a normal myelinated peripheral nerve, the axon is situated within a sheath of rolled myelin sheets, derived from extended processes of Schwann cells (Fig. 4-3). The myelin sheath is arranged in segments along the axon, with each segment consisting of the tightly rolled lamellae of a single Schwann cell. Between segments are nodes of Ranvier in which the myelin is absent. Surrounding the Schwann cells and the myelin sheath is the basal lamina of the Schwann cells. This basal lamina is continuous along the course of the axon, and in the regions of the nodes it represents the principal noncellular substrate of the axon. Nonmyelinated axons are embedded directly within Schwann cell cytoplasm.

During normal outgrowth of an axon, contact between the axon and its substrate (Schwann cells or the basal lamina) is actively mediated by receptors on the axons that attach to laminin in the basal lamina or to N-CAM, L1, or N- and E-cadherin in the Schwann cells (Ide, 1996). When a myelinated axon reaches a stable configuration, most of the active receptor–substrate interactions cease, but on damage and early

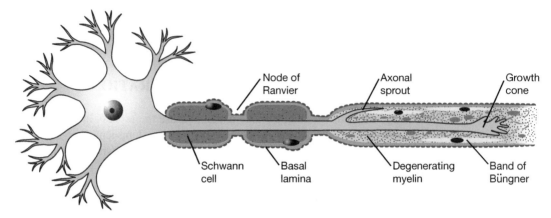

FIGURE 4-3 Regeneration of an axon in a damaged peripheral nerve. The regenerating axon (blue) sends out sprouts and a terminal growth cone into areas associated with dedifferentiating Schwann cells, which form bands of Büngner, and preexisting basal laminae.

regeneration, the appropriate molecules are reexpressed on the plasma membrane of the axon. In nonmyelinated axons, these connections are actively maintained even under equilibrium conditions.

Interruption of a myelinated axon sets in motion a specific sequence of events. Very early, the body of the Schwann cell separates from the myelin sheath, but the overlying basal lamina remains intact. This shedding of the myelin sheath has been described as a form of dedifferentiation of the Schwann cell, which appears to be driven by the Ras/Raf/ERK (extracellular signal–regulated kinase) signaling pathway (Harrisingh et al., 2004). Macrophages move in through the persisting basal lamina and begin to engulf the separated myelin sheath material, which is inhibitory to axonal extension. The macrophages also secrete mitogens, which stimulate the Schwann cells to enter the cell cycle and undergo mitotic division. After proliferation, the Schwann cells extend thin bands of cytoplasm, called *bands of Büngner* or *Schwann cell columns.* At one level, these bands serve as guides for regenerating axons. The axonal sprouts arise from nodal (Ranvier) regions, and several sprouts may arise from one axon. The axonal sprouts up-regulate their various integrins and adhesion molecules and begin to extend into the space between the Schwann cell columns and the basal lamina tube that originally surrounded the axon and its sheathing (see Fig. 4-3). During the extension phase, connections between the regenerating axon and the surrounding substrate (basal lamina and Schwann cell columns) are critical for the success of axonal regeneration. Although normally an important participant in axonal regeneration, Schwann cells are not absolutely necessary, because axonal regeneration can proceed in the absence of Schwann cells if the basal lamina remains intact (Fugleholm et al., 1994).

Another critical role for the substrate is the production or release, or both, of growth factors and trophic substances. Both the basal lamina and the Schwann cells play a role. One of the functions of the basal lamina is to sequester basic FGF on its intrinsic proteoglycans after production of FGF by the Schwann cells. FGF exerts trophic effects on the outgrowth of regenerating axons, and the use of the basal lamina as a physical substrate by regenerating axons ensures ready access to this growth factor. The Schwann cells themselves, in addition to producing Schwann cell columns to guide outgrowth, also produce greatly increased amounts of nerve growth factor (NGF) and brain-derived neurotrophic factor (BDNF). Such synthetic activity occurs after the Schwann cells are stimulated by the interleukin-1 of the invading macrophages to undergo pro-liferation. In the case of NGF, levels increase to 10 to 15 times normal within 24 hours after axotomy and remain elevated for at least 2 weeks (Heumann et al., 1987). Levels of BDNF, which support the regeneration of motor as well as sensory and sympathetic neurons, increase within 3 to 4 days and reach a maximum in about 4 weeks. After axonal elongation is completed, the production and availability of growth factors return to normal equilibrium values.

On the negative side, axonal regeneration, even in the PNS, is retarded by certain elements of the substrate. Although most myelin in the distal nerve segment is removed by macrophages, any remaining myelin can act as a local inhibitor of regeneration (see p. 84). Certain proteoglycans, especially the chondroitin sulfate or keratin sulfate side chains, also act as retardants of regeneration (Höke, 2005). Axonal regeneration, especially in peripheral nerve grafts, is facilitated if these side chains are removed enzymatically.

In vivo studies on crushed nerves have shown a remarkable specificity in the return of regenerating motor axons to their original end-plate regions on muscle fibers (Nguyen et al., 2002). In mice bearing yellow or cyan fluorescent genes linked to a fragment of mouse *thy1* gene that promotes neuron-specific expression, more than 95% fidelity in the accuracy of regeneration was noted after crush, as opposed to less than 5% in axons regenerating after transection. Such experiments show the importance of the mechani-cal integrity of the endoneurial tubes (basal laminae) in guiding the regenerating axons, but they also show great selectivity in choosing the correct branch point even at some distance from the muscle fiber.

When a peripheral nerve is transected, the two cut ends typically spread apart even after a simple lesion. Often after trauma, segments of the nerve are destroyed. In the clinical repair of nerve lesions, attempts to join the two cut ends directly have com-monly been unsuccessful because of the considerable tension that is imposed on the nerve. This has led to many attempts to devise suitable substrates, both natural and artificial, that could serve as conduits that would allow regenerating axons to bridge the gap created by the lesion. Implicit in the rational design of such substrates is the provision for suitable mechanical properties, as well as the addition of the growth and trophic factors that can maintain the momentum of the regenerating axons.

Regeneration of Axons in the Central Nervous System

For decades it had been assumed that axonal regeneration in the CNS was minimal because of intrinsic properties of the neurons themselves or because regeneration was blocked by a glial scar that formed at the site of injury. David and Aguyao (1981) added a new dimension to this field by demonstrating that central axons can regenerate well when provided with the environment of a peripheral nerve graft (Fig. 4-4). With that conceptual breakthrough, investigators began focusing on what there is in the environment of the CNS that inhibits the regenerative outgrowth of axons.

Persistence of inhibitory breakdown products of myelin was one of the foci of attention (Berry, 1982). Caroni and Schwab (1988) reported that neutralizing breakdown products of myelin by antibodies facilitates axonal regeneration. Ten years later, several groups identified the antigen, which was christened Nogo (rev. Filbin, 2003; Schwab, 2004). Subsequently, two other myelin-associated inhibitors of axonal extension, myelin-associated glycoprotein (MAG) and oligodendrocyte-myelin glycoprotein (OMGP), were identified (Spencer et al., 2003). All three of these myelin-associated inhibitors bind to the same receptor on the growth cone (Hunt et al., 2002) and cause a collapse of the growth cone (see Fig. 14-4). Nogo-A is absent in the spinal cord of fish and salamanders, which have good powers of spinal cord regeneration (Schwab, 2004). OMGP recently has been shown to prevent neurite outgrowth at nodes of Ranvier in the CNS (Huang et al., 2005), whereas in the PNS, most axonal sprouting occurs in the nodal region.

The myelin in the sheaths of peripheral nerves is also inhibitory to regeneration, but the massive infiltration of macrophages into a damaged nerve, together with phagocytic activities of Schwann cells, quickly removes the myelin. In the CNS, the influx of macrophages into regions of damage is less than that of peripheral nerves. In addition, oligodendrocytes both continue to produce myelin after injury and do not phagocytize myelin debris. These factors account in large part for the persistence of the myelin.

Myelin-associated inhibitory compounds are not the only substrate-related factors that discourage regeneration in the CNS. There are indications that the basal laminae in the CNS contain less laminin and interact less well with axonal adhesion molecules than those in the PNS. Reduced release of neurotrophic molecules (molecules, such as NGF and BDNF, that keep outgrowing axons alive before they have connected with their targets) by the substrate may also be a factor.

One issue that has received relatively little attention is why the presence of regeneration-inhibitory molecules in the myelin of the CNS has been preserved evolutionarily. One option is that once the various tracts have been set up in the CNS during

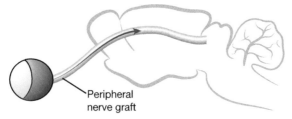

Peripheral
nerve graft

FIGURE 4-4   Regeneration of optic nerve fibers (red arrow) into a graft of peripheral nerve inserted into the central nervous system of a rat. (After David and Aguayo [1981].)

ontogenesis, these myelin-based inhibitory compounds promote anatomic and functional stability of the tract structure within the brain and spinal cord.

It is now apparent that earliest inhibition of axonal extension in the CNS occurs as the result of the effects of the myelin-associated inhibitors discussed earlier. A second level of inhibition occurs after a glial scar has formed. Glial scars are physical barriers, but molecules such as chondroitin sulfate proteoglycans and tenascin, which are produced by the reactive astrocytes and oligodendrocytes, are poor substrates for axonal extension.

A number of the research findings on physical factors that interfere with regeneration in the CNS have already been translated into experimental clinical procedures designed to improve axonal growth (Selzer, 2003). These include the use of peripheral nerve grafts as bridges across a lesion, the activation of macrophages, the enzymatic digestion of inhibitory molecules or antibodies against these molecules, and the addition of various trophic factors. Many older methods had concentrated on reducing the impact of the glial scar on regeneration. The rejuvenated field of CNS regeneration (see p. 290), which involves not only manipulation of the substrate but also the application of stem cells, remains in an embryonic state with respect to major human applications.

THE SUBSTRATE IN EPIMORPHIC REGENERATION

When a vertebrate appendage is amputated, the transection disrupts a number of cell–substrate relations at the plane of amputation. Immediately after amputation, the deep tissues of the limb are directly exposed to the outside medium. After epithelial wound healing, neither a basal lamina nor the dermis intervenes between the wound epidermis and the underlying tissues. Beneath the skin, the basal laminae covering any muscle fibers and nerve axons are transected, leaving cross sections of damaged axons and muscle fibers exposed to the fluids and clot material that collect at the amputation surface. Amputated blood vessels pour out blood until constriction of the vessel walls and clotting stem the flow of blood.

Once the amputation surface is covered by a wound epithelium, the next step in limb regeneration is characterized by the removal of many existing elements of the ECM, as the process proceeds toward dedifferentiation and blastema formation (Hay, 1970). Then, as tissue differentiation begins within the blastema, the dynamics change toward building up an ECM that is characteristic of the stable mature limb. Needham (1952, 1960) has divided epimorphic regeneration into two main phases: a phase of regression and a phase of progression. The regressive phase is associated with decreasing pH, reduced oxidation-reduction potentials, increased activity of lytic enzymes, and reduced activity of oxidative enzymes. As the regressive phase is taken over by the progressive phase (concomitant with growth of the blastema), these conditions are reversed and the overall metabolism gradually returns toward what is normal in the stable limb.

Throughout the periods of phagocytosis, dedifferentiation, and early blastema formation, the wound epidermis is not underlain by a basal lamina, thus facilitating interactions between the epidermis and the underlying tissues. Several studies suggest that the wound epidermis itself is a source of lytic enzymes (Eisen and Gross, 1965; Schmidt, 1966), and that the lack of a subtending matrix would facilitate diffusion of the enzymes. Another function of the early wound epidermis is physical removal of tissue debris through an encirclement mechanism (see p. 66).

Increases in the activities of a number of types of enzymes that degrade ECM during the regressive phase of limb regeneration have been noted (Kato et al., 2003; Schmidt, 1968; Stocum, 1995). These enzymes include the MMPs (e.g., collagenases, gelatinases, stromelysins; Table 4-3), serine proteases (e.g., plasmin), and acid hydrolases (e.g., cathepsin D, acid phosphatase, β-glucuronidase, carboxylic ester hydrolases). One of the earliest enzymes expressed (within 24 hours in the newt) is NvTIMP-1 (tissue inhibitor of metalloproteinase-1), which regulates the activities of the MMPs (Stevenson et al., 2006). This enzymatic activity removes many of the components of basement

TABLE 4-3

Matrix Metalloproteinases

MMP number	Enzyme	Main substrates
MMP-1	Interstitial collagenase	Fibrillar collagen
MMP-2	Gelatinase-A	Fibronectins, type IV collagens
MMP-3	Stromelysin-1	Nonfibrillar collagen, laminin, fibronectin
MMP-7	Matrilysin	Nonfibrillar collagen, laminin, fibronectin
MMP-8	Neutrophil collagenase	Fibrillar collagens (types I, II, and III)
MMP-9	Gelatinase-B	Type IV and V collagens
MMP-10	Stromelysin-2	Nonfibrillar collagen, laminin, fibronectin
MMP-11	Stromelysin-3	Serpin
MMP-12	Metalloelastase	Elastin
MMP-13	Collagenase-3	Fibrillar collagens
MMP-14	MT-MMP	Progelatinase A
MMP-15	MT2-MMP	Not defined
MMP-16	MT3-MMP	Progelatinase A
MMP-17	MMT4-MMP	Not defined
MMP-18	Collagenase-4 *(Xenopus)*	Collagen I
MMP-19	RAS-1	Collagens I and IV, fibronectin, tenascin
MMP-20	Enamelysin	Not defined
MMP-21	XMMP *(Xenopus)*	Not defined
MMP-22	CMMP (chicken)	Not defined
MMP-23	—	
MMP-24	MT5-MMP	Not defined
MMP-25	MT6-MMP	Not defined
MMP-26	Endometase, matrilysin	Collagen IV, fibronectin
MMP-27	—	
MMP-28	Epilysin	Not defined

MMP, matrix metalloproteinase.
Adapted from Rasmussen, H.S. 2000. Matrix metalloproteinase inhibitors in angiogenesis-mediated disorders with special emphasis on cancer. In: S.A. Mousa, ed. *Angiogenesis inhibitors and stimulators.* Georgetown, TX: Landes Bioscience. 124–133; and Sternlicht, M.P., and Z. Werb. 2001. How matrix metalloproteinases regulate cell behavior. *Annu Rev Cell Dev Biol* 17:463–516, by permission.

TABLE 4-4

Composition of Extracellular Matrix Components in Unamputated and Regenerating Urodele Limbs

Extracellular matrix molecule	Unamputated limb	Medium bud blastema
Fibronectin	+ Basal laminae; endoneurium and perineurium; dermal and loose connective tissue	++++ Throughout blastema
Laminin	+ Basal laminae	(−)
Reticulin	+ Basal laminae	?
Collagen I	+ Bone	+
Collagen II	+ Cartilage	(−)
Collagen IV	+ Basal laminae	?
Tenascin	+ Tendons, myotendinous junctions; periosteum; basal layers of epidermis	++++ Throughout blastema
Sulfated GAGs	+ Cartilage; dermis; cytoplasm of basal epidermal cells	+ Throughout blastema
Hyaluronate	+ Epidermis; basal laminae; dermal gland secretions; bone and cartilage matrix	++++ Throughout blastema

ECM, extracellular matrix; GAG, glycosaminoglycan; minus sign indicates absence; number of plus signs indicates relative amount.

Adapted from Stocum, D.L. 1995. *Wound repair, regeneration and artificial tissues.* Austin, TX: R.G. Landes, by permission.

membranes and other elements of the stable ECM and replaces them with molecules, such as tenascin, that have antiadhesive properties (Table 4-4). Osteoclasts and chondroclasts appear on the distal skeletal elements and secrete numerous enzymes that dissolve the matrix around the cells in the skeleton. A substantial increase in hyaluronic acid synthesis is noteworthy (Toole and Gross, 1971), because this molecule is associated with cell migration and rapid change in many developing systems.

As the regressive phase of regeneration comes to a close and the blastema builds up, the activities of many of the lytic enzymes are reduced, some through the activities of specific inhibitors, such as the TIMPs. Then newly organized ECM structures are re-formed (Tassava et al., 1996). The apical epidermis covering the blastema becomes underlain by a new basal lamina in a proximodistal sequence (Neufeld and Day, 1996), and regenerating muscle fibers begin to surround themselves with basal lamina material. As cartilage re-forms, the synthesis of its characteristic matrix components rapidly increases.

SUBSTRATE AS AN INDUCER

A long-recognized, but still controversial, function of the substrate in some regenerating systems is as an inducer. As introduced in Chapter 1 (p. 21), regeneration by induction has been claimed for a variety of tissues (Levander, 1964; Polezhaev, 1977b), but many of the former claims have not been substantiated. Nevertheless, little doubt exists that inductive influences emanating from the substrate are important in bone regeneration, and the induction of bone will serve as the basis for this section.

After an early report of bone induction by implants of urinary bladder mucosa (Huggins, 1931), attention turned toward bone itself as the inducer. During the 1950s, research results in Polezhaev's laboratory (rev. Matveeva, 1959a; Polezhaev, 1972b) showed that extensive defects in cranial bones can be filled in by new bone that was induced by the presence of bone shavings placed above the dura mater (see Fig. 1-9). Although the inducing activity was completely eliminated by autoclaving, it was preserved after lyophilization, thus eliminating live cells as the stimulus for new bone formation.

In the 1960s, Urist's laboratory extended this work by showing that bone matrix decalcified by hydrochloric acid (HCl; but not nitric or nitrous acid) is also a powerful inducer of new bone (Urist, 1965). Subsequent research led to the isolation of a "bone induction principle," which was later purified and christened *bone morphogenetic protein* (Urist et al., 1979). With the subsequent cloning of BMP (Celeste et al., 1990), it became apparent that BMP is a member of the large TGF-β superfamily. Members of that superfamily are often bound to molecules of the ECM, and BMP binds to collagen fibrils within the ECM. BMP is a generic term for a large family of molecules that play important roles in development, especially as inhibitors of other developmentally active molecules. Of the more than 30 members of the BMP family, BMP-2, -4, -6, and -7 have been shown to possess the greatest osteoinductive activity.

Because bone undergoes regular remodeling in response to local mechanical conditions, it is likely that the mechanisms underlying normal remodeling and regeneration are similar. In stable bone, BMP appears to be tightly bound to the organic matrix of the bone. After injury or during remodeling, osteoclasts accumulate on bony surfaces and enzymatically degrade the bone matrix beneath them. This releases the BMP molecules from their substrate and stimulates the induction of new bone in specific microenvironments. Osteoprogenitor cells in the periosteum or stem cells from the marrow are the main target cells in this system, and exposure to BMP sends them on the path of osteoblastic differentiation.

NATURAL SUBSTRATES IN GUIDED TISSUE REGENERATION

A variety of naturally occurring substrates have been used to facilitate tissue repair (Badylak, 2002; Hodde, 2002; Yannas, 2001). These substrates include acellular dermis, cadaveric fascia, amniotic membrane, small-intestinal submucosa (SIS), and urinary bladder ECM. Such substrates have been used both experimentally and clinically to repair defects in tissues as diverse as skin, blood vessels, esophagus, urinary bladder, and body wall. The basic principle is to provide a natural scaffold into which host cells can grow and restore the tissue architecture and physical properties of the original damaged tissue. In many cases, natural substrates have proved to be superior to implantable matrices of artificial materials.

Several features are common to the preparation and composition of natural tissue matrices, which can be either allogeneic or xenogeneic. First, all cells, whether epithelial sheets or individual cells embedded within the matrix, are removed to reduce the antigenicity of the matrix. Certain types of these tissues, such as SIS, are separated mechanically from adjacent layers of muscle; but in other cases, enzymatic separation of the matrix from nondesirable tissue elements is done. Whatever the treatment, an effort is made to maintain mechanical strength while retaining porosity and the presence of bound growth factors that will stimulate angiogenesis and the ingrowth of other host cells after implantation of the matrix. Table 4-5 lists major components of some of the more popular natural matrix scaffolds.

One of the most widely used natural tissue matrices is porcine SIS, which as of 2002 had been implanted in more than 100,000 human patients (Badylak, 2002). In experimental animals and in humans, SIS has been used in situations as diverse as arterial wall repair, urinary bladder reconstruction, esophageal reconstruction, skin grafting, dural repair, hernia repair and ligament reconstruction. An instructive example of the

TABLE 4-5

Components of Naturally Occurring Biopolymer Scaffolds

Scaffold material	Major components
SIS	Collagen types I, III, IV, V, VI
	Proteins: fibronectin, heparin sulfate proteoglycan
	GAGs: hyaluronic acid, heparin, heparin sulfate, chondroitin sulfates A and C, dermatan sulfate
	Growth factors: FGF-2, TGF-β, VEGF
Acellular dermis	Collagen types I, IV, and VII
	Proteins: elastin
	GAGs: not determined
	Growth factors: not determined
Bladder acellular matrix graft	Collagen types I and III
	Proteins: elastin, fibronectin
	GAGs: hyaluronic acid, heparin, heparin sulfate, dermatan sulfate, chondroitin sulfate A
	Growth factors: FGF-2, TGF-β, VEGF
Cadaveric fascia	Collagen types: I
	Proteins: not determined
	GAGs: not determined
	Growth factors: not determined
Amniotic membrane	Collagen types I, III, and IV
	Proteins: decorin
	GAGs: hyaluronic acid
	Growth factors: EGF, FGF-2, HGF, KGF, TGF-α and -β

SIS, small-intestinal submucosa; GAG, glycosaminoglycan; FGF, fibroblast growth factor; TGF, transforming growth factor; VEGF, vascular endothelial growth factor; EGF, epidermal growth factor; HGF, hepatic growth factor; KGF, keratinocyte growth factor.
Adapted from Hodde, J. 2002. Naturally occurring scaffolds for soft tissue repair and regeneration. *Tissue Eng* 8:295–308, by permission.

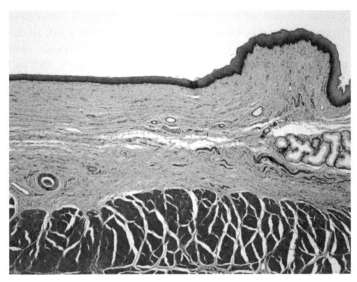

FIGURE 4-5   Regenerated esophageal lining 3 months after implantation of decellularized pig urinary bladder submucosa into an esophageal defect in the dog. The fidelity of tissue restoration is remarkable. (Courtesy S. Badylak.)

use of SIS grafts is the transplantation of porcine SIS or urinary bladder mucosal grafts into patch defects of the esophagus in dogs (Badylak et al., 2000). The typical natural reaction of the esophageal wall to damage is fibrosis and stricturing, rather than regeneration. When such natural matrix material is transplanted into the defects of the esophagus, inflammatory cells and blood vessels begin to invade the graft within a few days. As these events are occurring within the substance of the graft, squamous esophageal epithelium begins to migrate over the surface of the graft. Within a month, small clusters of cells with characteristics of skeletal muscle have appeared. Most traces of the SIS scaffold have disappeared by early in the second month, and a natural connective tissue matrix has taken its place. Later in the second month, the region of the graft becomes completely covered by epithelium, and the skeletal muscle has organized into prominent bundles (Fig. 4-5). One difference from normal in this model is the absence of submucosal glands in the regenerated esophageal segment. This is probably due to the absence in the regenerating tissues of substances that are capable of inducing glandular outgrowths from the healing esophageal epithelium.

WHAT ARE PROPERTIES OF A GOOD NATURAL REGENERATIVE SUBSTRATE?

To a certain extent, the properties of an ideal substrate for regeneration depend on the specific tissue or organ in question, but a set of generic characteristics defines what is

TABLE 4-6
Properties of a Good Natural Tissue Scaffold

1. Easy to obtain
2. Easy to prepare
3. Strong
4. Acellular
5. Sterile, or contains no pathogenic organisms
6. Has properties that allow stable implantation into the tissue defect
7. Permeable to cell and tissue ingrowth
8. Promotes rapid angiogenesis (exception: cornea)
9. Nonantigenic
10. Promotes controlled inflammation appropriate for the tissue
11. Retains intrinsic growth factors and cytokines
12. Contains attachment sites for cell-surface receptors
13. Permits cell–tissue interactions important for regeneration and maintenance of the tissue
14. Resorbable
15. Permits differentiation of cell types appropriate for the tissue
16. Recruits circulating progenitor cells
17. Does not interfere with tissue-specific morphogenesis
18. Nonshrinkable
19. Does not promote scarring response or fibrosis

usually considered to be a good natural substrate (Table 4-6). Virtually all natural regenerative substrates are derived from sources other than the recipient, often xenogenic sources. For human use, porcine materials have proved to be effective. The ideal substrate is easy to obtain and to prepare, the latter including separation from other tissues, removal of cells and other antigenic materials, and sterilization in a manner that does not destroy intrinsic active components such as growth factors. It should be strong and easily implantable.

After implantation, the substrate should allow the infiltration of appropriate inflammatory cells and the ingrowth of blood vessels early in the process. An exception to this is the cornea, where the objective is to discourage inflammation and angiogenesis. On a longer term basis, the implanted substrate should not shrink, but should be resorbable because it is being replaced by host tissues. It should permit the influx of cells and provide conditions for cellular differentiation and overall morphogenesis appropriate for the tissue to be replaced. It should not be replaced with fibrotic tissue.

The ideal substrate for tissue regeneration remains to be found. Currently, several natural substrates appear to possess a number of advantages over artificial ones (see discussion in Chapter 13), but there is room for improvement in all of them. In addition, much more remains to be learned about the specific requirements for successful regeneration of various tissues. This knowledge will, in turn, be instructive in the creation of new substrates and other conditions that are designed to promote regeneration.

SUMMARY

Essentially all regenerative activity occurs on a substrate, but the nature of the interactions between cells and substrates is as varied as the nature of the substrates themselves. In limb regeneration, one of the first activities is to remove any remaining elements of the original substrate at the distal end of the limb stump. Epidermal wound healing occurs through a broad expanse of clot and granulation tissue, whereas regenerating skeletal muscle fibers or peripheral nerve axons rely on the preservation of the basal laminae of the original cells for the most effective regeneration. Both the physical and molecular configurations of the substrate exert a major influence on a regenerative process, and substrates hold a key to the stimulation of cellular regenerative activity through the local release of a wide variety of growth factors that are bound to them. The nature of the substrate is a key component of many tissue-engineering approaches to regeneration, and tissue engineers use a variety of natural and artificial substrates. Natural substrates rely on the ingrowth of cells from the host to accomplish actual regeneration, and the level of restoration is often impressive.

CHAPTER 5

Tissue Interactions in Regeneration

Without a flow of information, very limited changes follow amputation. These are confined to wound healing and scar formation.

—S. Meryl Rose (1970)

Regeneration does not occur in isolation. A fundamental characteristic of regeneration is that the process, regardless of the nature of the structure to be regenerated, is intimately connected to the body of the individual that has sustained some sort of damage. This linkage of regeneration to the rest of the body is most strongly seen in the epimorphic regeneration of limbs, where interactions between the forming blastema and other elements of the limb stump are required to initiate the process. The regeneration of tissues, although not involving the level of interactions required for limb regeneration, nevertheless depends to a surprising degree on connections with surrounding tissues for progression and completion. This chapter highlights the role of tissue interactions in a variety of regenerating systems.

THE AMPHIBIAN LIMB

Tissue interactions in limb regeneration are important from the beginning of the process, starting with epithelialization of the amputation surface. In the absence of a wound epithelium in which epidermis directly abuts underlying mesodermal tissues, regeneration does not begin. This has been demonstrated in several ways. One approach is to prevent epidermis from covering the wound. Thornton (1957) amputated limbs of tiger salamander *(Ambystoma tigrinum)* larvae and removed the wound epidermis each day. Although epithelial covering of the amputation wound does occur within 24 hours, daily removal of the wound epidermis nevertheless prevented regeneration from taking place. Using a different strategy, Goss (1956a) and Poležajew and Faworina (1935) prevented epithelialization by inserting amputated limbs into the body cavity or nearby musculature, respectively. If the insertions were done immediately after amputation, regeneration was completely inhibited. However, if insertion was delayed until the amputation wound had undergone epithelialization, regeneration occurred (Goss, 1956b).

Another approach to the role of the epithelium has been to cover the amputation surface with a flap of full-thickness skin, with dermis intervening between epidermis and underlying damaged mesoderm (Godlewski, 1928; Tornier, 1906). As in the epidermal removal experiments, regeneration was completely inhibited. Subsequent experiments addressed whether epithelia derived from regions other than the limb could support limb regeneration. A variety of experiments consisted of removing the skin from the limb stump and replacing it with skin from other sources. Epithelium derived from skin transplants as diverse as flank and tail supported normal limb regeneration, but if the wound epidermis was derived from head skin, regeneration was inhibited (Efimov, 1933).

The nature of the early epidermal influence on the underlying tissues has still not been determined. For many years, investigators have suspected that through the direct production of lytic enzymes, the epidermis promotes the histolysis that characterizes the prededifferentiation stage of regeneration. Another possibility is that epidermal secretion of stimulatory factors is important in the earliest stages of limb regeneration, as well as during the later phases of blastema formation and outgrowth (Fig. 5-1).

The next critical interaction in limb regeneration involves innervation. Axons sprouting from the stumps of the severed nerves grow into the region of damaged tissue, and

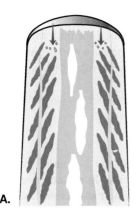

A. Epidermis-driven histolysis

B.
1. Neurotrophic influence on wound epidermis and blastemal cells

2. Blastemal cell influence on wound epidermis

3. Epidermal influence on blastemal cells

C. Blastemal regulation of dedifferentiation

FIGURE 5-1   Tissue interactions in the regenerating amphibian limb. *(A)* Just after initial wound healing, signals from the wound epidermis stimulate histolysis of the underlying mesodermal tissues. *(B)* During early blastema formation, influences from nerves (1) maintain both wound epidermis and blastemal cells. Influences from blastemal cells (2) help to maintain the wound epidermis, and influences from the apical epidermal cap (3) maintain the blastema. *(C)* The presence of the blastema regulates the amount of dedifferentiation in the stump.

sensory nerve fibers even penetrate the wound epidermis. Even before the elaboration of the cell theory, Todd demonstrated that if a limb is denervated, regeneration fails to occur (Todd, 1823). The reaction of an amputated limb to denervation varies, depending on the age of the animal. When limbs of urodele larvae are denervated at the time of amputation, not only do they fail to regenerate, but they undergo regression (uncontrolled dedifferentiation) as far proximally as the shoulder (Butler and Schotté, 1941). In contrast, when limbs of adult urodeles are denervated at the time of amputation, regeneration is completely inhibited, but the limbs do not regress (rev. Singer, 1952). Dedifferentiation either ceases quickly or fails to occur.

Early experiments on larval limbs suggest that the presence of the blastema itself places limits on the extent of dedifferentiation. Schotté et al. (1941) amputated denervated limbs of *Ambystoma* larvae and then transplanted young blastemas from other donors to the amputation surface (Fig. 5-2). The transplanted blastemas prevented the denervated limbs from regressing, leading to the conclusion that the accumulation of a blastema somehow checks the dedifferentiative process. The level of differentiation of the transplanted blastema is important, because if a differentiated blastema

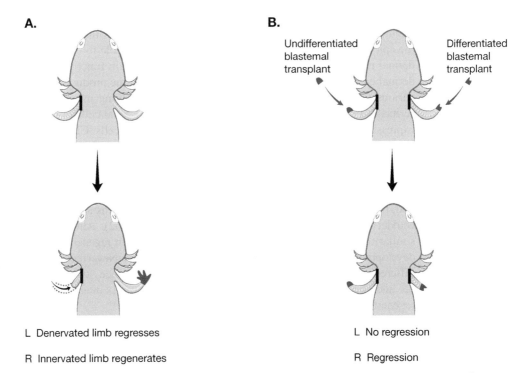

A.

B.

Undifferentiated blastemal transplant

Differentiated blastemal transplant

L Denervated limb regresses

R Innervated limb regenerates

L No regression

R Regression

FIGURE 5-2 Experimental studies on the regression of denervated limbs in larval *Ambystoma*. *(A)* After denervation, an amputated limb regresses. *(B)* If an undifferentiated regeneration blastema is transplanted to the end of an amputated, denervated limb, no regression occurs. In contrast, regression occurs if the blastemal graft is differentiated. (Based on findings by Schotté et al. [1941].)

is grafted to a denervated limb, it cannot suppress the excessive dedifferentiation of the limb.

Further evidence of the importance of nerves in regeneration is the stimulation of the formation of supernumerary limbs by deviating a nerve beneath a skin wound (Guyénot et al., 1948; Locatelli, 1929). The presence of a wound epidermis and attending trauma to underlying tissues is important for this method to succeed. From the types of experiments reported above, Rose (1962) formulated the concept of tissue-arc control of early events in regeneration, in which sprouts from regenerating nerves interact with the wound epidermis, causing the epidermis, in turn, to stimulate the mesodermal tissues beneath to begin the regenerative process. The trophic effect of nerves is not confined to their actions on the wound epidermis. There is considerable evidence for a direct action of nerves on the blastemal cells themselves. The mechanism of the neural influence is covered in detail in Chapter 6. Conversely, there is now evidence that blastemal cells produce a factor(s), different from known neurotrophic factors, that reciprocally stimulates the regeneration of nerve fibers into it (Bauduin et al., 2000; Tonge and Leclere, 2000).

The role of the wound epidermis does not cease with the initiation of dedifferentiation. During dedifferentiation, the wound epidermis greatly thickens, forming what is known as the apical epidermal cap (Thornton, 1956). This structure is required for the maintenance and outgrowth of the regeneration blastema. Although shaped differently, the apical cap in regeneration appears to play a similar role to that of the apical ecto- dermal ridge in the embryonic limb bud. One of the main functions of the apical epi- dermal cap, like that of the apical ridge, is likely to be the production and transport of various fibroblast growth factors (FGFs) to the underlying mesodermal cells. Tran- scripts of FGF-8 have been localized in the columnar basal cells of the apical cap, but expression of FGF-8 is considerably stronger in blastema cells (Christensen et al., 2002; Han et al., 2001). This could stimulate the ingrowth of capillaries into the developing blastema. FGF-4, which is an important secretion of the embryonic apical ectodermal ridge, is absent from the regenerating amphibian limb. However, FGF-10, which is an important mesodermal signaler from mesoderm to ectoderm in the embryonic limb bud, is expressed in blastemal cells, suggesting that it retains a role as a mesoderm-to- ectoderm signal in regeneration (Christensen et al., 2002). Although there are some similarities in morphology and gene expression between regeneration and embryonic development, there are also some significant differences. Nevertheless, despite possible differences in specific mechanisms, the available evidence still favors the existence of and necessity for mutual epidermal–mesodermal interactions during limb regeneration.

Another type of tissue interaction that is little understood is involved in morpho- genesis of the regenerating limb. The importance of morphogenetic tissue interactions is not apparent until the tissues of the limb are surgically manipulated. If certain tissues of the limb stump, such as muscle or dermis, are simply rotated by 180 degrees at the time of amputation, the disruption in tissue interactions results in the formation of grotesque multiple regenerates (Carlson, 1974a, 1975a; Lheureux, 1972). Similarly, if regeneration blastemas are rotated about the stump, multiple limbs of recognizable polarity regenerate (Iten and Bryant, 1975; Tank, 1978). This striking phenomenon is covered in greater detail in Chapter 7.

HUMAN FINGERTIPS AND THE MAMMALIAN LIMB

Fingertip amputations occur surprisingly commonly, especially in young children. For decades, the standard treatment was to graft full-thickness skin over the amputation surface to seal off the wound. This treatment results in a smooth digital stump without a nail. Attitudes toward treatment began to change when a case of distal digital amputation in a child was allowed to heal naturally. To the surprise of many, the fingertip not only healed, but it regenerated a new nail (Fig. 5-3). Although conservative treatment of distal digital amputations was advocated in several early reports (Douglas, 1972; Illingworth, 1974; Polezhaev, 1980), the surgical community was slow to accept this new mode of treatment (Martin and del Pino, 1998). Key parameters to successful fingertip regeneration are amputation past the most distal interphalangeal joint, retention of part of the nail bed, and a young age (regeneration is most successful in children younger than 12 years). Good regeneration is usually complete within 3 months, and regeneration of a new nail is the norm.

The essence of the success of the conservative method of treating such amputations is the allowance of a wound epithelium to grow directly over the damaged underlying tissues at the amputation surface. Treating such a wound with a skin flap essentially repeats in humans the skin flap experiments that had for years been known to completely inhibit limb regeneration in amphibians (see p. 93).

The mechanism by which a fingertip regenerates remains an open question. Singer and colleagues (1987) have investigated digital regeneration in rhesus monkeys. Despite the epithelialization of the wound and the fact that outgrowth occurred, the accumulation of cells beneath the wound epidermis did not resemble a regeneration blastema. In addition, that such regeneration cannot occur proximal to a joint suggests that this reparative process is not the same type that occurs in the amputated amphibian extremity. Nevertheless, it appears that some sort of epidermal–mesodermal interaction(s) is taking place, although there could be more than one tissue center of outgrowth in the fingertip.

Experimentation with digital regeneration in rodents has shed further light on this phenomenon. Essentially complete regeneration occurs if the digit is amputated through the last phalanx. Borgens (1982) confirms that complete regeneration fails to occur if the level of amputation is proximal to the last interphalangeal joint, and that in this case, healing occurs by scarring. Neufeld (1989) notes that at a more proximal level of amputation, the wound epidermis is not underlain by a basal lamina, but fibrin and clot material separates it from underlying cells. Contracture of the skin over the amputation surface after a few days further reduces the area of epidermal mesodermal contact.

If a rodent digit is denervated, regeneration after digit-tip amputation still occurs, in contrast to amphibian regeneration (Mohammad and Neufeld, 2000). Other experimentation by Zhao and Neufeld (1995) demonstrates a dependence of regeneration of

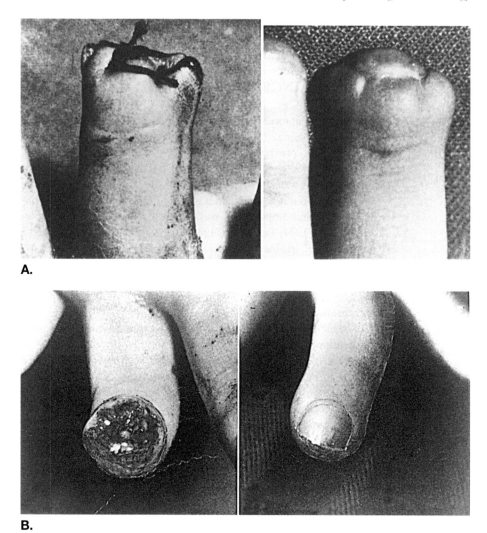

A.

B.

FIGURE 5-3   Human fingertip regeneration. *(A)* Treatment of amputated fingertip in 3-year-old girl by classic skin flap method (left). Four months after the operation (right), the finger has healed with a blunt stump. *(B)* Conservative treatment of an amputated fingertip in a 22-month-old boy. Good regeneration has occurred by 11 weeks after amputation. (Reprinted from Illingworth CM: (1974) Trapped fingers and amputated finger tips in children. *J Pediatr Surg* 9:853–858, by permission.)

the distal bone on an interaction with the nail bed. However, based on *in situ* hybridization studies on regenerating fetal mouse digits, Han and colleagues (2003) note expression of Msx-1 in a zone of fibroblasts beneath the nail bed and wonder whether this tissue, rather than the nail bed itself, is the active tissue element in digital regeneration. Msx-1 is expressed in the mesoderm beneath the wound epidermis of amputated digits that regenerate, and digital regeneration is greatly reduced in Msx-1 mutants (Han et al., 2003). In digital regeneration, Msx-1 regulates the downstream expression of bone morphogenetic protein 4 (BMP-4), without which digital regeneration does not proceed. Despite a significant increase in our understanding of digital regeneration in mammals, it is still premature to make a definitive determination of whether it is an example of epimorphic regeneration of the type that occurs in the amputated amphibian limb.

ANTLER REGENERATION

One of the most spectacular examples of natural regeneration of appendages is the annual replacement cycle of antlers (Fig. 5-4) (Bubenik and Bubenik, 1990; Goss, 1983). In animals with large antlers, such as elk and moose, daily elongation can be as great as 2 cm during the period of maximal growth in June.

Antlers are surprisingly complex structures that consist of two components: the pedicle and the antler itself. The first growth of an antler in a young deer is the formation of two bony pedicles from the periosteum of the frontal bones. This appears to be a process controlled intrinsically by the local periosteum without the need for significant tissue interactions. As the pedicle develops, a transformation occurs in the skin covering its tip. Instead of being covered with the usual scalp hairs, the growing tip becomes covered with fine, velvety hairs of the sort that eventually cover the entire growing antler. This marks the transition between growth of the bony pedicle itself and the formation of an actual antler. The growing antler has a core of well-vascularized bone, and under its cover of velvety skin, it is well innervated. The inner tip of the growing antler consists of a highly vascular zone of cartilage, and in contrast to the growth of horns, such as those of cattle or sheep, the antler grows by extension from its tip rather than being pushed out from the base. As rutting season approaches and testosterone levels increase, ossification of the antlers becomes complete and the velvet is shed.

During the winter, the antlers themselves are shed. This sets the stage for actual regeneration. Immediately after shedding, the remaining raw bony surface of the pedicle is exposed. The skin around it is already swollen at the time of shedding, and after shedding, it begins to cover the wound. The basis for antler regeneration remains controversial. Although both epidermis and cells from the underlying dermis appear to cover the wound, Goss (1983) believes that an epidermal covering with tongues extending into the underlying tissue is important for regeneration to proceed. He notes that if the exposed surface of the pedicle is covered by a full-thickness skin graft, at least in sika deer, antlerogenesis does not occur. As antler growth resumes, the underlying tissues induce the overlying skin to become velvet. Goss postulates a two-way interaction between skin (epidermis?) and the underlying skeletal tissues, whereby the overlying skin permits antler growth and the underlying tissues impart a specific velvet character to the skin covering the growing antler. In contrast to the need for periosteum from a specific territory on the frontal bone for antler regeneration to proceed, the source of skin does not appear to be as critical. Goss (1964a) surrounded an antler

FIGURE 5-4   Annual growth of antlers of a sika deer: *(A)* May 7; *(B)* May 13; *(C)* June 4; *(D)* June 15; *(E)* August 23; and *(F)* January 6. (Reprinted from Goss, R.J. 1983. *Deer Antlers: Regeneration, Function, Evolution.* New York: Academic Press, by permission.)

pedicle with skin from the ear and found that antlers, although slightly abnormal, regenerated.

A recent histologic study by Li et al. (2005) shows that the wound at the abscission surface of the fallen antler is healed by the ingrowth of full-thickness skin. They conclude that the process of early antler regeneration does not occur by the formation of a blastema, but through a process that is histologically similar to the postamputational healing of nonregenerating mammalian appendages. These authors conclude that actual antler growth occurs through the formation of anterior and posterior growth centers stemming from the periosteum of the pedicle. Clearly, one of the major remaining questions is what controls morphogenesis of the regenerating antler. Is the reasonably faithful replication of tine pattern during successive years of regeneration at all based on the same principles that control morphogenesis of a regenerating limb?

The growing antler is richly innervated with sensory nerve fibers, but in contrast to the regenerating amphibian limb, antler regeneration does occur after denervation (Suttie and Fennessy, 1985; Wislocki and Singer, 1946). Growth of the denervated antler, however, is less than that of the contralateral control antler. This response is reminiscent of that in mammalian tissue regeneration, in which innervation is not necessary for the process to begin, but nerves are necessary for the process to proceed to full completion.

MAMMALIAN EAR HOLE REGENERATION

One of the more unusual forms of mammalian regeneration is that of ear holes (Fig. 5-5). Markelova (1953) has described this phenomenon in detail, and it has since been

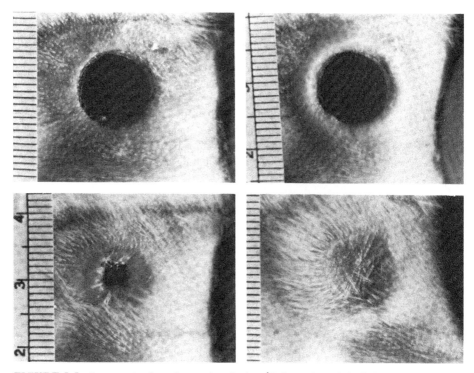

FIGURE 5-5   Regeneration from the margins of a 1-cm² hole cut through the full thickness of a rabbit ear. (top left, reading like a book) One day, 1 week, 4 weeks, and 8 weeks after surgery. (Reprinted from Goss, R.J. 1983. *Deer Antlers: Regeneration, Function, Evolution.* New York: Academic Press, by permission.)

studied by a number of investigators. For completely unknown reasons, ear hole regeneration has been found in only a few mammalian species, most notably rabbits, cats, and a certain variant of mice (MRL mice). Even in those forms that do regenerate ear holes, amputation of the tip of the ear is followed by only abortive attempts at regeneration. According to Goss (1983), the mammals that do not regenerate ear holes include normal mice, rats, guinea pigs, chinchillas, hamsters, gerbils, opossums, armadillos, Patagonian cavies, dogs, sheep and deer.

The rabbit ear consists of a sheet of hyaline cartilage closely opposed to skin attached on either side. Running along the interface are blood vessels and nerves. When a 1-cm^2 punch hole is made in the ear, a scab consisting of clotted blood and necrotic cartilage lines the wound. Over the course of 5 to 6 days, a wound epithelium cuts through the clot and necrotic cartilage and completely covers the circular wound surface (Fig. 5-6). Then a mass of cells closely resembling a blastema accumulates at the wound margin, and as cellular proliferation occurs, the wound begins to shrink and new cartilage forms, starting at the edge of the original ear cartilage. Within about 2 months, the hole has filled in entirely with tissue organized like that of the original ear. Even some new hair and sebaceous gland growth has been described in the area of regeneration.

Goss and Grimes (1972) have investigated interactions between the skin and cartilage after creation of a punch wound. When the central sheet of cartilage was removed, early regeneration began, but the process was not completed. When ear skin around the wound was replaced with belly skin, only limited ingrowth without cartilage formation occurred. Thus, ear skin possesses some special property that promotes the participation of cartilage in the regenerative process, but skin alone cannot substitute for cartilage. Goss and Grimes (1975) note the presence of prominent epidermal downgrowths from the wound epithelium and speculate that these might be the structural basis for interactions between the epidermis and underlying tissues. More recent studies on a "healer" mouse (MRL strain) have shown a histologic picture similar to that of the regenerating rabbit ear, including the presence of epidermal downgrowths into the mesodermal tissues (Heber-Katz, 1999). The availability of means of both genetic and molecular dissection of ear hole regeneration in the MRL mouse, combined with the existing knowledge of tissue interactions from the rabbit model, could greatly expand our understanding of this unusual regeneration model.

CATFISH BARBEL REGENERATION

Catfish are remarkable creatures that are uniquely adapted for sensing their external environment through modalities as diverse as sight, hearing, vibration sense, taste, smell, and even electromagnetic reception. Their large barbels are richly innervated, and the tips are covered with taste buds. These structures are one of the main means by which catfish, commonly swimming in muddy water, detect food or danger. Barbels are relatively simple structures, consisting of a central rod of cartilage, a nerve trunk, and an artery beneath a skin that is studded with taste buds.

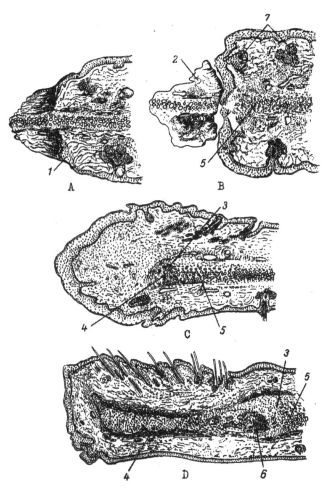

FIGURE 5-6 Histology of regeneration from the margin of a rabbit ear hole: *(A)* 3 days; *(B)* 5 days; *(C)* 20 days; and *(D)* 78 days after wounding. (1) Migrating wound epidermis; (2) scab; (3) proximal boundary of regeneration; (4) regenerating cartilage; (5) original cartilage; (6) focus of ossification within regenerate; and (7) sebaceous glands. (After Markelova, from Vorontsova and Liosner [1960], by permission of the publisher.)

As early as 1912, it was determined that not only do barbels regenerate after amputation (Beigel, 1912), but that the presence of the nerve is required for regeneration to proceed (Olmsted, 1920). Subsequent research (Kamrin and Singer, 1955b) has shown that after denervation, barbels of the bullhead regress spontaneously without additional damage (Fig. 5-7, B). This is in contrast to the regression of denervated limbs of larval *Ambystoma,* which require some type of distal trauma for postdenervation regression to occur (Schotté and Butler, 1941). After denervation, the taste buds also regress, and as is the case with taste buds in many other species, they recover on reinnervation.

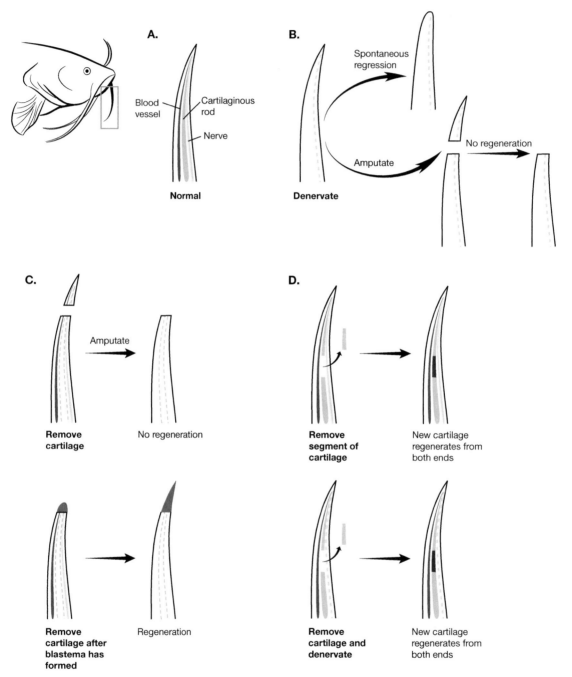

A.

Blood vessel

Cartilaginous rod

Nerve

Normal

B.

Spontaneous regression

Amputate

No regeneration

Denervate

C.

Amputate

Remove cartilage

No regeneration

Remove cartilage after blastema has formed

Regeneration

D.

Remove segment of cartilage

New cartilage regenerates from both ends

Remove cartilage and denervate

New cartilage regenerates from both ends

FIGURE 5-7 Tissue interactions in the regeneration of bullhead barbels. *(A)* Normal regeneration. *(B)* After denervation, the barbel regresses; after amputation, no regeneration occurs. *(C)* If the cartilaginous rod is removed at the time of amputation, regeneration fails to occur; if the cartilage is removed after a blastema has formed, regeneration takes place. *(D)* In both normal and denervated barbels, cartilage regenerates in the gap after removal of a segment of the cartilaginous rod. (Based on Goss's [1956c] experiments).

Another striking tissue interaction involves the central cartilaginous rod. After amputation and epithelial wound healing, a regeneration blastema appears beneath the apical epidermis. Removal of the cartilaginous rod at the time of amputation inhibits regeneration of the barbel (see Fig. 5-7, C), but removal of the rod a few days after the regeneration blastema begins to form does not inhibit further regeneration (Goss, 1954). Goss (1956c) proposed a model in which not only did the cartilage act upon the wound epidermis to allow regeneration to proceed, but the expanding wound epidermis permitted regeneration by providing space for the accumulation of blastemal cells derived from the cartilaginous rod.

A further experiment on regeneration of the cartilaginous rod alone (Goss, 1956c) illustrates well differences between the tissue regeneration of cartilage and epimorphic regeneration of the entire barbel. When a segment of the central cartilaginous rod is removed in the absence of amputation of the entire barbel, the gap is filled by the regeneration of new cartilage from both the proximal and distal cut ends (see Fig. 5-7, D). If the barbel is denervated, cartilage regeneration proceeds normally, in contrast to the inhibition of regeneration of the entire amputated barbel.

MAMMALIAN SKELETAL MUSCLE REGENERATION

A skeletal muscle consists principally of muscle fibers, connective tissue, blood vessels, and nerves. Either end of a typical muscle is connected to bone by a tendon consisting of regularly oriented collagen fibers. Muscle regeneration can be viewed at levels from a single cell to a whole muscle.

At the cellular level, the main interaction is between myogenic cells and the basal lamina that surrounds them. Satellite cells are highly responsive to a variety of growth factors for both their proliferation and their ultimate fusion into myotubes. These growth factors are largely bound to the basal lamina and the endomysial connective tissue until their release during the regenerative process.

At the tissue level, muscle requires a nearby vascular supply for regeneration to proceed (Hansen-Smith et al., 1980). The angiogenic response to muscle damage supplies the damaged muscle fibers with both the macrophages necessary for removal of the necrotic debris and the oxygen and nutrients needed for survival. In turn, damaged muscle possesses angiogenic properties that stimulate vascular ingrowth into ischemic areas (Phillips et al., 1991).

Early muscle regeneration proceeds normally in the absence of innervation, but full differentiation of regenerating muscle fibers requires the presence of motor innervation (Mong, 1977; Zhenevskaya, 1974). In some species, such as frogs and mice, regenerating myotubes break down and disappear within 3 weeks in the absence of innervation (Hsu, 1974; Mufti, 1977).

Muscle spindles represent tiny sense organs within muscles. Their formation in the embryo requires contact between the forming intrafusal fibers and sensory nerve fibers. In the absence of sensory innervation, spindleless muscles form (Zelená, 1957). Regeneration, however, sees a different set of interactions. Two different types of experiments that produced different results provided clues toward the nature of the interactions. In 1971, Zelená and Sobotková reported the absence of muscle spindles in minced muscle regenerates in the rat. However, in damage models (e.g., ischemia or local anesthetic damage) that do not involve physical disruption of the muscle, muscle spindles do degenerate and regenerate (Milburn, 1976; Rogers and Carlson, 1981). Even after complete denervation of the leg, muscle spindles regenerate (Rogers and Carlson, 1981). It appears that in postnatal life, the connective tissue capsule of the spindle provides sufficient information to the cellular contents of the spindle to allow regeneration of the intrafusal fibers, although their differentiation is typically incomplete. The absence of spindles in minced muscle regenerates is undoubtedly due to the physical destruction of the spindle capsule during the mincing process. In both intrafusal and extrafusal muscle fibers in adults, early regeneration can occur in the absence of innervation, but nerves are required for the complete differentiation of the regenerating muscle fibers.

At the level of a regenerating whole muscle, all of the interactions at the cellular and tissue level take place, but superimposed on that is a mechanical interaction that, to a large extent, determines both the gross form and internal architecture of the muscle regenerate. This interaction involves the regenerating tendons. A regenerating muscle in a functional environment is subjected to tension imposed on it through the tendons, as well as lateral pressure from surrounding muscles. As seen most prominently in minced muscle regenerates, the orientation of the regenerating muscle fibers and their surrounding connective tissue is strongly determined by mechanical forces that operate within the regenerate (Carlson, 1972b).

LENS REGENERATION IN NEWTS

One of the most remarkable and most extensively investigated systems of regeneration is that of the lens in the newt, in which after lentectomy a new lens regenerates from the dorsal iris. Under normal circumstances, the ventral iris is incapable of regenerating a lens. The process of dedifferentiation of the pigmented epithelial cells of the dorsal iris is outlined in Chapter 2 (p. 35).

Several features of the lens regenerating system involve tissue interactions (Fig. 5-8). It has long been known that the presence of the neural retina is required for lens regeneration to proceed (rev. Reyer, 1977; del Rio-Tsonis and Eguchi, 2004). Lens regeneration *in situ* is inhibited if an impermeable barrier is placed between the dorsal iris and the neural retina. If the neural retina is extirpated together with the lens, lens regeneration does not proceed until a new neural retina has regenerated from the retinal pigment epithelium. Similarly, if the dorsal iris is transplanted into many other sites, such as the peritoneal cavity, lens regeneration does not occur unless a piece of neural retina is implanted along with the piece of iris. Yet, when pieces of dorsal iris were implanted into regeneration blastemas of newts, lenses regenerated in a large percentage of cases (Reyer et al., 1973). In view of the importance of FGF in lens regeneration, it may be that the presence of numerous FGFs in the regeneration blastema is sufficient to support lens regeneration. Interestingly, when dorsal irises of newts are implanted into

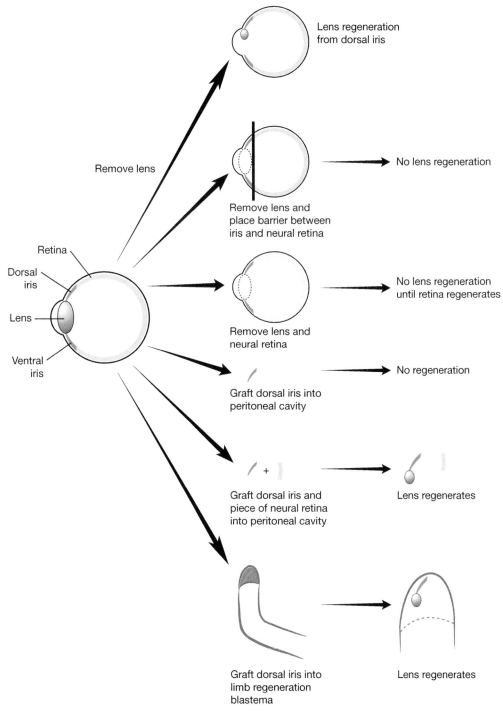

FIGURE 5-8   Tissue interactions in newt lens regeneration. (Based on experimental data from several investigations.)

lentectomized eyes of salamanders that lack the capacity to regenerate a lens, lens regeneration occurs from the implanted dorsal iris, but not from the dorsal iris of the host (Reyer, 1956).

Another interaction between retina and lens determines the orientation of the lens fibers. In the normal lens, the lens fibers form on the side of the lens facing the retina, and the opposite surface is covered by a cuboidal lens epithelium. Experiments on chick embryos have shown that if the lens is rotated in the eye, the retinal influence causes the lens epithelial cells, which are now closest to the retina, to elongate and form lens fibers (Coulombre and Coulombre, 1963). Stone (1954) rotated a portion of the dorsal iris by 180 degrees around its dorsoventral axis and found that the regenerated lens was normally oriented with respect to the retina and cornea, despite the rotation of the tissue that created the regenerated lens. In experiments that involve implantation of aggregates of dorsal iris cells into limb regeneration blastemas, Ito and coworkers (1999) have found that the regenerated lenses were often polarized with respect to the location of the thickened apical epithelium covering the blastema.

Remarkably little is yet known about the nature of the retinal influence. The evidence suggests that it is a positive stimulus, rather than something that inactivates an inhibitor. One set of molecules that has been implicated is the FGFs and the corresponding FGFRs in the tissues of the dorsal iris and regenerating lens (rev. del Rio Tsonis and Eguchi, 2004).

Another important interaction in lens regeneration is the action of the lens on the dorsal iris. The presence of the lens inhibits the formation of a new lens from the dorsal iris. The position of the lens in relation to the dorsal iris is also important, because under some circumstances, displacement of the lens allows a certain degree of lens regeneration to proceed. Two possible explanations for the inhibitory properties of the lens are that the lens produces a diffusible inhibitory factor that acts directly on the dorsal iris or that the lens somehow neutralizes a stimulatory factor from the retina. Another possibility is a matrix-mediated interaction between lens and iris. Definitive tests of these hypotheses remain to be done.

SUMMARY

It is apparent that to a greater or lesser extent, tissue interactions underlie most regenerative phenomena. The variety of interactions is remarkable. One of the common themes is that in the regeneration of essentially all vertebrate extremities, an interface between a wound epidermis and the underlying damaged mesodermal tissues is essential for regeneration to proceed. Two of the most commonly cited properties of the wound epidermis are the production of histolytic enzymes and the production of growth factors, especially members of the FGF family, which stimulate proliferation of the cells that will make up the bulk of the regenerate. Other stimulatory properties of the epidermal covering remain to be identified.

The role of skeletal elements in epimorphic regenerative phenomena is more complex. As noted earlier, the regeneration of both ear holes in mammals and barbels in catfish requires the presence of the normal cartilaginous element at the amputation surface. Yet, it has been known for almost a century that limb regeneration in salamanders proceeds normally even if the entire skeleton has been removed from the limb stump. One possibility is that, in the former two systems, the cartilage constitutes the major source of the blastema cells; another possibility is that some signal emanating from cartilage is required.

Virtually all tissues and organs require interactions with a local vascular supply for both the initiation and progression of regeneration. It is generally assumed that the basis of this requirement is the supplying of oxygen and nutrients by the blood. However, in view of recent findings (Mukouyama et al., 2002) that outgrowing sensory axons in the embryo closely follow the path of existing blood vessels, the possibility of other forms of interactions between tissues and the local vasculature should not be discounted.

As is discussed in the next chapter, the role of nerves in regeneration is complex. Most epimorphic regenerative processes require the presence of nerves, but the type of nerve appears less important than the amount. However, in digital regeneration, the nerve is not required for outgrowth. In contrast, most tissue-regenerative processes can proceed independently of nerves; but for some tissues, innervation is vital for the completion of regeneration. In these cases, the type of nerve (i.e., motor vs sensory) is important. Structures, such as taste buds, that rely on innervation for trophic support in the stable state maintain such requirements for their regeneration.

One question that remains to be thoroughly investigated is the degree to which tissue interactions in regeneration are mediated by diffusible molecules or through contact with molecules of the extracellular matrix. Because so many growth factors are bound to elements of the extracellular matrix, both may play an important role in tissue inter-actions. In the case of nonregeneration of a structure in one species that is related to one that can regenerate the same structure, one of the questions is whether an element(s) of a tissue interaction is missing or whether the ability of one of the components of a potential tissue interaction is incapable of responding to an existing signal.

CHAPTER 6

Role of the Nerve in Regeneration

If the sciatic nerve be intersected at the time of amputation, that part of the stump below the section of the nerve mortifies, reproduction following the cicatrix in the usual manner. If the division of the nerve be made after the healing of the stump, reproduction is either retarded or entirely prevented. And if the nerve be divided after reproduction has commenced, or considerably advanced, the new growth either remains stationary, or it wastes, becomes shriveled and shapeless, or entirely disappears.

—Tweedy John Todd (1823)

In 1823, 15 years before the elaboration of the cell theory, Tweedy John Todd published a remarkable article entitled "On the Process of Reproduction of the Members of the Aquatic Salamander." After a careful description of the gross features of a regenerating limb, he recounted experiments in which he transected the sciatic nerve. If this was done at the time of amputation, the limb stump healed by scarring. If denervation was done after wound healing, regeneration was either completely inhibited or was retarded. If a blastema had formed at the time of denervation, the regenerate remained static or completely regressed. Todd also observed that transection of the spinal cord at the time of amputation had no effect on regeneration of the tail. This was the first of a 180-year-long series of investigations designed to understand the effect and mechanism of the neural influence on regeneration.

EPIMORPHIC REGENERATION

Amphibian Limb

Todd's prescient experiments were not seriously followed up until the early 1900s, when several additional reports showed conclusively that limb regeneration fails to occur in the absence of nerves at the time of amputation. In attempts to understand the basis for the neural requirement in limb regeneration, a variety of early investigations removed various components of the nervous system at the time of amputation to determine their effect on the regenerative process. These experiments led to hypotheses that certain functional components of the nervous system (e.g., sensory, motor, or sympathetic) were the essential ones that supported limb regeneration (rev. Schotté, 1926).

It was not until the completion of a long series of detailed investigations that Singer (rev. 1952) proved definitively that it was the amount and not the type of innervation that was important for supporting amphibian limb regeneration. This concept was further refined to indicate that the important element was the density of innervation—that is, number of nerve fibers or cross-sectional axoplasmic area per unit area of amputation surface (Fig. 6-1) (Singer et al., 1967). Other experiments (Sidman and Singer, 1951; Singer, 1943) showed that the neural influence was not mediated by motor neural impulses. Singer (1959, 1965) then went on to elaborate a trophic theory of neural influence on regeneration in which the nerves, regardless of type, produce a trophic substance that is required in sufficient quantity for regeneration to proceed.

As proof of concept, Singer (1954) conducted one of the most influential experiments in the history of regeneration. The forelimb of the postmetamorphic frog does not regenerate, and its nerve supply does not meet the theoretic threshold requirements. Singer deviated the sciatic nerve from the hind limb under the skin of the forelimb and amputated the limb, thereby supplementing the normal nerve supply of the limb (Fig. 6-2). The result was substantially more regenerative outgrowth from the limb stumps with the deviated nerve than from control limbs. This experiment unleashed a wave of optimism concerning the possibility of stimulating limb regeneration in humans through a similar mechanism. However, a series of studies that applied this technique

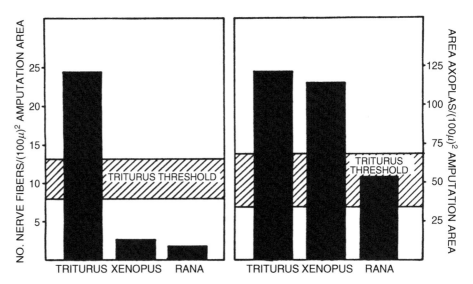

FIGURE 6-1 Relation between nerve fiber density (left) and axoplasmic area (right) and the ability to regenerate limbs in three amphibian species. The minimum threshold required to allow limb regeneration in the newt, *Triturus,* is indicated by the diagonal lines. Because of the greater axoplasmic area, limbs of *Xenopus* can regenerate, even though the number of nerve fibers is below the threshold. Innervation of nonregenerating *Rana* limbs falls below threshold in both categories. (Reprinted from Singer, M. et al., 1967. The relation between the caliber of the axon and the trophic activity of nerves in limb regeneration. *J Exp Zool* 166:89–98, by permission.)

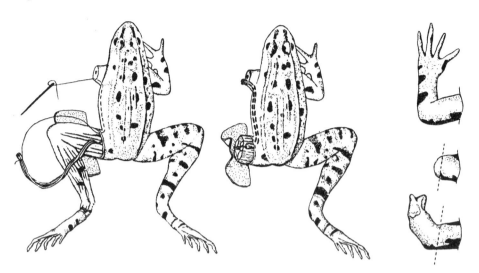

FIGURE 6-2 Induction of regeneration in the amputated frog limb after nerve deviation. *(A)* Operative scheme: The sciatic nerve is brought beneath the skin into the amputated forelimb. *(B)* Minimal regeneration from amputated upper arm in control (middle). The plane of amputation is indicated by dashed lines. More extensive regeneration from amputated arm containing additional innervation (bottom). (Based on Singer's [1954] experiments.)

to lizard and mammalian extremities (Bar-Maor and Gitlin, 1961; Mizell, 1968; Simpson, 1961; Singer, 1961) led to equivocal or negative results. With the benefit of time, a retrospective look at Singer's nerve deviation experiment in frogs leads one to wonder whether the regenerative response in that experiment was truly epimorphic regeneration.

Another study adds an additional level of complexity to the quantitative nerve threshold theory. Yntema had been conducting experiments on *Ambystoma* embryos in which he removed most of the neural tube, resulting in the outgrowth of aneurogenic or sparsely innervated limbs. As they grew, some of these animals bit off the limbs of their aquarium mates, and this trauma was followed by regeneration. A subsequent series of formal experiments (Yntema, 1959a,b) confirmed that if the limbs had never seen innervation, they were not dependent on nerves for regeneration.

The neurotrophic theory was modified on the basis of Thornton and Thornton's experiments (1970). They grafted aneurogenic limbs in place of normally innervated limbs of host larvae (Fig. 6-3). The brachial nerves of the host grew into the aneurogenic limbs. At various intervals after innervation, the grafted limbs were amputated. After almost 2 weeks of innervation, the formerly aneurogenic limbs did not regenerate after amputation. If the newly innervated limbs were then denervated and maintained in that state for 30 days, about half of the limbs regained their ability to regenerate in the absence of nerves.

These experiments led to a modified neurotrophic theory, according to which aneurogenic limb tissues themselves produce the trophic substance, so that after amputation

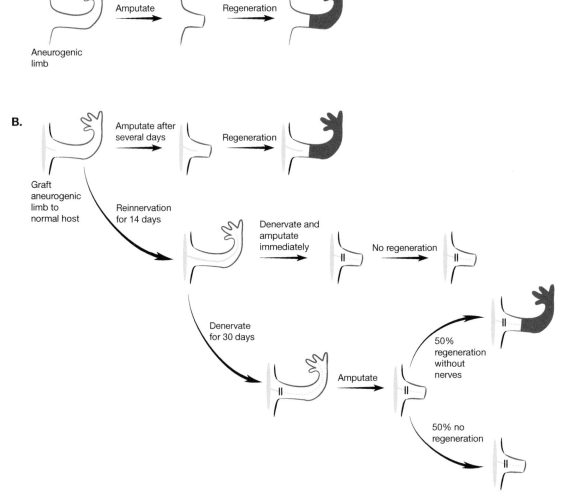

FIGURE 6-3 Experiments on the regeneration of aneurogenic limbs. *(A)* After amputation, an aneurogenic limb regenerates completely. *(B)* After grafting an aneurogenic limb onto a normal host, the grafted limb regenerates if amputated several days after grafting. If the grafted limb is allowed to become innervated for 14 days and is amputated immediately after denervation, it does not regenerate, whereas if it remains denervated for 30 days, it may recover the ability to regenerate. (Based on Thornton and Thornton's [1970] experiments.)

there is sufficient trophic substance to support regeneration. On innervation, the invading nerves suppress the production of trophic substance by the peripheral nonneural tissues, resulting in the limbs becoming "addicted" to nerves. Only after prolonged denervation of the transplanted limbs is the ability of the limb tissues to produce sufficient trophic substance restored in some of the limbs.

In other experimentation, Steen and Thornton (1963) grafted skin from normal limbs onto aneurogenic limbs; they found that, after amputation, regeneration failed to occur (Fig. 6-4, C). In contrast, grafts of normal mesodermal tissue beneath the skin of aneurogenic limbs were compatible with regeneration (see Fig. 6-4, E). The results of this experiment suggest that the wound epidermis, rather than the underlying tissues, is the target of the nerves. Later experiments by Fekete and Brockes (1987) show that the presence of a subpopulation of blastemal cells that react with the 22/18 antibody

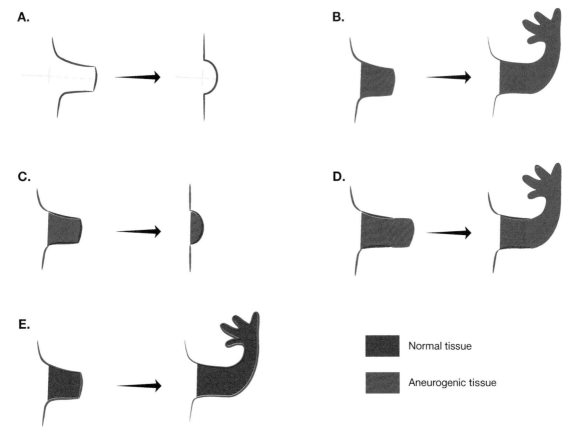

FIGURE 6-4 Exchange experiments between normal and aneurogenic limbs. *(A)* If a normal limb is denervated and then amputated, it regresses. *(B)* An amputated aneurogenic limb regenerates. *(C)* If an amputated aneurogenic limb is covered with normal skin and then amputated, the wound is covered by normal epidermis, and regeneration fails to occur. *(D)* If the proximal part of an aneurogenic limb stump is covered by normal skin, but a thin cuff of aneurogenic skin remains at the end of the stump, the wound epidermis is derived from aneurogenic epidermis, and limb regeneration occurs. *(E)* If the stump of an amputated aneurogenic limb is filled with mesoderm from a normal limb, then limb regeneration occurs. (Based on Steen and Thornton's [1963] experiments.)

is closely associated with the nerves growing into the blastema. Nerve-independent blastemas do not react with this antibody. Tassava and Olsen-Winner's (2003) more recent experiments suggest that aneurogenic limbs do not produce the neurotrophic factor and do not need it for regeneration. They also suggest that denervated neurogenic limb stumps or blastemas may actually produce a factor that inhibits regeneration, at least in aneurogenic limbs. Clearly, the conundrum of the aneurogenic limb raises more questions than we can provide answers for at this time.

The quest for determining the identity of the postulated trophic substance has been arduous, leading investigators down many blind alleys (Singer, 1978). The known neurotransmitters were eliminated early during the investigations. The standard assay for the trophic substance was to introduce a candidate substance into an amputated denervated limb and to determine whether a regeneration blastema forms. This assay has been largely supplanted by *in vitro* assays of blastemal growth. Because one of the major neural effects on regeneration is the stimulation of DNA synthesis or mitosis of blastemal cells (Boilly et al., 1985; Mescher and Tassava, 1975; Tassava and McCullough, 1978), putative neurotrophic effects have been quantified by calculating labeling or mitotic indices in cultured blastemas.

Many candidates for the neurotrophic substance have been proposed. The criteria for a successful trophic substance candidate should include: (1) presence in the blastema, (2) reduction after denervation, (3) possession of a mitogenic effect for blastemal cells, (4) ability to substitute for nerves in supporting regeneration, and (5) ability to block regeneration after its removal (Brockes, 1984). Among the numerous candidates for the trophic substance, fibroblast growth factor-2, glial growth factor, substance P, and transferrin have fulfilled many of the criteria (rev. Brockes, 1984; Globus, 1988; Mescher and Muniam, 1988; Mullen et al., 1996). Of these candidates, transferrin meets the criteria more completely than the others. Transferrin concentration in nerves increases and it is released from the growing ends of axons during early regeneration. It is lost from the distal blastema after axotomy (Kiffmeyer et al., 1991). In cultures of blastema cells, transferrin is both necessary and sufficient to maintain proliferative activity (Mescher et al., 1997). It is becoming apparent that the "trophic substance" may, in fact, be represented by a cocktail of factors with growth-promoting activity. Brockes (1984) has postulated that the ability of aneurogenic limbs to regenerate is based on the fact that their blastemas have a preponderance of a subpopulation of cells that are negative for a 22/18 antigen, which, if present on blastemal cells, makes them dependent on nerves.

Experiments conducted during the twentieth century confirm Todd's (1823) observation that denervation of a regenerating limb in the blastema stage causes regeneration to cease. Further experiments, however, showed that more advanced regenerates that were beginning to differentiate could continue to develop even though the limb bearing the regenerate was denervated (Schotté and Butler, 1944).

Experiments conducted early in the history of the study of neural influences on regeneration led some investigators to conclude that, in addition to their ability to support regeneration, nerves also play a role in determining the form of the regenerate.

Locatelli (1929) strongly put forth this viewpoint on the basis of her experiments involving the induction of regeneration by deviating nerves. Locatelli deviated limb nerves of newts into foreign areas and obtained the regeneration of a supernumerary limb. She believed that the nerve itself contained morphogenetic information that imparted a specific form to the regenerate. Not until other experiments (rev. Guyénot et al., 1948) showed that a deviated nerve could stimulate the formation of completely different structures (e.g., a deviated hind-limb nerve stimulating the formation of a new tail) was it accepted that, although nerves are essential for regeneration, they do not determine the morphogenetic nature of a regenerate. As will be discussed in Chapter 7, experiments such as this provided the basis for formulating the concept of regeneration territories and morphogenetic fields.

Amphibian and Reptile Tails

Regeneration of the amphibian or lizard tail involves an added degree of complexity with respect to the role of the nervous system, because both the peripheral and central nervous systems are present at the surface of an amputated tail. Therefore, attempts to determine the role of the nervous system in tail regeneration have had to take into account the presence of both central and peripheral nervous tissues.

Studies in both salamanders and lizards have clearly shown that regeneration of the tail depends on the presence of the spinal cord at the level of the amputation surface. When the spinal cord is destroyed in various ways, tail regeneration is inhibited (Holtzer, 1956; Kamrin and Singer, 1955a; Nikitenko, 1957; Roguski, 1957). Counts of peripheral nerve fibers in tails suggest that they are present in quantities too low to support regeneration (Cox, 1969; Mufti, 1971). However, the presence of the spinal cord is sufficient to allow regeneration to proceed. Simpson (1964) demonstrated experimentally that the ependyma, rather than nerve fibers, is the critical element for regeneration. This view was further supported when he found that grafts of ependymal cells induced the formation of supernumerary tails in lizards. Unfortunately, further experimental work on the nature of the neural influence in tail regeneration remains to be done before the role of innervation is completely understood.

Invertebrates

Invertebrates possess a variety of forms of nervous systems, ranging from generalized neural networks that are mainly neurosecretory in the most primitive phyla to well-defined nerve cords and pathways in the arthropods. Annelids, the most extensively investigated group, have a ventral nerve cord that contains ganglia for each body segment. Under normal circumstances, when part of an oligochaete worm is amputated, the nerve regenerates rapidly, and outgrowing fibers innervate the wound epidermis, as is the case in amphibian limb regeneration. If the nerve cord is surgically removed far enough from the plane of amputation so that reinnervation does not occur, amputated body segments do not regenerate (Fig. 6-5, A) (rev. Herlant-Meewis, 1964). More

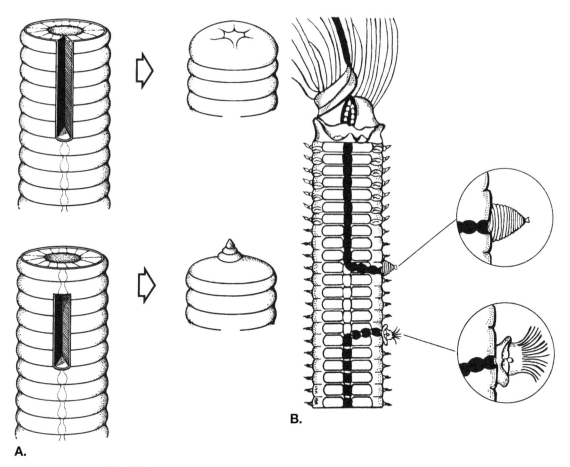

FIGURE 6-5 *(A*, top) Absence of regeneration in the worm *Allolobophora foetida* if the nerve cord at the amputation surface is removed. (bottom) Regeneration does occur if removal of the nerve cord does not extend to the level of the amputation surface. (Reprinted from Morgan, T.H. 1902. Experimental studies of the internal factors of regeneration in the earthworm. *Arch. f. Entw-mech.* 14:562–591, by permission.) *(B)* Formation of supernumerary head and tail after deviation of cut ends of the nerve cord in the marine annelid *Spirographis*. (Reprinted from Kiortsis, V., and M. Moraitou. 1965. *Factors of regeneration in* Spirographis spallanzani. In: V. Kiortsis and H.A.L. Trampusch, eds. *Regeneration in animals and related problems.* Amsterdam: North-Holland, by permission. 250–261.)

than a century ago, Morgan (1902), who first described the inhibition of regeneration caused by denervation in the worm, found that if the cut end of a nerve meets epidermis at a lateral site, a supernumerary head will form. In later experiments, Kiortsis and Moraitou (1965) deviated nerve trunks in a marine polychaete. They found that the posterior end of a deviated nerve induced the formation of a supernumerary tail, whereas a deviated anterior end of a nerve induced a new head (see Fig. 6-5, B). In discussing his experiments, Morgan (1902) noted that the presence of the digestive tract is not required for the regeneration of a head. Years later, Avel (1961), who was studying the effects of denervation on regeneration in worms, found that in the absence of nerves new head segments did not form, but the intestine nevertheless regenerated. Why the intestine is not needed for the regeneration of a head and does not require nerves for its own regeneration remains to be investigated.

Some annelids also use a neurosecretory mechanism to support regeneration of body segments. Certain species of polychaetes, in particular, require the presence of a brain (supraesophageal ganglion) for the regeneration of posterior segments. After amputation of a group of posterior segments, the cells in the brain accumulate and later release neurosecretory material. In the absence of a brain, posterior regeneration does not occur, but transplanted brains can compensate for the deficit and support posterior regeneration (Durchon and Marcel, 1962; Golding, 1967). Yet, other species of polychaetes can regenerate posterior segments without a brain (Hill, 1972). These species appear to rely more on the presence of the ventral nerve cord.

The presence of nerves is also required for the production of supernumerary parapodia (lateral projections) in the polychaete worm, *Nereis pelagica*. Boilly-Marer (1969) grafted parapodia to dorsal and ventral positions on the host and found that supernumerary parapodia formed at sites of dorsoventral discontinuity between graft and host. However, if the ganglion in the graft was removed, the production of supernumerary structures did not occur (Boilly-Marer, 1971).

Studies of the role of nerves in regeneration in arthropods have been conducted over the course of many years. In insects, it has been difficult to determine the role of the nerve in limb regeneration because of the difficulty in maintaining complete denervation for a sufficient period to obtain definitive results, but the early stages of limb regeneration appear not to require nerves (Nüesch, 1968). Regeneration of denervated limbs in cockroaches does occur, but it occurs more slowly and less completely than in normally innervated limbs (Penzlin, 1964).

One of the classic studies in the field of regeneration was that of Herbst (1896), who amputated eyestalks of prawns and reported that, in a high percentage of cases, antenna-like structures regenerated instead of eyes. Further experimentation showed that if the distal tip of the eyestalk was amputated, eyes regenerated, but if amputation was done through the base of the eyestalk, antennae formed (Fig. 6-6). Herbst (1902) finally determined that the presence of the optic ganglion (located in the base of the eyestalk) had to be present for an eye to form. Later studies (Maynard, 1965) show that the heteromorphic antennae were to some extent connected to the central nervous system. Some later investigators interpreted this phenomenon of heteromorphic regeneration as showing that the nervous system has the property of determining form in a regenerating structure. The experiments on nerve deviation that defined regeneration territories in amphibians (Guyénot et al., 1948), however, provide the most compelling evidence that, in vertebrates, at least, the role of the nerve in epimorphic regeneration is to stimulate and support regeneration, not to determine form directly. Nevertheless, several of the experiments that have been performed on invertebrates suggest that the nerves might have at least an indirect effect on the morphogenetic outcome.

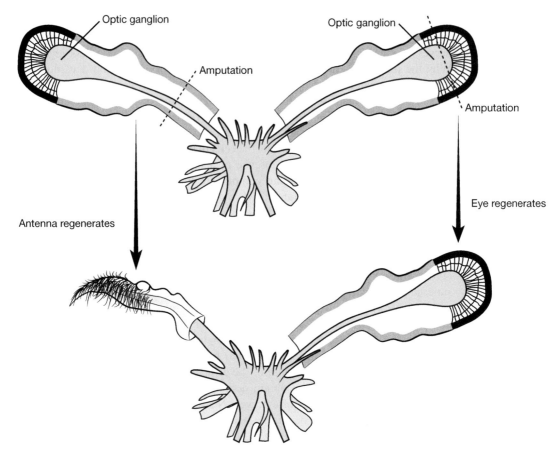

FIGURE 6-6 Regeneration of an antenna instead of an eye in the shrimp *Palinurus*. If the eyestalk is amputated distal to the optic ganglion (right), an eye regenerates. If the level of amputation is proximal to the optic ganglion (left), an antenna regenerates. (After Herbst [1896].)

TISSUE REGENERATION

Skeletal Muscle

Skeletal muscle is the end organ of motor nerves. In the absence of motor innervation, intact muscles lose their ability to contract voluntarily and undergo progressive atrophy over weeks and months. After damage, skeletal muscle fibers regenerate to nearly normal structure and function if they are adequately vascularized and innervated. In the absence of a nerve supply, the regeneration of skeletal muscle fibers is severely compromised (rev. Zhenevskaya, 1962, 1974).

In contrast to epimorphic processes, denervation has little observable effect on the initial stages of muscle regeneration. The degeneration of damaged muscle fibers, in both the intrinsic and phagocyte-mediated phases, typically occurs within the basal laminae of the damaged muscle fibers and independently of innervation. Similarly, the activation of satellite cells, their proliferation, and their fusion into multinucleated myotubes do not require the mediation of nerves. In fact, in most models of muscle transplantation, nerve fibers do not enter the grafted muscle until at least a week after transplantation, which is well after multinucleated myotubes have formed. A typical freely grafted muscle does not become restored to full mass or function; this is associated with incomplete innervation. Yet, in some transplantation models in which the motor nerve remains intact, it is possible to restore normal mass and nearly normal contractile function (Carlson et al., 1981). This does not mean, however, that nerves have no effect on satellite cells.

If an intact muscle is denervated, both the percentage and absolute number of satellite cells increase substantially over the short term. In the rat, denervation of a fast muscle results in a twofold to threefold increase in the percentage of satellite cells within a month (Viguie et al., 1997). In some of the early work on the regeneration and transplantation of entire muscles, Studitsky (1963, 1973) postulated that prior denervation for 2 to 4 weeks converts the muscle to a "plastic state," in which the muscle is better able to resist anoxia and to regenerate. Although at the time the focus was on metabolic changes (a greater anaerobic capacity) in the muscle, in retrospect it appears that a major basis for the concept of the plastic state is the increase in satellite cells that occurs during the weeks after denervation. Notably, Studitsky's concept of the plastic state was formulated (Studitsky, 1959) before the discovery of the satellite cell (Mauro, 1961). It may be that after the transplantation of a normal muscle, its separation from its motor nerve supply stimulates the proliferation of its satellite cells and facilitates early regeneration.

Zhenevskaya (1962) conducted a careful dissection of the neural influence on regenerating mammalian muscle. She found that after selective elimination of the sensory or autonomic innervation to a muscle, regeneration proceeded almost normally (Zhenevskaya, 1961), whereas if the motor innervation was disrupted, later stages of regeneration were unsuccessful (Zhenevskaya, 1960). These early results have been confirmed by subsequent investigators. As outlined in Chapter 5 (p. 105), the extent to which skeletal muscle regeneration can progress in the absence of nerves varies from species to species, but the effects of neural deprivation become noticeable first at the stages when the regenerating muscle fibers are passing from the myotube stage to that of early muscle fibers, with peripheral nuclei and cross striations. The ability of electric stimulation to maintain the mass of denervated muscle (Dow, 2002; Gunderson and Eken, 1992) strongly suggests that the main neurotrophic effect on regenerating skeletal muscles is the transmission of contractile impulses, rather than an effect mediated by chemical growth or maintenance factors.

The dependence of muscle spindles on sensory innervation for their formation in development and their relative independence from nerves during early regeneration is presented in Chapter 5. Although intrafusal fibers can certainly begin to regenerate in the absence of nerves, the differentiation of intrafusal fibers into both bag and chain morphologies and into histochemical fiber types is typically incomplete when an entire muscle regenerates. Soukup and Thornell (1997) have found that more than 85% of regenerated muscle spindles in muscle grafts contained the same fiber type, which was extrafusal in character. This may be related to the difficulty of restoring the complex pattern of sensory and motor innervation that characterizes a normal muscle spindle, because the type of intrafusal fiber is strongly influenced by its sensory nerve supply (Soukup and Novotová, 2000).

One important function of the nerve in mammalian muscle regeneration is to instruct the regenerating muscle fibers to become either fast or slow in character through the expression of either fast or slow myosins. In normal regeneration, the bulk of the muscle fibers in a slow muscle regenerate as slow fibers, and the majority of those in fast muscles regenerate as fast fibers. However, if a regenerating fast muscle is innervated by a slow nerve or vice versa, the fiber types of the regenerating muscle differentiate according to their new nerve supply rather than according to their original type (Fig. 6-7, A) (Carlson and Gutmann, 1974). Similarly, the differentiation of uniform extrafusal muscle fiber types within regenerating muscle spindles (Soukup and Thornell, 1997) is likely a reflection of the innervation of the regenerating intrafusal fibers by extrafusal motor axons.

Other Regenerating Tissues

A number of other regenerating tissues are influenced by nerves. For some, the neural effects are indirect; for others, a direct influence of the nerve on the regenerating end organ exists. An example of what is assumed to be indirect effects of nerves is the healing of skin wounds, which is slower and less complete in denervated areas. How much of this is a direct trophic effect and how much may relate to accompanying vascular disturbances remains to be sorted out.

One of the classic examples of a neurotrophic relation is that which exists between taste buds and the nerves that supply them. If the nerve supply to a taste bud is disrupted, the taste bud degenerates; when the nerve grows back, the taste bud is reconstituted. A denervated taste bud appears to have a specific requirement for gustatory nerve fibers, because if it is supplied *in situ* with motor or general sensory nerves, it will not reappear (Guth, 1958, Zalewski, 1969). However, when a graft of the vallate papilla (which contains taste buds) was placed into the anterior chamber of the eye together with a lumbar sensory ganglion, taste buds did form on the grafted papilla (Zalewski, 1972). This and later experiments reopened the possibility that under certain conditions some normally nongustatory nerves can produce what is needed to stimulate taste bud formation. Although it now appears that brain-derived neurotrophic factor produced in the area of the taste buds is important in the maintenance of their gustatory innervation, once that innervation is established, the chemical nature of the trophic influence from nerve to taste bud remains more elusive (Nosrat, 1998).

A cross-reinnervation experiment (Oakley, 1967) produced a result opposite to that of cross innervation in muscle. Taste buds in the front of the tongue respond to salt stimulation, whereas those in the back respond to quinine or saccharine. When the

A.

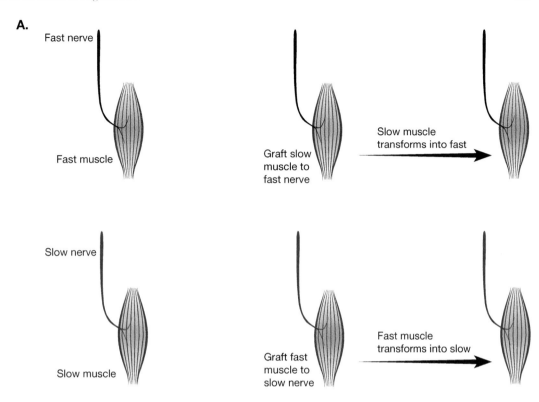

B.

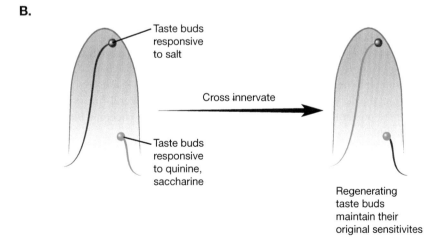

FIGURE 6-7   Cross-innervation experiments in mammals. *(A)* If a fast muscle is cross transplanted into the site of a slow muscle and is innervated by the slow nerve, it regenerates into a slow muscle and vice versa. (Based on Gutmann and Carlson [1975].) *(B)* In the rat tongue, if the nerves supplying anterior and posterior taste buds are switched, the original sensitivities of the taste buds are retained. (Based on Oakley [1967].)

TABLE 6-1

Comparison of Neural Requirements and Influences among Three Systems of Regeneration

Amphibian limb	Skeletal muscle	Mammalian taste buds
1. Nerves needed for the initiation of regeneration. (insect limbs possibly an exception)	1. Nerves not needed for the initiation of regeneration.	1. Nerves needed for initiation of regeneration.
2. Nerves not required for later regeneration, but growth is retarded.	2. Nerves required for final functional differentiation.	2. Nerves required throughout the regenerative process.
3. Quantity and not type of nerve is important.	3. Type of nerve is critical. (motor nerve needed)	3. Type of nerve is important (but some sensory can substitute for gustatory).
4. Neural effect mediated through some type(s) of trophic factor. Electric impulses ineffective.	4. Neural effect mediated by electric impulses.	4. Nerve effect mediated through putative trophic factor. Electric impulses ineffective.
5. Neural input can be replaced by defined molecules (e.g., transferrin).	5. Neural input can be replaced by defined electric impulses.	5. No substitutes found for direct innervation.

nerve that normally supplies the back of the tongue is crossed so that it now supplies the front of the tongue, the regenerated taste buds in that part of the tongue still respond to salt, and when the nerve that supplies the front of the tongue is deviated to the back, the taste buds in that area are still sensitive to quinine or saccharine (see Fig. 6-7, B). Therefore, the quality of the nerve does not determine the character of the taste receptor that it supplies; instead, it supports and maintains the development of regionally specified taste buds. In a somewhat similar transplantation experiment involving sensory (Grandry and Herbst-type) corpuscles in the skin, Dijkstra (1933) concludes that the characteristics of the regenerating sense organs are determined by the nature of the skin rather than by the nerve.

The lateral line in fish and aquatic amphibians consists of a string of vibration-sensitive lateral-line organs that contain both sensory and supporting cells supplied by a lateral-line nerve. Like mammalian taste buds, lateral line organs in fish regress after denervation and then become restored after reinnervation (Parker and Paine, 1934). In this system, the nerve supports the restoration of existing lateral-line organs, but it does not have the power to stimulate ordinary epidermal cells to become lateral-line organs (Wright, 1947).

SUMMARY

The examples cited in this chapter illustrate a wide range of dependencies of regenerative processes on nerves (Table 6-1). Clear differences exist between the requirement for a quantitatively sufficient supply of nerves of any type for the initiation of epimorphic limb regeneration and the ability of skeletal muscle to begin regeneration in the absence of nerves. In contrast, muscle has a type-specific requirement for motor inner-

vation for functional completion of the regenerative process, whereas the basic morphogenesis of the regenerating limb has only a general requirement of nerves for the
completion of growth. Once they have formed in the embryo, taste buds and lateral-line
organs are highly nerve dependent for their maintenance and regeneration, but recent
studies suggest less than total specificity in their requirements for neural support; for
example, other sensory, but not motor or sympathetic, nerves may be able to provide
partial trophic support. Currently, the broadest generalization is that if nerves are part
of a tissue, it is highly likely that they will be involved in some way in the regeneration of that tissue, but the manner in which this is done is highly tissue- or
organ-specific.

CHAPTER 7

Morphogenesis of Regenerating Structures

Carlson, stay away from morphogenesis. It's a bag of worms.

—C.S. Thornton (1966)

One of the remarkable features of regeneration is the fidelity with which many regenerating structures restore their original form. The typical result of an epimorphic regenerative process is almost anatomically perfect restoration of the structure that was lost. This is accomplished through a process that closely resembles the ontogenetic development of that same structure. On the other hand, many regenerating tissues come close to restoring their original form, but the result is not perfect. The regeneration of tissues follows a process that differs considerably from that seen in the embryonic development of the tissue at levels above that of the individual cell. In sharp contrast to these two examples, the hypertrophy of damaged internal organs often does not produce a structure that bears much resemblance to the gross form of the original organ. In this case, the bulk of the developmental energy appears applied to increasing functional mass, rather than to restoring external form, but internal architectural features may be restored to near-histologic normality. Of all aspects of both embryonic development and regeneration, morphogenesis—the development of form—remains the least understood. Numerous experiments have provided tantalizing clues to underlying mechanisms, but rarely can we claim to understand the cellular and molecular mechanisms that determine form.

TYPES OF MORPHOGENETIC PHENOMENA

Cataloging types of morphogenetic phenomena should be looked on as more than an exercise in classification. The final form of a regenerating structure can offer important clues toward understanding underlying mechanisms. This section introduces the types and vocabulary of shape in regeneration.

Absence of Regeneration

The absence of regeneration is usually attributed to a genetic deficiency or to the lack or deficiency of specific conditions (e.g., innervation) normally required for regeneration. Although this is all true, we now know that disturbances in morphogenetic control mechanisms in planaria or an amphibian limb, for example, can inhibit regeneration despite the presence of amputation, a wound epidermis, and adequate innervation (see pp. 153, 155). Therefore, missing or inappropriate distribution of morphogenetic information must be considered as an underlying cause in situations in which regeneration does not occur.

Perfect Restoration of Form

In most cases, the perfect restoration of form is the result of an epimorphic regenerative process, but there are also other circumstances that lead to the restoration of normal form. The ciliated protozoan, *Stentor,* shows remarkable fidelity in restoration of its original form after amputation of the oral region (see Fig. 1-7), but because this form of regeneration occurs within a single cell, it does not fall under the usual definition of an epimorphic process. Even certain regenerating tissues, such as bone, can be restored to almost perfect form, but the morphogenetic mechanisms are quite different from those that operate in a regenerating limb.

Hypomorphic Regeneration

Hypomorphic regeneration, the incomplete formation of regenerating structures, occurs commonly throughout the animal kingdom. Within the vertebrates, lizards and adult *Xenopus* (Fig. 7-1) typically form simple spikelike structures after the amputation of a limb. Among the invertebrates, hypomorphic regeneration is frequently seen in regenerating insect limbs.

The proper interpretation of hypomorphic regeneration is not always easy, because hypomorphism can be due to causes as diverse as nutritional deficiencies, delayed denervation, partial x-irradiation, or loss of regenerative power during metamorphosis. However, hypomorphic regenerates, like the absence of regeneration, can result from deficiencies of or inappropriate morphogenetic information at the end of an amputated structure. Although in some systems deficiencies in growth of the blastema lead to hypomorphic regeneration, planaria, for example, can regenerate perfectly normal parts of the body under nutritional conditions that cause the entire animal to shrink.

Supernumerary Regeneration

Under certain circumstances, regeneration results in the formation of duplicated or supernumerary structures. Supernumerary structures can also be stimulated to form without loss of the original structure, showing that amputation is not an absolute requirement for regeneration (Fig. 7-2). Supernumerary structures can be readily induced to form after amputation through a variety of means. A classic example is the formation of multiheaded planaria after amputation of the head and multiple incisions in the stump (Fig. 7-3). During the heyday of morphogenetic modeling in the 1970s,

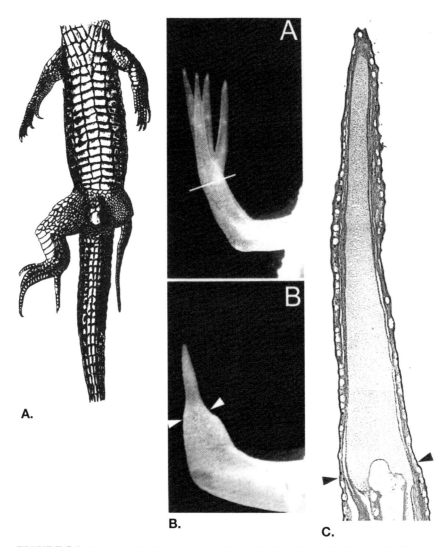

FIGURE 7-1   Hypomorphic limb regenerates in the lizard and frog. *(A)* Hypomorphic limb regenerate in *Lacerta muralis* found in nature. (After Marcucci, from Korschelt [1927].) *(B, C)* Spike regenerates from amputated forelimbs in *Xenopus*. *(B)* Gross photograph. (top) Normal limb. (bottom) Spike regenerate. Arrowheads indicate the level of amputation. *(C)* Photomicrograph showing a prominent rod of cartilage that occupies most of the spike. Arrowheads indicate the level of amputation. (B, C: Reprinted from Endo, T., et al. 2000. Analysis of gene expressions during *Xenopus* forelimb regeneration. *Dev Biol* 220:296–306, by permission.)

FIGURE 7-2   Complete supernumerary limb in the newt, induced by the implantation of a piece of kidney from the frog, *Rana pipiens,* beneath the skin of the upper arm.

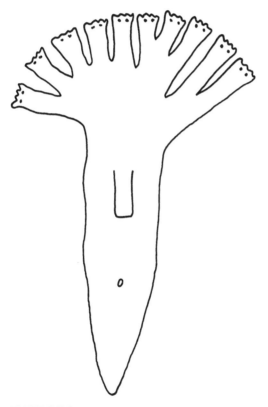

FIGURE 7-3   A ten-headed planarian *(Dendrocoelum lacteum)* that resulted from multiple cuts. (After Korschelt [1927])

the formation and analysis of supernumerary structures was an important tool in the study of morphogenetic mechanisms.

Heteromorphic Regeneration

Under most circumstances, a regenerated structure is a close copy of what was originally lost, but certain classic experiments have produced striking examples of the regeneration of structures that are totally different from those that were amputated. One of the earliest examples is the formation of two-headed planaria from either cut surface of a short segment of the body (see Fig. 7-8, A).

Another variety of heteromorphic regenerate has been called a *homeotic regenerate,* because an amputated extremity is replaced by one appropriate for another body segment. The best known example of this type of regeneration is the regeneration of an antenna in place of an eyestalk in *Palinurus* (see Fig. 6-6), but the regeneration of limbs in place of antennae has also been reported several times in insects (Korschelt, 1927). One of the most striking examples of heteromorphic (homeotic) regeneration is the regeneration of multiple pelvic girdles and hind limbs from amputated tails of tadpoles treated with retinoic acid (Maden, 1993; Mohanty-Hejmadi et al., 1992) (Fig. 7-4).

Morphallaxis

Morgan (1900) first used the term *morphallaxis* (see p. 27) to describe a regenerative phenomenon that he observed in planarians. He was studying regeneration from very small pieces and noticed that first the head and then the pharynx formed, but often the regenerated head and pharynx were proportionally too large in relation to the size of

FIGURE 7-4   Homeotic regeneration of a cluster of several hind limbs (right) from a regenerating tadpole tail after treatment with retinoic acid. (Courtesy M. Maden.)

the rest of the body. The worm responded by elongating and then remodeling the entire body so that a smaller, but better proportioned, individual resulted.

Morphallaxis occurs in a number of invertebrate phyla (Przibram, 1909). A classic example of morphallaxis is seen in *Tubularia,* a hydroid (Fig. 7-5). After amputation of the head, the remaining stalk becomes sealed. Then the distal end of the stalk undergoes reorganization, resulting in the formation of a new head region in the distal end of the stalk. Less profound degrees of morphallaxis accompany the regeneration of the head in certain polychaete worms (Berrill, 1978). When the head is amputated, epimorphic regenerative outgrowth occurs, but accompanying this is reorganization by morphallaxis of some of the abdominal segments near the amputation surface to form thoracic segments (see Fig. 1-12).

Somatic Embryogenesis

Tokin (1959) uses the term *somatic embryogenesis* to describe the formation of an entire individual from single cells or small parts of the body. This phenomenon is most prevalent in some of the lower, more poorly integrated animals, such as sponges, hydroids, and flatworms. Many of the animals most capable of somatic embryogenesis also reproduce by asexual reproduction. According to Tokin, an obligatory requirement for somatic embryogenesis is the disintegration of the normal organization, including loss of polarity and symmetry, of the portion of the body in which the process occurs.

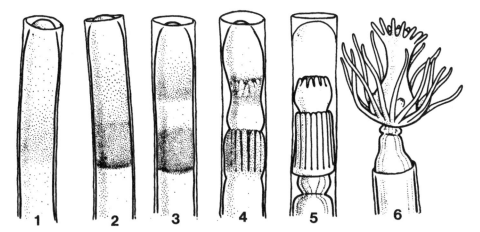

FIGURE 7-5   Anterior regeneration and morphallaxis in the hydroid *Tubularia:* (1) soon after amputation; (2) early formation of a pigment band presaging the formation of the proximal tentacles; (3) early formation of a second distal pigmented band above the proximal one; (4) and (5) early stages in the differentiation of proximal and distal tentacle rows; and (6) functional regenerated hydranth.

Intercalary Regeneration

As the name implies, *intercalary regeneration* is the filling in of gaps along the proximodistal or cross-sectional axes of the limb. It occurs when two nonadjacent tissues are placed next to one another. First described in insects (Bohn, 1970), intercalary regeneration has also been shown to occur in the regenerating amphibian limb (Iten and Bryant, 1975; Stocum, 1975) (Fig. 7-6, B).

In insects, the nature of the proximodistal discontinuity determines not only whether intercalary regeneration will occur, but also the polarity of the intercalated segment. If a distal part of a limb is grafted onto a more proximal surface, the gap is filled in by intercalary regeneration of the same polarity as that of the original missing segment (see Fig. 7-6, A). In contrast, if a proximal limb segment is grafted to a distal segment, intercalary regeneration again occurs, but in this case, the polarity of the regenerate is opposite to that of the original limb. A significant difference between intercalary regeneration in amphibians and insects is that, in the former, the intercalated segment arises from the proximal stump (Pescitelli and Stocum, 1980), whereas in insects, grafting experiments that involved both bristle patterns and color markers showed that the intercalary portion comes from the distalmost segment (Bohn, 1976).

Growth as a Morphogenetic Phenomenon

Most embryonic and regenerating structures grow as they take shape. In some cases, growth, defined as an increase in mass or number of cells, is a critical factor in the expression of the normal morphogenetic pattern or blueprint. The regenerating limb provides a good example of this. If the growth of an embryonic limb bud or a regeneration blastema is inhibited or retarded, a reduction in the number of digits is a common result. Yet, growth is not obligatory for normal morphogenesis, as is seen especially in morphallactic phenomena, such as the reorganization of pieces of planaria. During early stages of regeneration, growth can be an important factor in permitting the expression of morphogenetic information, but growth commonly occurs after the overall form of the regenerate has been fully established. In allometric growth, proportions of structures change as overall growth is occurring.

MAJOR CONCEPTS IN MORPHOGENESIS

Morphogenesis continues to lack a central theory; as a result, the field is littered with observations, concepts, and hypotheses. Nevertheless, many of these apply to specific and often important morphogenetic phenomena that have been uncovered over years of research on many different species. The level and depth of experimental backing for these concepts vary considerably, but most have proved to have predictive value in their own realms. In several cases, more than one mechanism has been proposed as the basis. One of the few incontrovertible facts in the study of morphogenesis is that morphogenetic phenomena are usually much easier to describe than to explain.

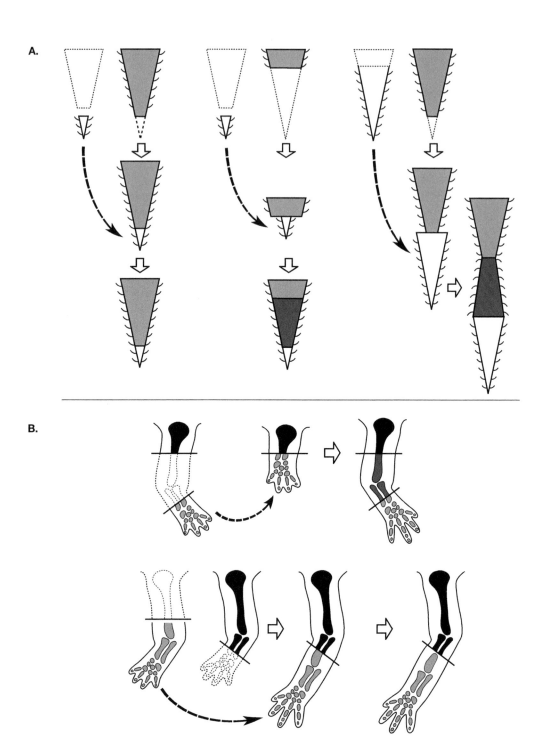

FIGURE 7-6 Intercalary regeneration in cockroach and amphibian limbs. *(A)* In the cockroach, if the tip of a limb is grafted onto a corresponding level of a host limb (left), the graft heals in place, but no intercalation occurs. If a distal tip is grafted onto a proximal level of the host (middle), intercalation with normal polarity (red) takes place. If a proximal segment is grafted onto a distal segment (right), intercalation (red) also takes place, but the polarity is reversed. (Based on Bohn [1970].) *(B)* In the amphibian limb, intercalation (red) occurs if a hand is grafted to an upper arm (top), but if an upper arm segment is grafted to a forearm, no intercalary regeneration takes place. (Based on data and experiments by French et al. [1976].)

Organ Specificity

One of the fundamental observations in almost every system of regeneration that has been studied is that when an amputated structure regenerates, the resulting regenerate is a replica of the original. That is, an amputated forelimb is replaced by a forelimb and not by a hind limb or a tail. Although (with the exception of retinoic acid effects) this rule is now taken for granted, the topic of organ specificity in regeneration was subject to considerable experimentation and debate in earlier years.

A historical example (rev. Polezhaev, 1979) will suffice to introduce some of the issues. One of the early questions asked in the study of epimorphic regeneration was whether the early blastema is essentially an empty vessel into which tissues of the stump pour morphogenetic information. In the older literature, the early blastema was considered to be undetermined (i.e., morphogenetically "empty"), and as it develops it becomes supplied with morphogenetic information from the stump (or becomes "determined"). This concept evolved from experiments in which early blastemas from a forelimb, for instance, were transplanted onto an amputated hind limb. In this case, the result was usually regeneration of a hind limb. If a later blastema was transplanted, the regenerate typically became a forelimb. The main problem with these older experiments is that grafts of early blastemas were usually resorbed and replaced by blastemas arising from the tissues of the host limb, whereas grafts of older blastemas resisted total resorption. When multiple early blastemas were transplanted into a heterotopic site, enough cells survived to produce a regenerate whose form was that of the donor extremity and not that of the recipient. This showed that even early blastemas contain sufficient morphogenetic information to replicate the structure that was lost. Subsequent experiments in which blastemas were transplanted into neutral sites, such as the dorsal fin or the orbit, showed that, like the embryonic limb bud, the blastema is a self-differentiating system. How the blastemal cells get their morphogenetic information and what it is remain the central problems in current investigations of morphogenesis.

Tissue Specificity

Many of the earliest investigators in the field postulated that the regeneration of complex structures is a tissue-specific phenomenon; for example, muscle creates muscle, epidermis creates epidermis, and so forth. Tracing cell origins and migrations has been a difficult proposition in almost all regenerating systems, but according to contemporary understanding, there are cases of firm tissue specificity, for example, epidermis-to-epidermis in the regenerating amphibian limb or several tissue types in the regenerating tail of *Xenopus* (Slack et al., 2004). In certain cases, however, other tissues can substitute for one that has been removed; thus, if the humerus is removed from an amputated upper arm, a regenerate replete with a full skeleton grows out. In dealing with morphogenesis, this concept is important when considering sources and locations of morphogenetic information.

Distalization (Distal Transformation)

One of the cardinal rules of regeneration is that only more distal, rather than proximal, structures form from an amputation surface. One of the earliest attempts to explain this phenomenon was made by Weismann (1892, p. 136), who postulated that the cells at any given level of a limb possess the information to form more distal structures, but not more proximal structures. In this pre-Mendelian era, Weismann elaborated the concept of determinants to account for proximodistal growth. At that time, it encompassed the known phenomenology of regeneration. According to this concept, a total of 35 determinants (information units), for example, would be sufficient to account for the proximodistal organization of a limb (Fig. 7-7). The cells of the most proximal

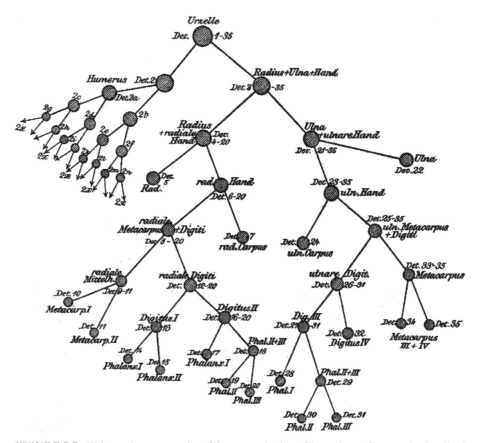

FIGURE 7-7   Weismann's representation of the progressive loss of determinants from proximal to distal levels in the limb. The original cell *(Urzelle)* contains determinants 1 to 35. As development proceeds distally, proximal or cross-sectional determinants are lost, so that the radial metacarpus, for example, contains only determinants 8 to 20, and the tip of phalanx II contains only determinant 15. (Reprinted from Weismann, A. 1892. *Das Keimplasma. Eine Theorie der Vererbung.* Jena, Germany: G. Fischer Verlag, by permission.)

starting point would contain all 35 determinants. As the limb grows out, more proximal determinants are lost, so that cells in the elbow would contain determinants 3 through 35, but the wrist would contain only determinants 23 through 35. Finally, the tips of the digits would contain only one (determinant 35). Remarkably, this way of organizing the phenomenology stood the test of time until the discovery in the 1980s (see p. 145) that the application of retinoic acid could cause distal structures to produce more proximal structures than already exist.

Polarity

The term *polarity* implies an inherent directionality in a structure without specifying beyond mere description the mechanisms that underlie that directionality. In a bilaterally symmetric animal, the craniocaudal axis is a reflection of an inherent polarity in the system, as is the proximodistal axis of the limb. In most regenerating systems, the polarity has already been set during embryonic development through the graded expression of members of gene families, such as the *Hox* genes, and the influence of signaling centers, such as those in the limb bud or the developing brain. Regeneration may reactivate some of these embryonic mechanisms, but because most regeneration is based on existing polarity in the remaining structure, normally there is no need to establish polarity *de novo*. Historically, polarity in regeneration has been linked to metabolic gradients (Child, 1941) or differences in electric potential or current flow (Athenstaedt, 1974; Borgens et al., 1989a), but it is not easy to determine whether any property—molecular, metabolic, or electric—is a cause or a result of polarity.

Polarity can be reversed in regenerating systems. Some of the most striking examples are the formation of two-headed regenerates from small segments of planaria or annelids (Fig. 7-8). One of the explanations for the reversal of polarity in short segments of worms is that inherent metabolic gradients are so truncated that they are ineffective regulators of the basis for polarity. This viewpoint is strengthened by experiments in which reversal of polarity follows treatment of segments with inhibitors, such as colchicine compounds (Kanatani, 1958) or chloramphenicol (Flickinger, 1959). Reversal of polarity has also been accomplished by exposing segments of animals to externally applied electric fields. By this means, Lund (1925) reversed the polarity of hydranth regeneration in the hydroid *Obelia*. Marsh and Beams (1952) embedded cut pieces of planaria in agar and obtained bipolar regenerates or even a total reversal of polarity when reversed electric fields of varying strengths were applied to body segments (see Fig. 11-5).

Gradients

Gradients have been hypothesized to explain many developmental phenomena, and regeneration is no exception. In some cases, it is easy to demonstrate the existence of a gradient, but it has been difficult to make a causal connection between the gradient and a mechanism that controls regeneration. In other cases, it has been convenient to

A. Double-headed planaria

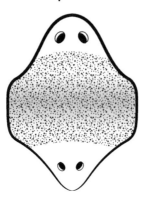

B. Double-headed sabellid

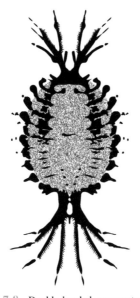

FIGURE 7-8   Double-headed regenerates from short body segments. *(A)* Planaria. *(B) Sabella.*

postulate the existence of gradients as the basis for certain morphogenetic phenomena, but it has not been possible to demonstrate that such a gradient actually exists.

One of the early champions of gradients in regeneration is Child (1941), who studied metabolic gradients mainly in annelids. Despite the accumulation of a large amount of experimental data, it was never possible to determine whether the gradients were parts of a regenerative mechanism for establishing polarity or whether they merely reflected the existence of a polar organization. Perhaps the best evidence for the existence of gradients that play a fundamental role in regeneration is seen in *Hydra,* in which a pronounced gradient is related to the existence of a head organizer (see p. 158) (Bode, 2003).

Regulation

The concept of regulation is one of the first ones to be developed in the field of experimental embryology more than a century ago. It arose from experiments originally designed to test the theories of preformation versus epigenesis. These experiments consisted of removing parts of embryos and observing whether the remaining parts developed with defects corresponding to the parts that were removed or whether somehow the defect was restored. In many types of embryos, the latter scenario occurred, and the concepts of regulative versus mosaic development were formulated.

Fundamental to a regulative system are the following properties: (1) If a structure is divided into two parts, each part can reconstitute the whole; (2) if two regulative structures are combined, the cells will merge and form one harmonious structure rather than two structures; and (3) a regulative system can be dissociated into fragments or individual cells, and these will come together to form a unified, whole structure.

Regeneration blastemas often exhibit these regulative properties. Some authors have even considered that epimorphic regeneration itself is a variety of regulation, but in studies on the postamputational restitution of embryonic extremities, it is important to distinguish between true embryonic regulation and a regenerative process that builds on the existence of fixed body parts.

The fundamental question in regulation is how an organism recognizes that some part of itself is missing. Or, in the case of the remaining blastomere of a two-cell embryo, does this cell even "know" that anything is missing? This type of question may not fully apply to regeneration in adults, but the answers to questions such as this are likely to provide a key to understanding both what initiates a regenerative process and what controls the development of form in an epimorphic regenerative process.

Regeneration Territories

The concept of regeneration territories arose from nerve deviation studies, which were first conducted in the 1920s. Locatelli (1929) deviated nerves from newt limbs into tissues outside the limb. If the end of the deviated nerve was placed close to the limb, a supernumerary limb of the same type regenerated. When the end of the nerve was deviated to the flank some distance from the limb, no regeneration occurred. If the end of the nerve was placed near the tail, a supernumerary tail appeared. Schotté (1926) removed the entire tail (or limbs) in newts and found that a new tail regenerated from tissues surrounding the original tail. However, if he removed too much tissue from around the base of the tail, no regeneration occurred.

Such experiments allowed the construction of a map that showed the areas, or territories, that were competent to regenerate an extremity (Fig. 7-9). Territory maps have also been constructed for some invertebrate species. Importantly, these territories are larger than the extremities themselves.

Morphogenetic Fields

The concept of morphogenetic fields, which also arose during the 1920s, is a broad one, of which regeneration territories represent a subset. Based on experiments on regeneration and embryonic regulation, a morphogenetic field is defined as that body of information (e.g., within a regeneration territory) that guides the formation of a complex structure. Weiss (1939), who formulated the field concept most specifically, defined a field district as the structural area that contains the morphogenetic field. He listed (Weiss, 1939), a number of characteristics of morphogenetic fields, most of which apply to both regulative and epimorphic systems. Weiss's list includes the following characteristics:

1. Field activity is invariably bound to a material substratum (i.e., the field district), rather than being some vitalistic force.
2. A field is not a mosaic, but rather an entity whose integrity can be restored after a disturbance.
3. In complex organs and organisms, a field is heteraxial and heteropolar; that is, the structures are asymmetric in three-dimensional space, and there are differences from end to end in any of the three Cartesian axes.
4. Within a field, none of the structural entities can be identified with any specific component of the field. For example, in the lower arm, the information that determines that the ulna will develop does not reside only in the ulna.
5. When the dimensions of a field are shrunk, nothing is lost and the proportions remain unaltered. This is the case in the morphallactic reorganization of a starved and shrunken planarian.

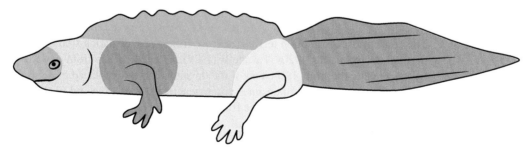

FIGURE 7-9 Map of regeneration territories in the crested newt. All colors except gray represent regeneration territories. (After Guyénot, from Lender [1974].)

6. If a field is split, each half retains an entire set of field characteristics. A good example of this is human identical twinning, in which each half of a split embryo retains the information required to form a complete individual.

7. If two field districts are merged with coinciding axes, their individual fields will merge into a single field. If their axes do not coincide, they will produce complex, multiple structures. (Weiss, pp. 293–294)

The preceding properties also describe those of a regulative field in embryonic development. A type of experiment that reflects field properties in regeneration is the removal of tissue components of the stump, either bone or muscle, and then allowing the limb to regenerate (Weiss, 1925; Carlson, 1972a). In both cases, the regenerate contains an anatomically normal skeleton or musculature, indicating that the regeneration can compensate for structural defects in the stump, a phenomenon often called *double assurance* in the regeneration literature (see p. 308).

The field hypothesis has often been criticized for its inability to account for specific mechanisms, but it does provide a good framework for describing and predicting the behavior of cells and tissues within a territory or field. Although use of the term *morphogenetic field* has gone out of style with the rise of the molecular era, the concept and its properties remain a useful basis on which to view the often confusing literature on morphogenesis in epimorphic systems.

Positional Information and Positional Memory

About a century ago, Driesch (1908) inferred from some of his experiments on sea urchin embryos that, in some manner, cells are able to recognize their relative positions within the embryo (cited in Wolpert [1971].) This idea did not strike a responsive chord with Driesch's contemporaries, and it was not pursued further until Wolpert (1969, 1971) formulated a more rigorous and testable theory of positional information in development. In essence, Wolpert's theory stresses that, within a developing organism or a field within the organism, the cells are able not only to recognize their positions, but they can respond to this information by differentiating into structures appropriate for their positions. According to Wolpert, boundaries and simple coordinate systems relating to them could provide the basis for the positional information that the cells within the system receive. From positional information, the cells are assigned positional values, which could be viewed as analogous to giving a house a street address. This is assumed to be the foundation for pattern formation. Once a cell is assigned a positional value, it interprets it in relation to its own genetic background, which determines the specific nature of its response to the positional information to which it was exposed.

Epimorphic systems appear to have maintained an active system of positional information, which has been called *positional memory* (Carlson, 1983). According to this concept, the cells in a system capable of regeneration retain a memory of their originally assigned position even if they are moved to some other location within the regeneration territory. Positional memory, which can be viewed as the long-term persistence of positional values, is the basis for intercalation when two dissimilar regions are brought together experimentally.

Although abundant experimental evidence exists for positional information, positional values, and positional memory, the molecular basis underlying the phenomenology represented by these concepts remains almost totally unknown. To complicate matters even further, Frankel (1974) has shown that many of the properties of positional information that have been described for multicellular fields can also be demonstrated within unicellular organisms.

Bateson's Rule

More than a century ago, Bateson (1894) collected data on a large number of duplicated and supernumerary structures that arose through either embryogenesis or regeneration. He found that duplicated asymmetric structures, such as limbs, are normally mirror images of one another. This observation has been called *Bateson's rule,* and after many years of neglect, it has proved to be valuable in the interpretation of experiments on limb morphogenesis. The formation of mirror-image structures during both embryogenesis and regeneration has often been postulated to be due to the actions of opposing gradients.

Barfurth's Rule

Also more than a century ago, Barfurth (summarized in Morgan, 1901, pp. 44–52) noted that if a structure is amputated obliquely, the regenerate grows out perpendicular to the amputation surface (Fig. 7-10). Later, during the period of growth, the regenerate begins to straighten out, so that it ultimately becomes normally aligned with the stump. In a number of cases, the straightening out can be at least partially explained by differential rates of growth between proximal and distal regenerates. In general, the growth rate of proximal regenerates is greater than that of distal ones on an equivalent extremity.

EXAMPLES OF MORPHOGENETIC CONTROL IN REGENERATING SYSTEMS

This section outlines the elements of morphogenetic control and the types of experimentation used to define it in two dramatically different systems of regeneration: the amphibian limb and the head of *Hydra*. Morphogenesis is not confined to tightly controlled systems such as these. Morphogenesis also occurs in regenerating mammalian tissues, but in this case, the control of morphogenesis typically depends on extrinsic environmental factors, as well as intrinsic factors. Examples of morphogenetic control in mammalian regeneration are given in Chapter 8.

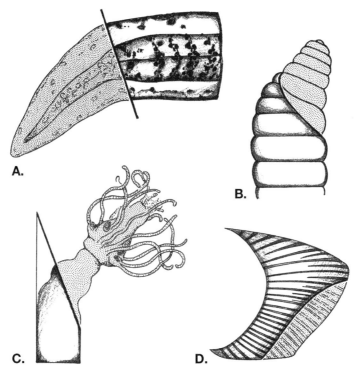

A.

B.

C.

D.

FIGURE 7-10   Examples of Barfurth's rule in regeneration. Regenerates grow out at right angles from an oblique surface. (After Morgan [1901].) *(A)* Tadpole tail. *(B)* Worm: *Allolobophora. (C)* Hydroid: *Tubularia. (D)* Caudal fin of fish: *Fundulus.*

Amphibian Limb

The epimorphically regenerating limb of urodele amphibians becomes restored to essentially normal morphology both externally and internally. The main deviation from normality is the pattern of innervation of the regenerate, which is characterized by highly abnormal branching from more normal major nerve trunks (Piatt, 1957).

Proximodistal Axis

A major mystery of morphogenesis is how an outgrowing structure knows where to start and where to stop growing. A regenerating limb begins to take shape at the level of amputation, and the process normally continues until the most distal parts of the digits have formed. One of the cardinal rules of regeneration, sometimes called the *law of distal transformation,* is that the regenerate produces more distal structures to the plane of amputation in a polarized fashion. One of the clearest demonstrations of this was an experiment by Butler (1951), in which he implanted the distal end of a sala- mander limb into a depression made in the body wall, and then amputated the limb through the level of the humerus (Fig. 7-11, A). After a period of regression, the transplanted limb, which had reversed polarity, produced a regenerate that contained all structures normally distal to the level of the amputation surface, regardless of the fact that many of them were present in the stump of the reversed transplanted limb. In a variant of this experiment, conducted for another purpose, the front limbs of newts were joined together at the wrist and the right limb was denervated, allowing nerve fibers from the left limb to grow into the right (see Fig. 7-11, B). A couple of weeks later, the right limb was amputated at the shoulder. The proximodistally reversed right limb stump regenerated a left limb of normal polarity in a manner similar to that in Butler's experiment (Carlson et al., 1974).

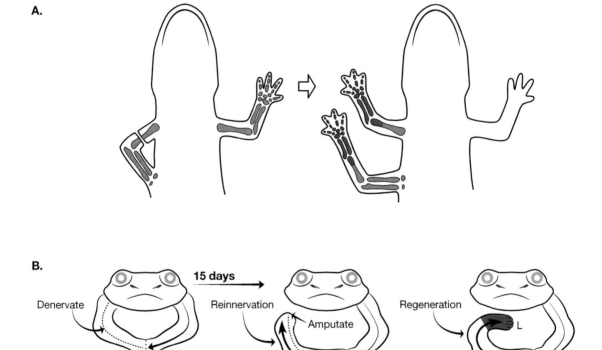

A.

B.

Denervate   **15 days**   Reinnervation   Amputate   Regeneration

R   L   R   L   R   L

FIGURE 7-11   Experiments illustrating the law of proximodistal transformation. *(A)* After amputation of the hand, the wrist was embedded in the flank. On healing, the limb was amputated through the level of the humerus. Both segments regenerated, but the regenerate arising from the distal segment produced everything distal to the level of the amputation surface, thus duplicating the distal segments already present. (Based on Butler's experiments [1951].) *(B)* The same principle. Newt forelimbs amputated at the wrist were joined, but the right limb was denervated. Nerves from the left limb regenerated into the right limb, which was later amputated. A left limb with reversed proximodistal polarity regenerated from the amputation surface of the right limb. (Based on Carlson and colleagues' experiments [Carlson, B.M. et al. 1974].)

Another fundamental property of epimorphic regeneration is intercalation, which is well illustrated by level-shift grafting (Iten and Bryant, 1975; Stocum, 1975). If a distal (e.g., wrist level) blastema is grafted onto a proximal (e.g., midhumeral) stump, the result is not a wrist growing out of the upper arm; rather, dedifferentiation and cellular proliferation between the grafted blastema and stump result in filling the morphologic gap, so that the regenerate contains the segments (i.e., the distal upper arm and forearm) that were originally missing between stump and blastema (see Fig. 7-6, B). In the converse experiment—a proximal blastema grafted to a distal stump—intercalation does not occur, and the regenerate consists of an almost complete arm growing out from the wrist of the host limb stump.

The grafting experiments described earlier strongly suggest that there are differences along the proximodistal axis that form the basis for the phenomenon of intercalation. Experiments on cellular behavior lend further support to this notion. Nardi and Stocum (1983) investigated surface adhesive properties of blastemal cells from different proximodistal regions of the limb. Combining blastemal masses from proximal and distal levels of amputation, they found that cells from proximal blastemas invariably engulf those of distal blastemas (Fig. 7-12, A). In contrast, if blastemas from the same level are approximated, Nardi and Stocum found no evidence of engulfing behavior.

An *in vivo* assay also indicated proximodistal differences in cell-surface properties. Crawford and Stocum (1988) grafted forelimb blastemas from upper arm, elbow, and wrist levels onto the blastema-stump junction in the hind limb (see Fig. 7-12, B). As the hind-limb blastema grew out, the grafted forelimb blastemas also developed and were carried along with the outgrowing hindlimb regenerate. The location of the forelimb/hindlimb regenerate interface depended on the origin of the forelimb blastema, so that an upper arm–level blastemal graft remained at the midthigh level, whereas a wrist-level graft was carried out to the ankle level of the host regenerate. This experimental model was called the *affinophoresis assay*.

One of the most influential findings in the field of regeneration was the discovery that retinoic acid can cause a regenerating limb to violate the law of distal transformation. Niazi and Saxena (1978) amputated the limbs of toad tadpoles through the shank, and then exposed the tadpoles to vitamin A suspended in the water. Several of the animals regenerated a full limb, starting with a thigh, despite the fact that the original limb was amputated at a more distal level. Shortly thereafter, Maden (1982) repeated the experiment on axolotls with exposure to a variety of retinoids and provided unquestioned evidence of proximal limb structures regenerating from distal stumps. Retinoic acid proved to be the most effective of the retinoids (Fig. 7-13). The retinoic acid model provided the first concrete tool for dissecting the basis for proximodistal polarity.

For several years, studies on the effects of retinoic acid on regeneration explored the phenomenology induced by exposure of regenerating systems to retinoids. Later studies began to delve into mechanisms. A conclusion stemming from the earliest experiments is that retinoic acid respecifies the positional values of cells along the proximodistal axis of the limb, causing more distal values to become proximalized. A clear experimental demonstration of proximalization by retinoic acid was made by the experiment of Crawford and Stocum (1988) in a follow-up of their affinophoresis assay

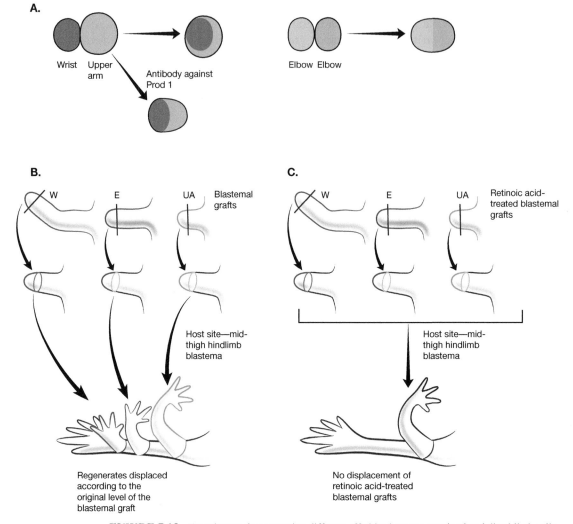

FIGURE 7-12 Experiments demonstrating different affinities between proximal and distal limb cells. *(A)* If proximal (upper arm) and distal (wrist) blastemas from regenerating axolotl limbs are juxtaposed in culture, cells of the proximal blastema surround those of the distal blastema. If blastemas from the same level are juxtaposed, their cells do not mix. Treatment of the blastemas with an antibody against Prod 1 (a proximal determinant) results in no engulfment of the distal blastema by the proximal one. (Based on Nardi and Stocum's [1983] and Morais da Silva and colleagues' [2002] experiments.) *(B)* If axolotl forelimb blastemas from different levels are grafted to thigh-level hind-limb blastemas, the grafted forelimb blastemas are carried out to their appropriate proximodistal level as the hind limb regenerates. *(C)* In contrast, if the blastemas are treated with retinoic acid, they become proximalized and remain at a proximal site when grafted as in *B*. (B, C: Based on Crawford and Stocum's experiments [1988].)

FIGURE 7-13   After exposure to retinoic acid, this axolotl regenerated complete upper arms, lower arms, and hands from wrist-level amputation surfaces of both arms.

system. They took blastemas from upper arm, elbow, and wrist levels of retinoic acid–treated *Ambystoma* larvae and grafted them onto midthigh hind-limb blastemas (see Fig. 7-12, C). In this case, the positional values of all of the transplanted blastemas had been proximalized, and the regenerates from all levels of blastemal grafts arose from the thigh of the host.

One of the first theoretic attempts to explain proximodistal transformation was an averaging model, which postulated a linear array of positional values along the proximodistal axis of the limb (Maden, 1977; Stocum, 1978b). According to averaging models, the proximodistal axis of the salamander limb is represented by a series of positional values from the base to the tip of the limb. The distalmost epidermis is considered to be a fixed reference point, and beneath it, positional values increase in a proximal direction. When a limb is amputated, a discontinuity is created between the wound epidermis, which is assigned a zero value, and the distalmost remaining underlying tissues. Recognition of the discontinuity leads to cell dedifferentiation and later proliferation and the reassignment of positional values so that the distal cells or their progeny in the limb stump will gradually fill in the missing positional values as regeneration proceeds (e.g., see Fig. 7-14).

Considerable recent progress has been made at understanding the differences between proximal and distal cells at a concrete molecular level. As in limb development, expression of *HoxA* and *D9-13* genes occurs in a general proximal-to-distal pattern (Gardiner et al., 1995; Savard et al., 1988; Simon and Tabin, 1993). After treatment of distal blastemas with retinoic acid, expression of key *Hox* genes adopts a more proximal pattern (Gardiner et al., 1995), as does expression of the T-box gene *NvTbox1*, which is normally expressed more strongly in proximal regions of the regenerating limb (Simon et al., 1997).

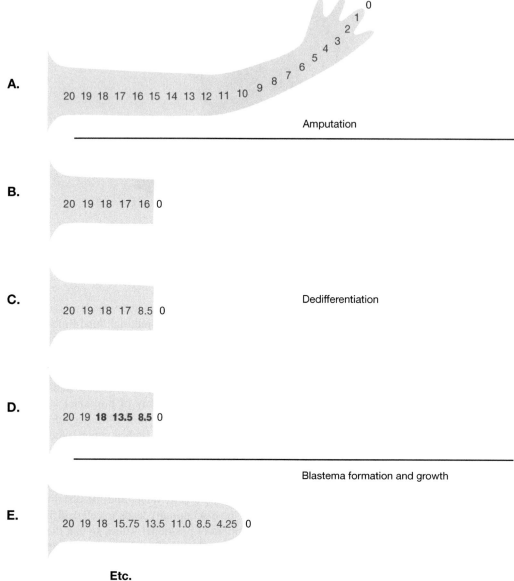

Etc.

FIGURE 7-14   Example of a proximodistal averaging model. *(A)* An array of cellular positional values in a stable limb. The distal epidermis has a constant boundary value of zero. *(B)* After amputation and epidermal wound healing, a discontinuity exists between the epidermal boundary value (0) and the distal-most underlying tissues (16). Assume that during dedifferentiation cell 16 will average its positional values without dividing. *(C)* The new positional value of cell 16 is 8.5 (between 17 and 0). Next, cell 17 will average its positional value between 18 and 8.5 as nonmitotic dedifferentiation comes to a close. *(D)* As the blastema forms, the cells indicated in bold begin to divide. One daughter cell will retain its original positional value, and the other will adopt a positional value between that of the parent and the next most distal cell. *(E)* As the blastema grows and the cells continue to divide, their progeny will continue to fill in the missing positional values until the original array (1–20) is restored. (After Maden [1977].)

Prod 1 is a small cell-surface protein that is related to the mammalian CD59 protein, which is involved in inhibiting the terminal phase of complement activation. Prod 1 is expressed in a proximodistal gradient in the normal salamander limb (Morais da Silva et al., 2002), and it is much more highly concentrated in cells of proximal versus distal blastemas (Fig. 7-15, A). Corresponding to the proximalization of blastemas by retinoic acid, *Prod 1* was up-regulated approximately 15-fold over control values in distal blastema cells after treatment with retinoic acid.

The power of Prod 1 in influencing cell behavior was shown in an experiment by Echeverri and Tanaka (2005), who electroporated Prod 1 together with a fluorescent marker onto distal blastemas and found that these cells became situated in the proximal regions of the regenerate (see Fig. 7-15, B). In contrast, when only dye was introduced into the same area of comparable blastemas, the labeled cells were found in the regenerated hand. These experimental data strongly suggest that Prod 1 may be a component of the elusive system of positional memory, which guides much of morphogenesis in the regenerating limb. Meis1, 2, and 3, homeodomain transcription factors in the axolotl, are downstream targets of retinoic acid and also act as proximalizing factors during limb regeneration (Mercader et al., 2005). When they are overexpressed in distal blastemal cells, for example, after retinoic acid treatment, these distal cells relocate to more proximal regions of the regenerate. Interestingly, *Meis* knockouts do not affect the course of regeneration after proximal amputation. Available evidence suggests that the Meis proteins may act as effectors of the system of positional memory, rather than as part of the memory system itself.

These newer molecular data suggest a modification of Maden's (1977) proximodistal averaging model (see Fig. 7-14). In its original form, the distal wound epidermis was postulated to be a reference point that was used in averaging positional values with the mesodermal cells immediately underlying it. In view of Echeverri and Tanaka's (2005) results, it now appears that if the wound epidermis plays a role in actual proximodistal patterning, it may instruct the cells directly underlying it to assume the most distal positional values, and that averaging would occur between distal and more proximal blastemal cells, rather than between blastema and wound epidermis. At a superficial level, this would appear to represent a convergence with Rose's (1970) ideas of polarized distal inhibitory control of regenerative processes. In both cases, prospective distalmost components of the structure would appear first, but in Rose's model, the distal cells inhibit the formation of more proximal structures. However, the accumulating molecular evidence suggests that, with respect to positional memory molecules, proximal cells are more information-rich than are distal cells. Much more information about the dynamics of the cells that form between the most distal and proximal cells of the regeneration blastema is needed before proximodistal morphogenetic control models can be fully refined.

At a broader level than the amphibian limb, a principal question in proximodistal pattern formation is whether the new pattern is laid down in a sequential or simultaneous fashion. If sequential, directionality (proximodistal or distoproximal) of determination is an important consideration. Research on planaria (Romero and Bueno, 2001) and *Tubularia* (Rose, 1970) suggests a distoproximal determination, whereas the reverse appears to be true in vertebrate systems.

A.

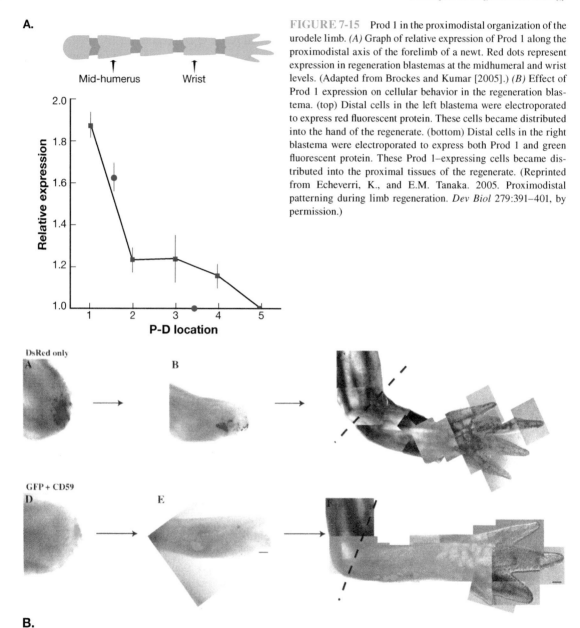

FIGURE 7-15   Prod 1 in the proximodistal organization of the urodele limb. *(A)* Graph of relative expression of Prod 1 along the proximodistal axis of the forelimb of a newt. Red dots represent expression in regeneration blastemas at the midhumeral and wrist levels. (Adapted from Brockes and Kumar [2005].) *(B)* Effect of Prod 1 expression on cellular behavior in the regeneration blastema. (top) Distal cells in the left blastema were electroporated to express red fluorescent protein. These cells became distributed into the hand of the regenerate. (bottom) Distal cells in the right blastema were electroporated to express both Prod 1 and green fluorescent protein. These Prod 1–expressing cells became distributed into the proximal tissues of the regenerate. (Reprinted from Echeverri, K., and E.M. Tanaka. 2005. Proximodistal patterning during limb regeneration. *Dev Biol* 279:391–401, by permission.)

Transverse Axes

Evidence also exists for the presence of positional memory in each of the transverse axes (anteroposterior and dorsoventral) and for the ability of retinoic acid to affect each of these axes. The contemporary study of morphogenetic control along the transverse axes began with experiments involving rotation or transplantation of either blastemas or stump tissues.

Lheureux (1972) and Carlson (1974a) continued a line of earlier experiments in which skin was rotated about the underlying tissues of the stump of an amputated limb. They found that after such rotation, the regenerates formed complex multiple structures with as many as four times the normal number of digits (Fig. 7-16). If either the skin or the underlying tissue was exposed to high doses of x-radiation, regeneration proceeded, but multiple structures did not form. Despite the fact that normal skin rotated around irradiated underlying tissues could produce normal-looking regenerates (although without muscles), altering the topographic relation of the normal skin graft profoundly affected the regenerative process. If the nonirradiated skin was turned 90 degrees around irradiated underlying tissues, so that all dorsal or all ventral skin abutted the wound surface (Carlson, 1974a; Samarova, 1940), or if small pieces of purely dorsal or ventral skin were aligned around the entire circumference of the limb (Lheureux, 1975b), regeneration did not occur. These experiments showed that the juxtaposition of morphogenetic opposites resulted in supernumerary regeneration, and that in the absence of morphogenetic opposites, regeneration did not occur at all. A later experiment (Carlson, 1975a) showed that positional displacement, rather than actual angular rotation, is the important manipulation, and that dermis and muscle possess the qualities necessary to produce multiple regenerates, whereas epidermis, bone, and nerve do not. The experiments suggest that dermis and muscle possess

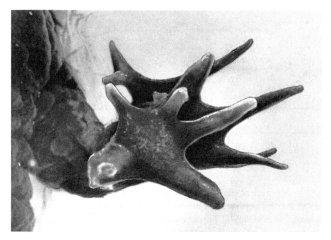

FIGURE 7-16   A complex supernumerary regenerate that formed in the axolotl after 180-degree rotation of stump skin and immediate amputation.

positional memory, and that after rotation, the original memory remains. Further experimentation (Carlson, 1975a,b) shows that positional memory is retained in stable rotated tissues for at least several months, and that it persists in rotated muscle that has undergone both degeneration after mincing and a tissue regenerative process (Fig. 7-17). Thus, positional memory is difficult to erase.

Other experiments involved the rotation of blastemas around the transverse axes of a limb stump (Iten and Bryant, 1975; Bryant and Iten, 1976). Like the stump-tissue rotation experiments, this manipulation also produced multiple regenerates; but in contrast with the grotesque regenerates produced by stump-tissue rotations, well-formed limbs appeared at the sites of axial discontinuities (Fig. 7-18).

The formation of supernumerary structures after rotation of tissues around the transverse axes is not confined to amphibian limb regeneration. It has also been documented in insect limbs (Bart, 1971; Bohn, 1972; French, 1976). These observations led to the elaboration of the polar coordinate model, which provided a unified basis for

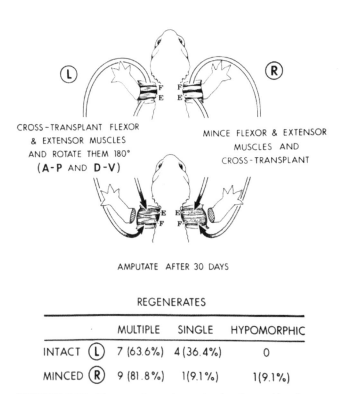

REGENERATES

		MULTIPLE	SINGLE	HYPOMORPHIC
INTACT	(L)	7 (63.6%)	4 (36.4%)	0
MINCED	(R)	9 (81.8%)	1(9.1%)	1(9.1%)

FIGURE 7-17   Diagram of experiment showing that positional memory persists for a long period and even after complete tissue regeneration of muscle in the forelimb of the axolotl. The formation of supernumerary regenerates is evidence of preservation of positional memory. On the right, positional memory survived both degeneration and regeneration of the muscle tissue. (Reprinted from Carlson, B.M. 1975b. Multiple regeneration from axolotl limb stumps bearing cross-transplanted minced muscle regenerates. *Dev Biol* 454:203–208, by permission.)

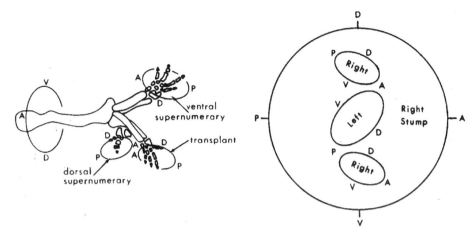

FIGURE 7-18 Supernumerary regenerates formed in the newt after grafting of a left forelimb blastema onto a right forelimb stump in the newt. Two right supernumerary regenerates are formed at the graft–host junction. A, anterior; D, dorsal; P, posterior; V, ventral. (Reprinted from Bryant, S.V., and L.E. Iten. 1976. Supernumerary limbs in amphibians: Experimental production in *Notophthalmus viridescens* and a new interpretation of their formation. *Dev Biol* 50:212–234, by permission.)

interpreting the results of many of the extirpation and rotation experiments that had been reported (French et al., 1976). The polar coordinate model views an epimorphic field as a group of concentric circles, with positional information embedded in both the circular and radial dimensions (Fig. 7-19). The center of the circles represents the most distal part of a three-dimensional epimorphic field. Two fundamental rules characterize the model. The first is the shortest intercalation rule, which states that in the event of a discontinuity around the circumference, missing positional values will be filled in by intercalation through the shortest route. That is, if a discontinuity is created that brings cells together with arbitrary 3 and 6 values, intercalation will fill in the discontinuity in a 3,4,5,6 fashion, rather than 6,7,8,9,10,11,12,1,2,3. The second rule is the complete circle rule for distal transformation. If, after simple amputation or transposition of tissues, a complete circle of positional values exists or is created by intercalation, distal outgrowth (regeneration) will occur. In the absence of a complete circle, regeneration will not occur.

The shortest intercalation rule functioned well to account for the formation and handedness of supernumerary regenerates in a wide variety of circumstances (e.g., see Fig. 7-20). In a test of the complete circle rule of distal transformation, Bryant (1976) created double anterior forelimbs in newts and then amputated them. Such limbs would not have on the amputation surface the complete array of positional values required for distal transformation, and they did not regenerate. However, in an apparent contradiction to this rule, Stocum (1978a) produced symmetric double posterior regenerates from double posterior hind-limb stumps at thigh level, but double anterior stumps produced only truncated regenerates. Such results produced a modification of the polar coordinate model to suggest that distal transformation may not be an all-or-none

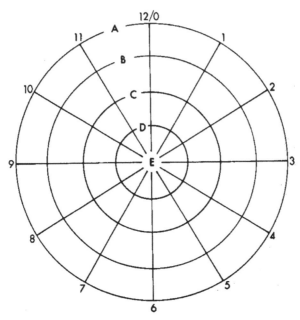

FIGURE 7-19 Diagram illustrating the principle of polar coordinates in an epimorphic field. Each cell in the field is considered to possess information relative to its position along the radius (A–E) and its position around the circle (1–12). E represents the tip of the limb, and A represents its base. (Reprinted from Bryant, S.V., and L.E. Iten. 1976. Supernumerary limbs in amphibians: Experimental production in *Notophthalmus viridescens* and a new interpretation of their formation. *Dev Biol* 50:212–234, by permission.)

response, but rather a graded one that takes into account the number of positional values around the circumference of the stump and how the values interact (Bryant et al., 1981).

In further tests of the distalization model, Stocum and Thoms (1984) treated newts with double anterior forelimb stumps with retinoic acid and obtained the unexpected result that many of the regenerated forelimbs not only underwent distal transformation, but were complete along the anteroposterior axis. This showed that, in addition to proximalizing, retinoic acid also posteriorizes limb tissues during regeneration. Molecular evidence for posteriorization was provided by experiments tracking expression domains of sonic hedgehog (shh), a signaling molecule normally expressed in the posterior edge of a developing of regenerating limb. After retinoic acid treatment, the normal endogenous expression of shh is soon down-regulated, and a new site of expression appears near the anterior margin of the regeneration blastema (Torok et al., 1999).

Other experimentation showed that retinoic acid also ventralizes, but clear-cut molecular correlates are lacking. Despite considerable research on retinoids, their receptors, and cytoplasmic binding proteins (rev. Stocum, 1995, pp. 163–172; Maden

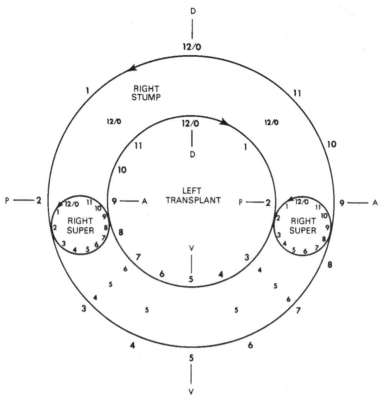

FIGURE 7-20 Diagram showing how the handedness of supernumerary regenerates can be predicted from the polar coordinate model. In this case, a left blastema is grafted onto a right stump. Where the circumferential disparity between graft and host is greatest, a new circle of values forms through intercalation, and a supernumerary limb regenerates. (Reprinted from Bryant, S.V., and L.E. Iten. 1976. Supernumerary limbs in amphibians: Experimental production in *Notophthalmus viridescens* and a new interpretation of their formation. *Dev Biol* 50:212–234, by permission.)

and Hind, 2003), the means by which retinoic acid affects morphogenesis during regeneration remains little understood.

The requirement for the meeting of morphogenetic opposites (e.g., anterior and posterior tissues) as a precondition for regeneration is not confined to the amphibian limb. Chandebois (1976, 1979) suggests that the meeting of dorsal and ventral tissues is essential for the initiation of head regeneration in planaria. Later experimentation involved the transplantation and 180-degree rotation of pieces of planaria so that dorsal and ventral tissues were juxtaposed (Kato et al., 1999). Blastemas and regenerative outgrowth appeared in the regions where dorsal and ventral tissues came together. This is similar to the requirement for the intersection of anterior and posterior skin at the site of a deviated nerve to obtain good morphogenesis of nerve-induced supernumerary structures in amphibian limbs (Lheureux, 1977). In planarian regeneration, the regenerative outgrowth is believed to arise from stem-cell–like neoblasts, which migrate to the wound surface and form the regeneration blastema. In planaria, x-radiation kills neoblasts and prevents regeneration (Dubois, 1949). When Kato and colleagues (2001) combined irradiated and nonirradiated tissues in dorsoventral combinations, blastemas did form, and the cells arose from neoblasts residing in the nonirradiated tissue at the dorsoventral interface (Fig. 7-21). This experiment suggests that, in planaria at least, positional memory resides in radiation-resistant differentiated cells, which may then transmit positional information to the morphogenetically blank neoblasts. These results in planaria differ from those obtained in salamander limbs (Carlson, 1974a; Lheureux, 1975b). If irradiated skin is rotated over nonirradiated underlying tissues or the reverse combination, only single regenerates arising from the nonirradiated tissues form. Multiple regenerates are rarely seen. This suggests that there may be fundamental differences in morphogenetic control between the epimorphic regeneration driven by dedifferentiation in amphibians and epimorphic regeneration driven by neoblasts in planaria.

The Head of *Hydra*

The experimental science of regeneration biology began in 1740 when Abraham Trembley first bisected a *Hydra* in an effort to determine whether these creatures were animals or plants (Lenhoff and Lenhoff, 1991). When he found that the pieces regenerated whole individuals, he began a systematic study of their regeneration that stimulated a flowering of eighteenth-century research on the phenomenon of regeneration. Head regeneration in *Hydra* differs greatly from limb regeneration in amphibia. Whereas the distal part of an amputated limb reorganizes to form a regeneration blastema through an epimorphic process, head regeneration in *Hydra* makes use of cellular and molecular mechanisms that operate in the ongoing maintenance of its normal body form. Cell division is not necessary for head regeneration in *Hydra* (Cummings and Bode, 1984; Hicklin and Wolpert, 1973), making it a morphallactic, rather than an epimorphic, regenerative process.

Cellular Organization and Structure of *Hydra*

Hydra are tiny creatures, usually measured in millimeters. Their overall structure is quite simple, composed of a tubular body, a basal disk (foot) that holds them to the substrate, and a head, consisting of a hypostome (mouth) and a circumferential row of tentacles forming a ring below the hypostome (Fig. 7-22). The cellular construction of a *Hydra* is relatively simple. The body tube consists of two concentric epithelial layers, ectoderm and endoderm, with a thin basement membrane-like layer of extracellular matrix, called *mesoglea,* interposed between them. The cells of both the ectodermal and endodermal layers are mitotically active and are in a state of constant movement. From the body tube, they are displaced in a steady stream toward both the head and the foot until they are finally shed—in much the same way that epithelial cells move and are shed along the length of intestinal villi in mammals. Once in the tentacles or

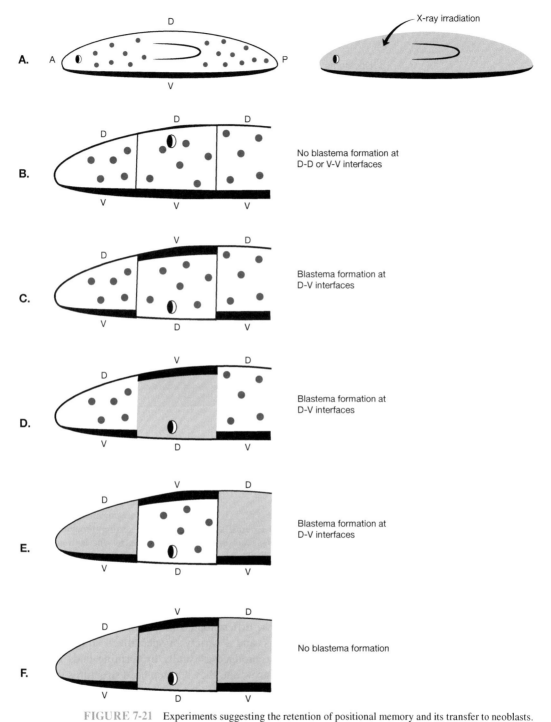

FIGURE 7-21 Experiments suggesting the retention of positional memory and its transfer to neoblasts. *(A)* Neoblasts (red) in a normal planarian and their absence in an irradiated planarian (blue). *(B)* If a piece of head is removed and grafted in normal dorsoventral (D-V) orientation, no blastemas form. *(C)* If a segment of a normal planarian is inverted along the D-V axis, blastemas form at the D-V interfaces and projections appear. *(D, E)* If either graft or host is irradiated, neoblasts from the nonirradiated tissue form blastemas at the rotated D-V interfaces, suggesting that the irradiated tissue can still provide positional information to the morphogenetically naive neoblasts. *(F)* After grafting a D-V–rotated irradiated head segment in an irradiated host, no regeneration occurs at the areas of D-V interface. (Based on Kato and colleagues' experiments [2001].)

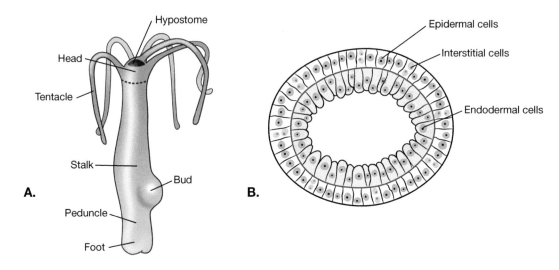

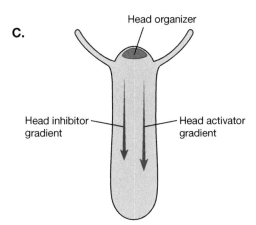

FIGURE 7-22 Structure and organization of a normal *Hydra.* *(A)* External structure. *(B)* Cross section through stalk. *(C)* Schematic representation of head organizer and activator and inhibitor gradients.

the foot, the cells leave the cell cycle (G2 arrest). Multipotent stem cells located in the epidermis, called *interstitial cells,* differentiate into a variety of cell types, such as gametes, secretory cells, neurons, and a variety of specialized stinging cells, called *nematocysts. Hydra* commonly reproduce asexually by forming buds that grow out from the body tube. As the buds grow out, they develop head structures and ultimately break off from the body tube of the parent. One of the fundamental issues in *Hydra* biology and regeneration is understanding the nature of the extracellular signals that provide both morphogenetic and differentiative information to the cells migrating out of the body stalk. An equally important set of questions is how these cells receive, interpret, and respond to this information.

Regeneration in *Hydra*

If a *Hydra* is bisected midway through the body stalk, the cut surface of the head end will regenerate a foot, and the cut surface of the foot end will regenerate a new head (Fig. 7-23). Short segments (as little as 5% of the body stalk) from most regions of the body stalk will regenerate both a head and a foot end (Bode and Bode, 1980). Even pellets of disaggregated cells will regenerate a complete animal (Gierer et al., 1972). Regeneration is rapid, with a foot regenerating in 30 hours and a head within 72 hours (Bode, 2003). After amputation of the head, the two epithelial sheets stretch over the amputation surface within 3 to 6 hours. This is followed by the reorganization of some of these cells to form the rudiment of a hypostome and the formation of the first traces of new tentacles within 30 to 36 hours. The differentiation of specific cell types occurs within the regenerating tentacles.

A large number of experiments, mainly of the transplantation type and conducted over several decades, led to the conclusion that the overall structural organization of a normal *Hydra* depends on the existence of organizing centers and gradients of morphogenetically active substances (Bode, 2003; Burnett, 1962; Gierer and Meinhardt, 1972; Wolpert et al., 1974). These same processes were found to play critical roles in regeneration. MacWilliams (1983a,b) has outlined four types of experiments that demonstrate both head activation and inhibition:

1. If grafts from the body stalk of donor *Hydra*s are placed midway down the stalks of intact *Hydra*s, only a low percentage of heads will form in the grafts (Fig. 7-24, A). In contrast, if the heads of the recipient *Hydra*s are removed, a greater percentage of grafts form heads. This suggests that the presence of a head inhibits the formation of new heads elsewhere along the body.
2. If grafts from the same region of donor *Hydra*s are placed close to the head region of host *Hydra*s, the percentage of grafts that form heads is less than if the grafts are placed farther away from the heads of the hosts (see Fig. 7-24, B). This shows that the inhibitory influence is stronger closer to the head, suggesting the existence of a gradient of the inhibitory influence.
3. The closer that tissue is taken to the head of the donor, the greater the percentage of heads that are formed when this tissue is grafted to similar sites in a recipient (see Fig. 7-24, C). This suggests the presence of an activator in a gradient that is stronger closer to the head.
4. If grafts are derived from the headmost area of a donor *Hydra* after the original head has been amputated, the head-forming activity is stronger as the donor *Hydra* is regenerating its own head (see Fig. 7-24, D). This suggests that the presence or development of the head strengthens the head-activating influence.

Gradients of the sort demonstrated in the preceding list imply a source and sink for the substance(s) that constitutes the gradients. For both the head activator and head inhibitor, transplantation experiments assumed that the head was the source of each. To account for the different effects of the transplantation experiments, he also had

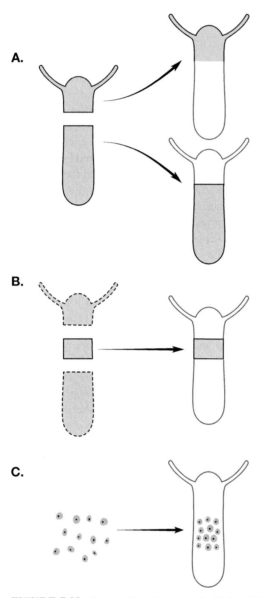

FIGURE 7-23 Regenerative phenomena in *Hydra*. *(A)* Regeneration (red) from head or stalk pieces. *(B)* Regeneration (red) of head and base from a middle segment. *(C)* Regeneration of an entire *Hydra* (red) from disaggregated cells.

A. Existence of Head Inhibition

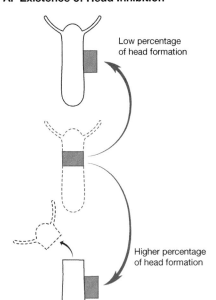

Low percentage of head formation

Higher percentage of head formation

C. Existence of a Head Activation Gradient

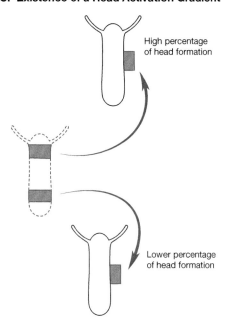

High percentage of head formation

Lower percentage of head formation

B. Existence of a Gradient of Head Inhibition

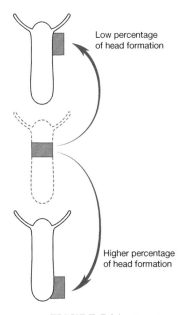

Low percentage of head formation

Higher percentage of head formation

D. Increase in Head Activation Activity with Regeneration

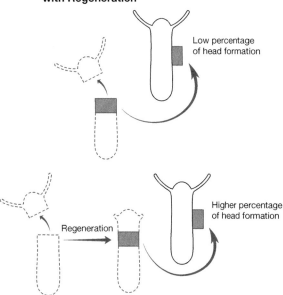

Low percentage of head formation

Regeneration

Higher percentage of head formation

FIGURE 7-24 Experiments demonstrating activating and inhibitory influences in head formation in *Hydra*. *(A)* If a midstalk segment is grafted onto the same level in a host, a greater percentage of head formation will occur in the grafts if the head of the host is removed. *(B)* If grafts from the same segmental level of the stalk are placed close to the head, the percentage of head formation in those grafts will be less than that in equivalent grafts placed close to the foot. *(C)* If stalk grafts taken from near the head or the foot are placed in the same location on a host stalk, those from nearer the head will form a greater percentage of heads than those from near the foot. *(D)* A segment of stalk taken from just beneath a regenerating head will form a greater percentage of heads than one taken from the same location in a nonregenerating *Hydra*. (Based on MacWilliams's experiments (1983a,b].)

to assume that there were different rates of breakdown of the active molecules (the sink) or that there were different concentration curves for some other reason. Without specific molecular candidates, it was not possible to get more specific. More recent research (rev. Bode, 2003) suggests that head activation consists of two components: (1) a head organizer (much like an embryonic organizer) that is centered in the hypostome, and (2) a head-activator gradient that operates principally in the body stalk. It has been assumed that in the head region the effective concentration of head activator exceeds that of the inhibitor, and also that far down along the body stalk the inhibitor becomes less effective than the activator. Otherwise, the natural budding process could not occur. The regeneration of a head was explained on the basis of a shorter half-life of the inhibitor than that of the activator. Once the concentration of the inhibitor has declined relative to that of the activator, the regeneration of a head can begin. A mutant strain (reg-16), which regenerates heads in only 10% to 20% of cases, has been shown to have a higher level of inhibitor, which does not decline as rapidly after amputation as that in normal animals (Achermann and Sugiyama, 1985). In normal regeneration of the head, a first step is to reconstitute the head organizer, and then begin to produce the head activator and inhibitor gradients.

More detailed knowledge of the dynamics of head activation and inhibition has had to await the identification of specific molecules that constitute the gradients. Understanding mechanisms at the molecular level is still quite incomplete, but some patterns are beginning to emerge (Steele, 2002).

To date, the head-activation system is best understood. As early as 1973, Schaller identified what was later found to be an 11–amino acid peptide that fulfilled many of the characteristics of the head activator. The head activator is produced by nerve cells and can also be produced by epithelial cells in nerveless *Hydra*. The head activator, which is mitogenic to several cell types in *Hydra*, not only accelerates head regeneration, but also stimulates budding. Subsequent to the identification of the head activator as a peptide, an international consortium has identified more than 800 peptides within adult *Hydra* (Steele, 2002). What they all do and whether the presence of this enormous number of peptides is the norm for higher animals or is, in contrast, an evolutionary experiment remains to be determined.

Two other molecules are expressed early after removal of the head and appear to be important in head regeneration. One is a 12–amino acid peptide, called HEADY, which is strongly up-regulated by 6 hours after head amputation and already declines by 8 hours (Lohmann and Bosch, 2000). It acts as a head inducer. In contrast, its RNA does not increase after amputation of the foot.

Even more rapid is the expression of the signaling molecule Wnt, which appears within 1 hour of head amputation (Hobmayer et al., 2000). *Hydra* express almost all components of the Wnt pathway, indicating that this pathway is truly functional. Bode (2003) believes that the activity of the Wnt pathway is closely associated with head-organizing activity. Expression of the Brachyury homologue *HyBra* 1 in the hypostome correlates with changes in head activation.

Other peptides are involved in the positional gradient leading to foot formation (Bosch, 2003). The peptides pedibin/Hym-346 and Hym-323 are present in decreasing basoapical gradients and favor foot formation by lowering the positional value gradient along the basal body column. Hym-323 acts on basal epithelial cells, whereas pedibin is expressed in endodermal cells and is coexpressed with *CnNK-2,* a homeodomain factor involved in translating the positional values reflected in the gradient to cellular behavior.

What is not currently understood is whether the regenerative controls that have been demonstrated in *Hydra* can be extrapolated to higher animals, and if so, in what context. Certainly it would be of interest to determine whether simple elongated structures in vertebrates, such as intestinal villi, which also exhibit cell streaming in their epithelial layer, are served by any of the types of molecular mechanisms that guide morphogenesis in *Hydra*.

SUMMARY

Morphogenesis remains one of the last major frontiers in the study of epimorphic regeneration. Until recently, our understanding of morphogenetic phenomena has come largely from the interpretation of various deletion and transplantation experiments, but despite the existence of some elegant formal models, we still know amazingly little about the cellular or molecular basis for pattern formation and the shaping of the regenerate. It is becoming increasingly clear that morphogenetic boundary conditions are important even in the initiation of epimorphic regeneration. This is seen in the principle of opposites, whereby opposite tissue qualities (dorsal and ventral or anterior and posterior) must be present at the amputation surface for regeneration to proceed. The greatest understanding of the cellular and molecular control of morphogenesis has come from experiments that involve the proximodistal axis. The serial duplications caused by retinoic acid and demonstrations of different adhesive properties by blastemal cells at various levels produced data leading to the conclusion that there is a gradient of surface properties from proximal to distal in the regeneration blastema. Recent studies of the surface molecule, prod 1, are now providing a molecular basis for these differing properties. We continue to know almost nothing regarding morphogenetic control about the anteroposterior or dorsoventral axes, and we do not understand the basis for positional memory, which appears to represent the embodiment of that morphogenetic information. Why the regenerating tail, which histologically looks much like a regenerating limb, reacts so differently from the limb to various tissue manipulations is also a mystery. In some invertebrate systems, such as *Hydra,* considerable evidence favors the existence of molecular gradients that can both stimulate and inhibit regeneration.

CHAPTER 8

Reintegrative Processes
in Regeneration

It can be shown, I think, with some probability that the forming organism is of such a kind that we can better understand its action when we consider it as a whole and not simply as the sum of a vast number of smaller elements.

—Thomas Hunt Morgan (1901)

One of the fundamental characteristics of regeneration is that the regenerating structure takes shape in association with an already existing remainder of the body. Reintegration entails the harmonious coupling of the regenerate with the body, and it can develop into a variety of forms.

At one extreme, reintegrative processes can be vital for the initiation of a regenerative process. One example is the epithelial healing of the amputation surface of an amphibian limb. The simple healing of a skin wound in this case is a requirement for epimorphic regeneration to proceed (see p. 68). Another example of early reintegration is the ingrowth of blood vessels into an ischemic area of tissue. Without a functional vascular connection, the ischemic tissue would remain in that state until it is sloughed or converted into a nonfunctional scar or calcified body.

Especially in tissue regeneration, morphogenesis is accomplished largely through mechanical interactions between the regenerating tissue and the parts of the body with which it interfaces. In this case, reintegration occurs almost continuously as the regenerating structure takes shape, with modeling and remodeling of the regenerating tissue occurring as the basic tissue structure is being laid down. One of the best-studied examples of this is regenerating bone, whose gross and internal structures reflect amazingly closely the mechanical environment in which regeneration is occurring. Although not so well studied, regenerating muscle responds similarly to its mechanical environment.

Reintegration can also begin near the termination of a regenerative process. This is particularly true for neural and sensory structures. Only after a regenerating motor nerve reaches its target muscle fiber can true structural and functional integration be established. Similarly, the axonal processes that emanate from a regenerating retina undergo final functional reintegration when they have reached and connected with their central targets. Functional reintegration in regenerating internal organs, such as the liver and kidneys, involves restoring a metabolic balance throughout the body; but even before that, the regenerating internal structure of these organs must assume a configuration that allows such a functional interchange.

Even in structures that regenerate epimorphically, reintegration is necessary for the proper connection of the regenerate with the stump. In the regenerating amphibian limb, there is a zone located just outside the sphere of influence of the regeneration blastema where tissue regeneration of bone and muscle forms a link between the perfect morphogenesis of the regenerate and the surviving tissues of the stump. This is vital for the reestablishment of normal function of the limb. In some worms, the process of morphallaxis remodels the tissues closest to the regenerate so that there is a smooth transition between the regenerate and the remaining body segment.

Reintegration involves many different mechanisms, some partially understood and some hardly recognized. This chapter describes how a number of regenerating systems achieve structural and functional reintegration.

SKELETAL TISSUES

The classic case of skeletal tissue regeneration is the healing of a fracture of a long bone. Although at a microscopic level, there is considerable similarity in the healing of virtually all types of fractures, some reintegrative processes are most apparent in the healing of a nontreated displaced compound fracture. The most important principle of reintegration was enunciated more than a century ago by Wolff (1892). In essence, what is now known as Wolff's law states that mechanical forces acting on bone are accompanied by predictable changes in the internal architecture and external form of the bone. This law, which was based on mechanical engineering principles, applies to growing bone, mature bone, and regenerating bone. In practice, it means that regenerating bone will be most heavily deposited in areas subjected to the most pressure; thus, if a fractured long bone is set at an angle, the most prominent site for new bone formation will be at the concave surface of the break (Fig. 8-1).

The healing of a fracture can be broken down into several steps (Table 8-1). During the fracture process, blood vessels become severed and a clot forms within the fracture site. Within hours after the fracture, the regions of bone closest to the fracture plane become necrotic because of a disruption of the local microvasculature. During the long healing process, the necrotic bone is gradually removed by dissolution of the inorganic salts and the enzymatic breakdown of the organic components of the bone matrix. A few days after the fracture, the clot is replaced by a mass of granulation tissue that forms the first connection between the two ends of the broken bone (Fig. 8-2). A short time later, cells of the inner endosteal and outer periosteal linings of the bone proliferate to form a provisional callus that joins the two ends of the broken bone (Fig. 8-3). The extent of the contribution of marrow-derived stem cells in normal fracture healing remains to be determined. The interior of the bone and the far reaches of the external part of the callus become vascularized quite quickly, and cells in these regions of the callus form tiny spicules (trabeculae) of bone. In contrast, the external callus often takes shape before it becomes completely invaded by regenerating capillaries. In this

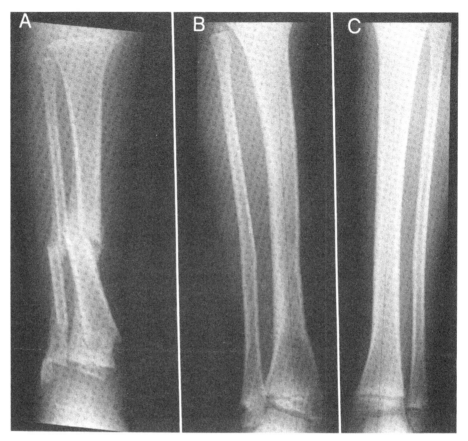

FIGURE 8-1 X-ray images showing postfracture remodeling of a young patient with severe fractures of one leg. *(A)* Early healing, showing new callus material filling in gaps between displaced edged of the fractured tibia and fibula. *(B)* Two years after the injury, the fractures had healed and remodeling had straightened out the contours of both fractured bones. *(C)* Normal bones in the contralateral leg. (Reprinted from Frost, H.M. 1973. *Orthopaedic biomechanics.* Springfield, IL: Charles C. Thomas, by permission.)

TABLE 8-1

Stages in the Healing of a Fractured Long Bone

1. The fracture itself, with accompanying hemorrhage
2. Formation of a blood clot between the ends of the fractured bone
3. Formation of a temporary bridge of soft granulation tissue over the fracture site
4. Replacement of granulation tissue by a harder callus of cartilage and bone spicules
5. Replacement of all cartilage in the callus by spongy bone
6. Conversion of the spongy bone in the callus to lamellar bone
7. Final remodeling of the regenerated bone to meet the functional demands of the site

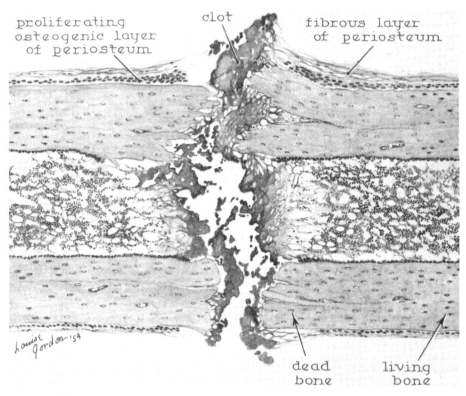

proliferating
osteogenic layer
of periosteum

clot

fibrous layer
of periosteum

Louise
gordon '54

dead
bone

living
bone

FIGURE 8-2   Drawing of an early stage in the healing of a fractured rib in a rabbit 48 hours after injury. Most of the fracture site is occupied by a blood clot, and the bone at the immediate margins of the fracture is necrotic. Early activation of the periosteum is evident. (Reprinted from Ham, A.W. and W.R. Harris. 1956. Repair and transplantation of bone. In: G.H. Bourne, ed. *The biochemistry and physiology of bone.* New York: Academic Press, by permission. 425–506.)

metabolically deficient environment, cartilage first forms within the avascular portion of the callus (Fig. 8-4). Over time, blood vessels penetrate the cartilaginous callus and facilitate its conversion to spongy bone.

While the callus first becomes ossified, the forming trabeculae do not appear to respond to local mechanical forces; rather, their orientation appears to be almost random. According to Frost (1989), the orientation of the first-formed trabeculae corresponds to the orientation of the ingrowing capillaries that support the formation of new bone. Later remodeling of the woven bone in the bony callus involves the laying down of lamellar bone in superficial regions of the callus and the formation of tubular secondary osteons (the fundamental functional unit of compact bone) that become aligned parallel to the lines of mechanical tension and compression across the fracture site (Ashhurst, 1986). Resorption of the original trabeculae of the early callus is part and parcel of the remodeling process. In humans, it takes from 1 to 4 months to get to this stage in the development of the callus. The overall dimensions of the callus shrink

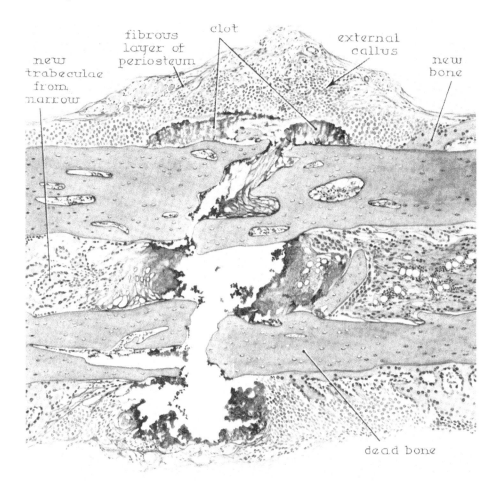

new trabeculae from marrow

fibrous layer of periosteum

clot

external callus

new bone

dead bone

FIGURE 8-3   Early stages in callus formation 1 week after fracture of a rib in a rabbit. Around the fracture site a collar of newly forming cartilage is creating the external callus, and within the callus, as well as the marrow cavity, a few spicules of new bone are forming. Internally, much of the fracture site continues to be occupied by clotted blood. (Reprinted from Ham, A.W. and W.R. Harris. 1956. Repair and transplantation of bone. In: G.H. Bourne, ed. *The biochemistry and physiology of bone*. New York: Academic Press, by permission. 425–506.)

as remodeling is occurring. Final replacement of the bony callus by typical arrays of well-formed and well-oriented osteons and lamellae typically takes from 1 to 4 years in humans. Although the resulting gross form and internal architecture of the healed bone are quite predictable, considerable uncertainty remains about the mechanism by which this structural reintegration occurs.

   One of the fundamental reintegrative processes in bone regeneration is the remodeling of the regenerating bone to provide the best structural fit to the mechanical demands placed on the bone. At a gross level for a fractured long bone, this means that the

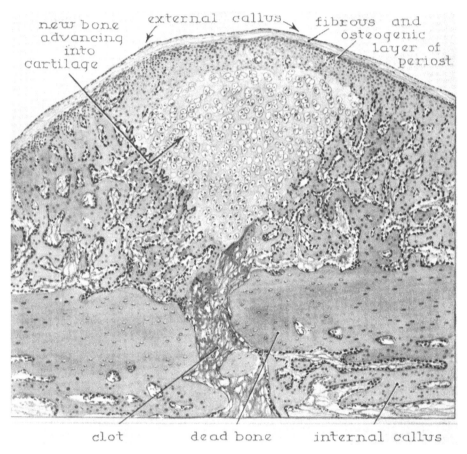

new bone advancing into cartilage   external callus   fibrous and osteogenic layer of periost

clot   dead bone   internal callus

FIGURE 8-4 A well-developed callus 2 weeks after fracture of a rabbit rib. Around the edges of the callus where blood vessels have penetrated, newly forming bone spicules have replaced the cartilage of the avascular callus. An internal callus is also forming in the marrow cavity. (Reprinted from Ham, A.W. and W.R. Harris. 1956. Repair and transplantation of bone. In: G.H. Bourne, ed. *The biochemistry and physiology of bone.* New York: Academic Press, by permission. 425–506.)

regenerated bone should be restored to a basically linear configuration. Fundamental aspects of regenerating bone remodeling can be readily visualized by examining a fractured bone that was set at an angle (Fig. 8-5). At the gross level, the general principle is that new bone will be deposited in areas of pressure (the concave side of a bent bone) and removed in areas of tension (the convex side). Over time, the original cross-section of the bone at the site of the fracture can migrate over a substantial distance through selective resorption and deposition of bone (see Fig. 8-5). At the microscopic level, a characteristic configuration of remodeling bone is the deposition of new bone by osteoblasts on one side of a trabecula and the removal of existing bone matrix on the opposite face by marrow-derived, multinucleated osteoclasts.

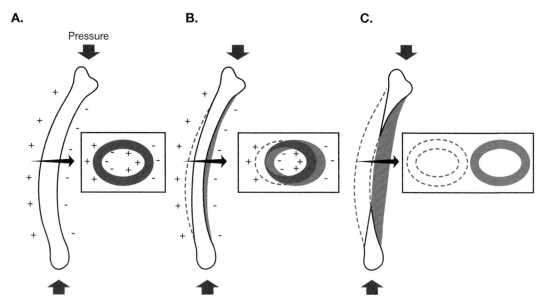

FIGURE 8-5   Diagrams illustrating the remodeling of a bent bone in relation to its mechanical environment. *(A)* The concave side of the bone has a net negative electrical charge, and that of the convex side is positive. *(B)* New bone (red) is deposited on the electronegative concave surface of the bone. Net resorption of bone (dashed lines) occurs on the positively charged convex face. As remodeling continues *(B, C),* the original shaft (blue) is completely resorbed, and the new shaft (red) is composed of newly regenerated bony material.

Gross remodeling of regenerating bone, such as that discussed earlier, does not occur until the regenerated bone has developed a lamellar or osteonal tissue structure. Although this has been recognized for many decades, only recently has there been some consolidation of thought concerning the cellular mechanisms underlying this phenomenon. It required an understanding of the cellular structure of compact bone.

A classic histologic section through compact bone is dominated by the osseous intercellular matrix that is periodically punctuated by small lacunae containing the osteocytes (Fig. 8-6, A). A ground bone preparation shows that emanating from the lacunae are numerous tiny canaliculi (see Fig. 8-6, B). These lacunae contain fine processes that extend from the cell bodies of the osteocytes, and it is now known that processes from neighboring osteocytes are not only in contact, but are coupled by gap junctions. The result is a cytoplasmic meshwork, resembling a diagram of a neural network, which extends throughout the compact bone. Through this meshwork, the osteocytes are afforded a means of rapidly communicating with one another.

To translate gross mechanical forces, such as footsteps or muscle contractions, into cellular phenomena within regenerating bone, there needs to be a recognizable stimulus, a means of sensing that stimulus, the transduction of the stimulus into cellular behavior, and the final response by the cells. Although details are still sketchy, a general model is beginning to take shape (Cowin and Moss, 2000).

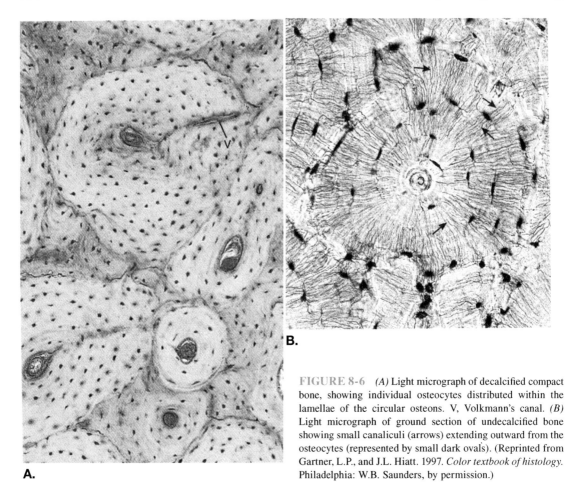

A.

B.

FIGURE 8-6 *(A)* Light micrograph of decalcified compact bone, showing individual osteocytes distributed within the lamellae of the circular osteons. V, Volkmann's canal. *(B)* Light micrograph of ground section of undecalcified bone showing small canaliculi (arrows) extending outward from the osteocytes (represented by small dark ovals). (Reprinted from Gartner, L.P., and J.L. Hiatt. 1997. *Color textbook of histology.* Philadelphia: W.B. Saunders, by permission.)

After mechanical stress in a bone, the strain (deformation) of the bone structure (only 0.2–0.3% *in vivo*) results in the propagation of microfluid pulses through the affected canaliculi. Pulsatile waves at low frequency (tenths to tens of Hertz) appear to be more effective stimuli than steady fluid pressure. Of the cells in compact bone, osteocytes are the most sensitive to these fluid-based signals, which could be received by the cells as either mechanical or electrical stimuli (Burger, 2001), the latter in the form of streaming potentials (Pollack, 2001). Because of the distances involved and the timing of responses, many investigators have ruled out intracellular chemical signaling through second messengers in the propagation of the stimulus. It is known that osteocytes respond to fluid displacement by producing nitric oxide, which stimulates the release of prostaglandin E_2 (Klein-Nulend et al., 1995; Rawlinson et al., 1991).

The osteocytic canalicular network extends to the surface of the bone matrix and connects to both osteoblasts and periosteal fibroblasts, but neither cell type is as sensitive as osteocytes to fluid shear stresses. Nevertheless, these are the cells that become activated in response to mechanical stimuli. The osteocytes appear to be the primary sensors and relayers of the mechanical information to actual bone-producing cells, but exactly how they accomplish this remains a matter of conjecture. Investigators have postulated the local release of growth factors or a general stimulation of metabolism by the released prostaglandins. What is even less well understood is how these signals are translated into the resorption of bone matrix, because in this instance, some signal must attract the monocyte-derived osteoclasts to the specific sites where bone resorption is slated to occur.

Distraction Osteogenesis (the Ilizarov Method)

An excellent example of the clinical application of knowledge concerning the reintegrative powers of bone is the Ilizarov technique (Green, 2003); this technique was devised in the early 1950s by the Siberian surgeon Gavriil Ilizarov. The essence of the technique is the external fixation of a fracture site by a series of rings placed outside an extremity and pins passing through the bone above and below the fracture site (Fig. 8-7). The rings are supported by metal rods, which allow the rings to be moved in a manner required for adaptation of the regenerating bone. This arrangement stabilizes the fracture site, and by judicious manipulation of the rings through the supporting rods, the healing fracture can be molded in a wide variety of manners, ranging from no displacement to partial rotation to lengthening.

One of the most dramatic applications of the Ilizarov technique is its use in lengthening short bones. A critical element in the success of this technique is the timing of the application of traction by use of the Ilizarov apparatus. Initially, a long bone, such as the femur, is divided in a manner to preserve the marrow, endosteum, and periosteum as much as possible. Normally, the early stages of fracture healing (collectively called the *latency period*), involving inflammation and the early formation of a type of granulation tissue, are allowed to proceed until the start of the next stage, that of the soft callus. The period of the soft callus, which lasts about 3 weeks in humans, occurs when the granulation tissue becomes replaced by a bed of vascularized fibroblasts. In some peripheral nonvascularized areas, cartilage forms (Fig. 8-8). This is the time when the Ilizarov apparatus is put to work. During the period of the soft callus, the metal support rods in the Ilizarov apparatus are adjusted in a way that gradually moves the two ends of the fractured bone apart (sometimes 1 to 2 mm/day depending on the need). As traction is applied by the apparatus, the fibroblasts in the gap between the two cut ends of the bones become highly oriented parallel to the lines of traction (Fig. 8-9). The traction also stimulates a much greater growth of blood vessels than would be the case in the healing of a normal fracture. This results in the expansion of the fracture callus and the ultimate deposition of sufficient bone in the fracture site to increase its length by several centimeters.

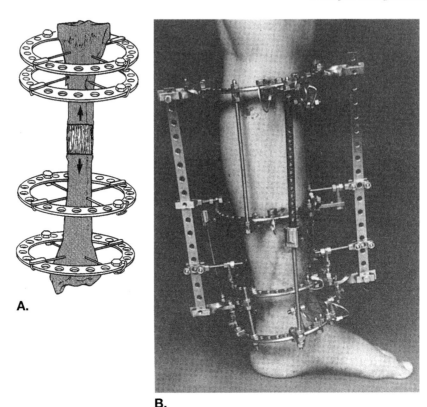

FIGURE 8-7 *(A)* Apparatus used for the Ilizarov method of lengthening a bone. The circular rings *(B)* are fixed to the bone, and the metal rods are used to gradually force apart the open ends of the fractured bone. *(B)* Diagram of the positioning of the metal rings connected to the bone in the Ilizarov apparatus. When the top two rings are gradually pushed away from the bottom two rings, the soft tissues at the fracture site elongate and will ultimately lengthen the bone. (Reprinted from Green, S. 2003, In: B.D. Browner et al., eds. *Skeletal trauma*. 3rd ed. Philadelphia: W.B. Saunders, by permission.)

Interestingly, although the Ilizarov technology was new in the 1950s, the concept of lengthening bones by distraction was not. As early as 1923, Bier illustrated a technique of dividing femurs and increasing the length of the fracture callus in an operation designed to lengthen the legs of a dwarf (Fig. 8-10).

Electrically Mediated Fracture Healing

When bones are mechanically deformed, they become electrically polarized. The mechanism of polarization is still not completely clear. The role of streaming potentials in the adaptation of living bone (see earlier) relates to the displacement of fluid in the canaliculi of compact bone matrix. Another source of electrical potential resides in the piezoelectric properties of bone collagen (Fukada and Yasuda, 1957). Regardless of

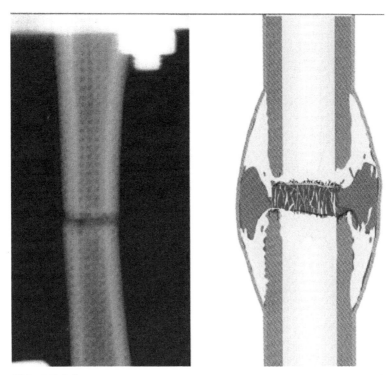

FIGURE 8-8 Radiograph and drawing showing the soft callus stage in a healing fracture. The radio-lucent granulation tissue that has formed in the fracture site is converting to fibrous tissue within the bone and cartilage within the external callus. (Reprinted from Samchukov, M.L., et al. 2001. *Craniofacial distraction osteogenesis*. St. Louis: Mosby, by permission.)

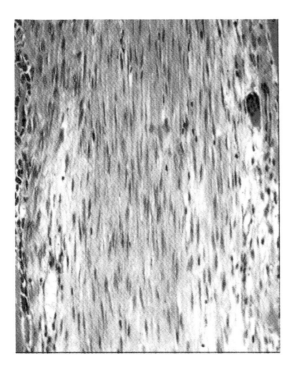

FIGURE 8-9 Photomicrograph showing the parallel arrangement of newly forming collagen fibers and associated fibroblasts located in the distraction gap between two ends of a fractured bone undergoing elongation by the Ilizarov method. (Reprinted from Samchukov, M.L., et al. 2001. *Craniofacial distraction osteogenesis*. St. Louis: Mosby, by permission.)

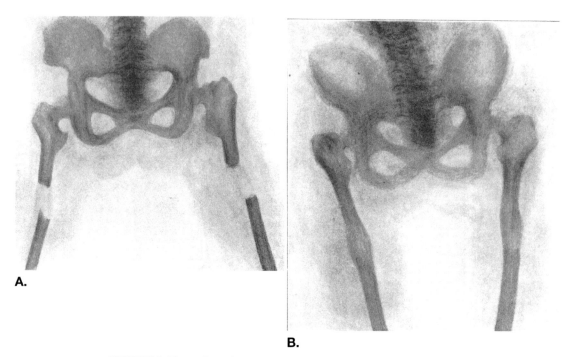

A.

B.

FIGURE 8-10   Radiographs showing lengthening of the femurs in a dwarf by division, and then lengthening the limb by 4.5 cm. *(A)* Gap between the divided bones. *(B)* Regenerating bone is filling in the spaces between the ends of the divided bones. (After Bier [1923], from Korschelt, E. 1927. *Regeneration und transplantation.* Vol. 1. Berlin: Verlag Gebrüder Borntraeger, by permission.)

the mechanism, a net negative charge builds up on the side of a bone that is subjected to compression (e.g., the concave face of a misaligned fracture), and a positive charge is found on the side exposed to tension (the convex surface). As a generalization, there is a net deposition of bone on a negatively charged surface and a net resorption of bone at an electropositive surface. After Yasuda's (rev. 1974) demonstration that new bone forms at the negative pole (cathode) of an artificially applied electrical field (see Fig. 11-6), orthopedic surgeons have applied a variety of devices designed to accelerate fracture healing or to stimulate bony healing of nonhealing fractures (pseudoarthroses) (McGinnis, 1989). Typical success rates are about 80%. The variety of electrical healing devices is large and beyond the scope of this text, but they have been applied successfully to hundreds of thousands of patients. Although all do not borrow directly from the reintegrative bioelectrical mechanisms normally found in bone, they nevertheless illustrate the importance of identifying and exploiting such mechanisms.

SKELETAL MUSCLE

Reintegration of regenerating skeletal muscle occurs in several steps. At the first level, damaged areas of muscle that have been disconnected from the local vasculature must become revascularized before the regeneration of new muscle fibers can begin. If a muscle has been severely damaged, the mechanical environment that acts on the region of regeneration works to align the regenerating muscle fibers with the lines of local mechanical stress. Finally, when regenerating nerve fibers make their way back to the regenerating muscle fibers, neural reintegration must occur.

In areas of devascularized damaged muscle, the normal sequence of events is a short period of intrinsic breakdown of the necrotic muscle fibers, followed by an influx of macrophages that phagocytize the remains of the necrotic muscle fibers (Carlson et al., 1979a). The appearance of macrophages is closely accompanied by the ingrowth of capillary sprouts. At this point, satellite cells within the persisting muscle fiber basal laminae become activated, and the regeneration of new muscle fibers begins. By the time muscle fiber regeneration has reached the late myotube stage, the regenerating capillaries have already begun to align themselves parallel to the regenerating muscle fibers. Over time, the alignment of capillaries with muscle fibers becomes more ordered, and ultimately, the normal mosaic pattern of capillaries is reestablished (Grim et al., 1986).

Similar to a number of other tissues, damaged skeletal muscle exhibits angiogenic properties (Phillips and Knighton, 1990), which foster vascular reintegration early in regeneration. Assays with partially purified muscle extract show evidence of both monocyte and endothelial cell chemotaxis (Phillips et al., 1991).

Once revascularization has been initiated and muscle fiber regeneration has begun, the next major reintegrative event is the alignment (or realignment) of the regenerating myotubes along lines of mechanical tension. This is best illustrated in the model of minced muscle regeneration (Carlson, 1972b), in which fragments of the originally implanted mince are oriented randomly. Within 2 weeks after implantation, regenerating muscle fibers at the periphery of the mince begin to be aligned parallel to the long axis of the muscle (see Fig. 4-2). This alignment depends on the reestablishment of connections between the tendon stumps of the host and the implanted mince of muscle fragments. Over time, most of the regenerated muscle fibers assume a parallel orientation. In the absence of tendon connections, the regenerating muscle fibers in a mince retain their random orientation (Fig. 8-11), but if directed tension is applied artificially to such a system, good orientation of muscle fibers results (Fig. 8-12).

Directed mechanical tension also is important in the growth of regenerating muscle. This becomes apparent in experiments in which most of the rat gastrocnemius muscle, together with all of the plantaris, is removed (Carlson, 1974b; Litver et al., 1961). In some cases, a 5-mm proximal stump of the muscle attaches to nearby tissues and does not grow further. However, if the regenerating Achilles tendon makes a connection with the distal end of the gastrocnemius muscle stump, the tension applied through the regenerated tendon results in an up to fivefold growth in length of the regenerating muscle stump.

Subsequent research shows that the same principles apply to artificial muscles constructed *in vitro*. Vandenburgh and Kaufman (1979) have devised a model for stretch-induced hypertrophy of cultured muscle, and Vandenburgh (1982) has shown that directed tension in the *in vitro* stretch model results in the alignment of developing chicken muscle fibers along the lines of mechanical tension. Later, with a different *in*

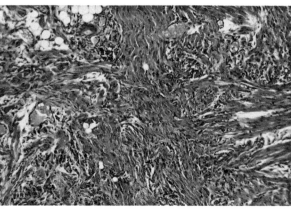

B.

A.

FIGURE 8-11   Regeneration of minced muscle fragments from the gastrocnemius muscle placed under the abdominal skin in the rat. *(A)* Fourteen days after mincing, the regenerating muscle (arrow) has rounded up into a button-shaped mass. *(B)* Microscopic section through the same regenerate. The regenerated muscle fibers are randomly oriented in three dimensions. (Reprinted from Carlson, B.M. 1972b. *The regeneration of minced muscles.* Basel, Switzerland: S. Karger, by permission.)

vitro model, Kosnik and colleagues (2001) showed the importance of a fibroblastic matrix in the orientation of the developing muscle fibers. Perrone and coworkers (1995) have shown that stretch influenced the secretion of insulin-like growth factor-1 in their *in vitro*–cultured avian myotubes, and Tatsumi and coworkers (2001) have found that quiescent satellite cells subjected to stretch rapidly released hepatocyte growth factor, resulting in their activation. Each of these growth factors could be a means by which a mechanical stimulus is translated to cellular behavior in regenerating muscle.

Another reintegrative force is well illustrated in minced muscle regeneration. Minced muscle regenerates develop proximal and distal tendon connections, and they assume the external form of a generic muscle. However, both external form and internal architecture typically differ substantially from that of the original muscle. It has been established experimentally that gross form of a regenerating minced muscle can be accounted for by external mechanical forces (Carlson, 1972b), rather than the morphogenetic field influences that guide the shaping of an epimorphically regenerating muscle to near-perfect morphology. An implant of the surgical sponge material Gelfoam (Pharmacia & Upjohn, Bridgewater, NJ) into the bed of a removed gastrocnemius muscle becomes connected to the host through integration with the proximal and distal

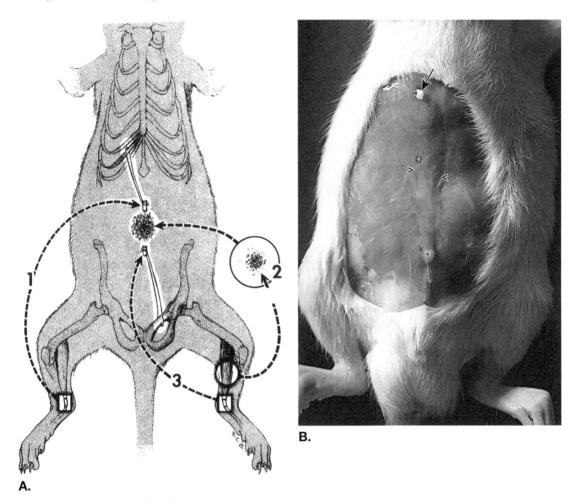

A.

B.

FIGURE 8-12 Experiment demonstrating the effect of applied tension on the external morphogenesis of regenerating muscle in the rat. *(A)* Experimental design. (2) Half of the gastrocnemius muscle is minced into 1-mm^3 fragments and implanted beneath the abdominal skin. (1 and 3) Pieces of Achilles tendon are attached to fixed skeletal points, and the tendons were positioned at the edge of the implanted mince. Within days, the tendon pieces become attached to the implanted muscle fragments by connective tissue. As the animal grows, the tendon stumps are pulled apart, creating continuous tension on the regenerating minced muscle. *(B)* Gross photograph of a 14-day abdominal mince that regenerated under tension. The cranial tendon piece (arrow) broke free, and that end of the regenerate shows radially oriented connective tissue adhesions (arrowheads). The caudal tendon piece (asterisk) remained connected to the mince and pulled it into a tongue-shaped structure. Within it, the muscle fibers were oriented in a parallel fashion. (Compare with Fig. 8-11.) (Reprinted from Carlson, B.M. 1972b. *The regeneration of minced muscles.* Basel, Switzerland: S. Karger, by permission.)

tendon stumps of the host. Its overall form closely approximates that of a regenerating minced muscle, even though the Gelfoam implant contains no muscle cells (see Fig. 9-3).

A final form of reintegration of skeletal muscle is reinnervation. In the case of severe injury or after transplantation, regenerating nerve fibers must grow back to the regenerating muscle fibers and make functional contacts. Reinnervation involves both motor and sensory nerves. After certain types of ischemic injury, the nerves are not mechanically damaged, and regenerating axons can follow existing pathways. Under these circumstances, the regenerating motor axons typically return to the sites of the original neuromuscular junction. It has been clearly established that, under these circumstances, information inherent in the junctional basal lamina of the muscle fibers provides a preferential end point for the regenerating axons (Sanes et al., 1978). After severe trauma, where the regions of the original neuromuscular junctions are not available, the regenerating motor axons are capable of forming neuromuscular junctions *de novo* (Womble, 1986). Shortly after the establishment of neuromuscular junctions, the regenerated nerves are able to send contractile signals to the regenerating muscle fibers (Carlson et al., 1981). Although muscle spindles re-form in regenerating muscles if the original spindle capsule is left intact, their innervation is incomplete (Rogers, 1982). How well sensory reintegration occurs will not be known until further functional studies are conducted.

NERVOUS SYSTEM

Any time regeneration occurs in the nervous system, both structural and functional reintegration must necessarily occur for regeneration to be complete. Of the many examples of neural reintegration, one of the best studied is that of regenerating retinal neurons and the connections that they make with the visual areas in the brain. Retinotectal regeneration has been studied for more than a half century, and much of our understanding of neural reintegration has been based on analysis of this phenomenon.

In most vertebrates, there is a strong, point-for-point correspondence between the position of neurons located in the retina and the sites of termination of their axons in the tectal area of the midbrain. In lower vertebrates, the axons of the optic nerves decussate almost completely to the contralateral side; by the time they reach the optic tectum on the opposite side, the orientation of the tectal map typically is rotated to some extent (90 or 180 degrees), even though the basic map remains the same as that of the retina (Fig. 8-13).

One of the early breakthrough experiments in the field of retinotectal regeneration is that of Sperry (1943, 1944), who cut the optic nerve in newts and frogs and rotated the eyeball 180 degrees about the axis of the optic nerve. The optic nerve regenerated, and vision was restored. However, the animal's response to a visual stimulus was thereafter 180 degrees opposed to normal, so that when striking at a target located dorsal to the eye, the action was actually directed ventrally. The interpretation of this

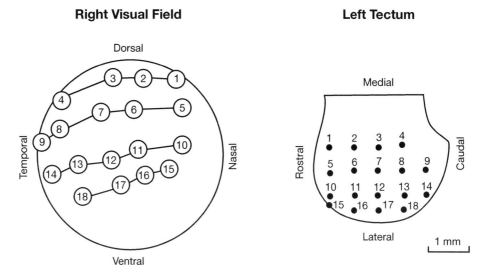

FIGURE 8-13   Maps of correspondence between stimulated electrophysiologic points on the right retina of a goldfish (left) and their representation on the left optic tectum (right). (After Purves and Lichtman [1985], by permission.)

experiment was that the regenerating retinal axons retained their original specificity, and that no relearning occurred. Thus, these axons made synaptic connections with their original tectal neurons in correct map order, but because there was no postregeneration respecification within the array of retinal neurons, the visual reflex was reversed. Although some form of chemical specificity was postulated, decades passed before significant progress was made on identifying the nature of the molecular basis for the retinotectal map.

Other, less widely known experiments shed light on an important property of this system, namely, that it is able to compensate for defects through a form of regulation. Yoon (1971) ablated a portion of an optic tectum of goldfish and severed the contralateral optic nerve. When the nerve regenerated, a complete retinotectal projection could be mapped on the incomplete tectum, which did not itself regenerate (Fig. 8-14, A). In the converse experiment, Schmidt and coworkers (1978) excised part of the retina of one eye and severed the optic nerve. In this case, a complete retinotectal projection ultimately took shape, indicating that, again, the system compensated for the defect in the retina (see Fig. 8-14, B). In addition to demonstrating considerable powers of regulation in both major components of this system, these experiments also showed that the retinotectal map is not based on point-for-point correspondence from specific qualitative chemical signatures, or visual defects would have persisted after regeneration was completed. These experiments lent support to Sperry's (1963) chemoaffinity hypothesis, which includes the proposition that quantitative matching within chemical gradients could be the basis for spot-to-spot matching within the tectum.

A.

Compression of retinotectal map

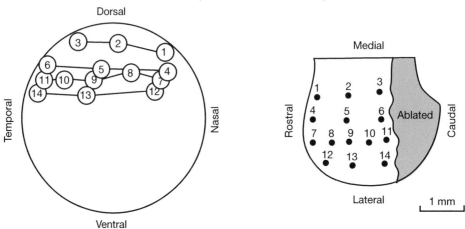

Right Visual Field

Left Tectum

B.

Expansion of retinotectal map

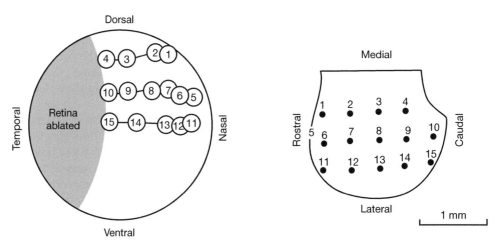

FIGURE 8-14 Functional visual pattern regulation within the visual system. *(A)* After partial removal of the tectum (green area), the retinal field is compressed, but after several months, the entire retinal field is represented in the remaining tectum. *(B)* Several months after partial ablation of the retina (green area), the retinal visual field is restored. (After Purves and Lichtman [1985], by permission.)

This hypothesis was borne out in the 1990s, when a greater understanding of the molecular basis for neural interactions allowed further productive experimentation, principally in birds. In the chick, temporal retinal neurons project onto the anterior tectum, whereas nasal retinal neurons project onto the posterior tectum. *In vitro* experiments (Stahl et al., 1990; von Boxberg et al., 1993) involved culturing neurons from nasal or temporal regions of the retina against a background of strips coated with membrane preparations of anterior or posterior tectum (Fig. 8-15). Given the choice, the temporal retinal axons preferentially selected the strips coated with anterior tectal membrane, whereas the nasal retinal axons preferred the posterior strips.

With the recognition of the importance of ephrin ligands and Eph receptors in neural guidance and mapping phenomena (rev. Holder and Klein, 1999), it was established that the chicken homologue of ephrin-A5 is distributed in an anterior (low) to posterior (high) gradient in the tectum (Drescher et al., 1995). Conversely, the Eph-A3 receptor is distributed in a nasal (low) to temporal (high) gradient in the retina (Fig. 8-16). Ephrin-A2 in the tectum repels temporal (high Eph-A3) retinal axons, causing the most temporal of the axons to terminate in the anterior part of the tectum. In contrast, nasal axons are not repelled by posterior tectal cells. Although this answers part of the puzzle, more remains to be learned about why nasal axons project to anterior tectum and whether attractive forces, as well as repulsive forces, may be involved in constructing and maintaining the overall tectal map.

The tectum is not the only location in the brain that contains gradients. Topographic specificity exists in the connections between the entorhinal cortex and the hippocampus and from the hippocampus to the lateral septum of the brain (Zhou and Black, 2000). A gradient of ephrin ligands exists in the lateral septum, with a countergradient of Eph receptor in the hippocampus, and the distribution of connections operates according

Neurons from temporal half of retina   Neurons from nasal half of retina

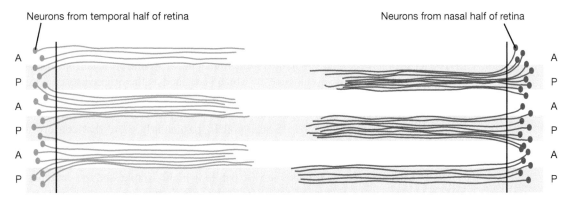

FIGURE 8-15   An experiment demonstrating selectivity of retinal axons as they interact with tectal membranes. In culture, the substratum has been coated with alternating strips of membrane from anterior or posterior tectum. Retinal neurons are then plated onto the dish. Axons from neurites derived from temporal halves of the retina selectively grow along strips of anterior tectal membrane (left), whereas axons from nasal halves select strips of posterior tectal membrane. (Based on von Boxberg and colleagues' experimental data [1993].)

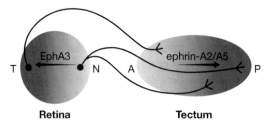

FIGURE 8-16   Role of ephrins in specifying the retinotectal map. Ephrins A2 and A5 are expressed as a concentration gradient in the tectum, and the ephrin receptor, Eph3, forms a gradient in the retina. Neurites from the high point in the retinal Eph3 gradient (red) seek out tectal locations at the low point (yellow) of the ephrin A2/A5 gradient. Conversely, neurites from the low Eph3 gradient in the retina connect in areas of the tectum characterized by high points in the ephrin A2/A5 gradient. A, anterior; N, nasal; P, posterior; T, temporal. (Based on Zhou and Black's experimental data [2000].)

to principles similar to that seen in the tectum. Topographic relations, such as the preceding example, may be a feature underlying much of the organization of the central nervous system.

ANGIOGENESIS

Almost every variety of injury leading to regeneration involves a disruption of the local vasculature. The establishment or reestablishment of vascular connections between the body and the regenerating structure is of paramount importance. For many types of tissue regeneration in mammals, revascularization and regeneration progress hand in hand. In cases that involve massive ischemia, early revascularization is a precondition for the initiation of regeneration. Limb regeneration in urodeles is unusual in that the distal limb stump remains poorly vascularized during the early stages of epimorphic regeneration. Even more extreme, regeneration of a lens from the dorsal iris of urodeles involves no direct disruption of a local vascular supply.

The term *angiogenesis* refers to the production of new vascular sprouts from existing vessels; this differs from vasculogenesis, which is defined as the original formation of a primitive vascular network in an embryo or a formerly avascular area. Despite the importance of angiogenesis in regeneration and healing processes, most contemporary studies have been conducted on disease models, particularly those that concern the vascularization of tumors and attempts to inhibit tumor angiogenesis. Nevertheless, one can extract highly relevant information from that area of the literature.

Under normal adult conditions, the vasculature is highly stable, and the endothelial cells may remain mitotically quiescent for years. After injury or the development of certain pathologic agents, an "angiogenic switch" is turned on. This results in the local production of angiogenic factors, as well as a change in the properties of the endothelial cells so that they are receptive to the effects of the angiogenic factors. The blood vessels themselves must become destabilized for angiogenesis to occur.

The general principle of regenerative angiogenesis is that injured or hypoxic tissues produce angiogenic factors (ligands) that, upon acting on specific receptors on endothelial cells, stimulate their ingrowth into the injured area. Hypoxia in the injured tissue activates a hypoxia-inducible factor (HIF) system, which has emerged as a major regulator of the vasculature (Maxwell and Ratcliffe, 2002). Under normal conditions of oxygenation, HIF subunits are destroyed intracellularly by proteasomes. The HIF system induces the production of vascular endothelial growth factor (VEGF) by cells in the region of damage. VEGF induces the proliferation, chemotaxis, and differentiation of endothelial cells through binding to a specific receptor, VEGFR2, which is expressed principally on endothelial cells. Like so many developmental and regenerative processes, there appears to be a backup system. In this case, a number of growth factors (e.g., basic fibroblast growth factor, epidermal growth factor, transforming growth factor, tumor necrosis factor), bound by local connective tissues, have for years been known to have angiogenic properties (DiPietro and Nissen, 1998). The activation of endothelial cells in mature blood vessels appears to be facilitated by a destabilizing influence, namely, angiopoietin 2 (ang2), which inhibits the signaling of Tie2, a receptor that normally acts to stabilize the association between endothelial cells and the cells plus matrix that normally surrounds them (Yancopoulos et al., 2000).

Within a couple of days, the VEGF-stimulated, unstable endothelial cells begin to proliferate and form minute, solid endothelial cell buds, which then grow out into the region of tissue damage (Fig. 8-17). With continued outgrowth, the angiogenic buds form a network of endothelial cell cords, which undergo considerable pruning and remodeling (Risau, 1997). These vessel precursors begin to canalize, allowing blood to enter. Initially, the blood is stagnant, and the new vessels become highly dilated. As this is occurring, another angiogenesis factor, angiopoietin 1 (ang1), binds to the endothelial Tie2 receptors and stimulates the process of building up and stabilizing the elements of the vascular wall (Carmeliet, 2000). As perivascular smooth muscle differentiates, the lumen of the new arterial vessel constricts. Once connections are made between the arterial and venous components of the local circulation, vigorous blood flow resumes.

Local perivascular cells appear to supply tissue-specific information to the regenerating vessels, resulting in the development of specific morphologic and functional characteristics of the endothelial cells (LeCouter and Ferrara, 2002). Thus, capillaries invading sites of damage in the central nervous system reconstitute the blood–brain barrier, whereas those in secretory structures may become fenestrated.

In addition to local endothelial cells, circulating marrow-derived precursor cells may also participate in the vascular healing process (Rafii et al., 2002). It has been estimated that these VEGFR2-positive precursor cells constitute less than 0.01% of all circulating mononuclear blood cells. Like resident endothelial cells, these precursor cells respond to the locally produced VEGF and undergo a transformation into reparative endothelial cells (Takahashi et al., 1999). They also appear to be a main source of the endothelial cells that ultimately line the inner surface of vascular prostheses (Rafii et al., 1995).

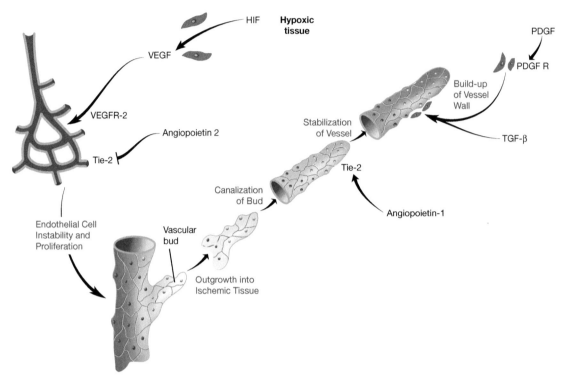

FIGURE 8-17   Scheme of regenerative angiogenesis. The process is stimulated by hypoxia-inducible factor (HIF) originating in ischemic tissues and working through vascular endothelial growth factor (VEGF) to stimulate vascular budding. PDGF, platelet-derived growth factor; TGF, transforming growth factor.

In most cases of tissue regeneration, the vasculature adapts remarkably well to the structure and needs of the regenerating tissues. When the course of regeneration has reached stability, the distribution and functional characteristics of the regenerated blood vessels are often almost normal for the region.

REGENERATING LIMB

Although not greatly stressed in limb regeneration studies, reintegration plays an important role in epimorphic regeneration. From a purely morphologic standpoint, the covering of the amputation wound surface by epithelium, the ingrowth of nerves, and the ingrowth of the vasculature into the regenerate are all examples of reintegration. Another level of reintegration is seen at the interface between the epimorphically produced regenerate and the remaining tissues of the limb stump (see Fig. 1-4, E). This is a zone where tissue regenerative processes provide continuity between damaged stump tissues and the tissues arising from the blastema. In the case of the skeleton, subperiosteal cartilage forms a collar around much of the remaining bone in the stump and connects with the most proximal skeletal element formed from the blastema. A similar situation applies to muscle, where myotubes generated through the tissue mode of regeneration cross into the area under the influence of the epimorphic field and grow into the regenerating part of the limb. In the case of both subperiosteal cartilage and regenerating muscle fibers, these cells regenerating by the tissue mode have begun to differentiate into easily recognizable cartilage and muscle fibers well before any differentiation of these tissues occurs within the blastema itself.

The importance of the tissue regeneration interface between stump and epimorphic regenerate is seen particularly well in experiments that involve removal of tissue components of the stump. When either skeletal elements or muscles are removed from the stump, tissue regeneration of these structures does not occur within the stump, and the epimorphically regenerated limb remains structurally and functionally disconnected from important corresponding tissue elements of the body (Carlson, 1972a). Similarly, in the amputated urodele tail, unilateral removal of the stump musculature is followed by the regeneration of a tail with normal musculature, but muscle is not restored in the side of the tail stump where muscle was removed (Dinsmore, 1981a).

Under conditions where either tissue or epimorphic regeneration could occur in the amputated limb, epimorphic regeneration takes precedence. This was demonstrated in an experiment in which the limb muscles of the axolotl were minced, and the limb was amputated at a level transecting the minced muscle (Fig. 8-18). In the proximal parts of the limb stump, the muscles regenerated by the tissue mode; within the blastema, muscles regenerated by the epimorphic mode. At the distal 2-mm of the limb stump,

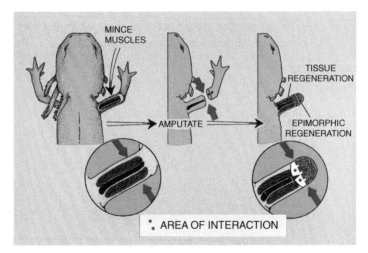

FIGURE 8-18 Interactions between tissue and epimorphic regenerative processes in the axolotl. The upper arm muscles are minced, and the limb is amputated through the level of the minced muscle. Proximally, the minced muscles regenerate by the tissue mode. In the distal end of the limb stump, epimorphic regeneration takes precedence over the tissue regeneration of the minced muscles.

the minced muscle was swept into the dedifferentiative reaction and into the orbit of the regeneration blastema. This shows that, in this case at least, epimorphic regeneration is dominant over tissue regeneration.

MORPHALLAXIS

One of the most spectacular examples of reintegration is seen in processes that involve morphallaxis. In these processes, large parts of the body literally reorganize themselves to accommodate the regenerate, which is often produced by an epimorphic process. An excellent example of this is seen in the regeneration of the head in sabellid worms after amputation through the level of the abdominal segments. As head regeneration is occurring, the abdominal segments close to the amputation surface undergo a pronounced transformation into thoracic segments (see Fig. 1-12). This involves the resorption of abdominal setae (bristles) and their replacement with thoracic appendages in a different location on the segments.

SUMMARY

Reintegration is a relatively neglected facet of regeneration, but particularly in mammalian tissue regeneration, reintegrative forces are critical in returning the regenerating tissues to a functional relationship with the body. Many reintegrative mechanisms are tissue specific, and they may involve factors as different as physical forces or chemical growth factors. With the increasing interest in regenerative medicine, whether involving natural regenerative processes, tissue engineering approaches, or stem cell therapy, a greater understanding of reintegration between the regenerating tissue and its surroundings will be critical for human applications of regeneration.

CHAPTER 9

Regeneration and Embryonic Development

A brief consideration of regenerative processes will disclose that they are fundamentally of the same nature and follow the same principles as the ontogenetic processes.

—Paul Weiss (1939)

Regeneration, especially that of the limb, shares many similarities with embryonic development. This inevitably leads to the question of the relation between the two processes. One of the most important aspects of this question is whether a regenerating structure uses the same genetic program of developmental control as its embryonic counterpart, or whether regenerating structures have evolved their own unique mechanisms that lead to restoration. Especially for structures in which pattern-forming mechanisms are involved, the issue becomes more complex because regeneration in this case involves both the reestablishment of a pattern and cellular differentiation. In tissue regeneration, a significant question is whether any aspects of morphogenesis reuse embryonic mechanisms, or whether morphogenesis is accomplished through different mechanisms.

AMPHIBIAN LIMB

The most obvious feature of limb regeneration is that whatever is going to regenerate is, by necessity, connected to an already existing body. The cut surface of an amputated limb or the site at which a supernumerary limb is induced to form is vastly different from the part of the body wall of the embryo that creates a limb bud. From the standpoint of morphology, it is not until a regeneration blastema forms that there is an apparent similarity between a regenerating and an embryonic limb. Nevertheless, one of the most important questions addresses what is happening in the preblastemal phase of limb regeneration and what relationship, if any, exists between the events of this phase and those underlying the initiation of limb development.

The period that determines whether limb regeneration will occur begins with epidermal wound healing of the amputation surface and ends when dedifferentiation is in full swing (Polezhaev, 1950). Whereas initially a developing limb in the embryo arises solely from lateral mesoderm and its overlying ectodermal covering, an amputated adult limb will regenerate on a diverse cellular foundation. Beneath the wound epidermis are a fibroblast-rich dermis, transected muscle and skeletal tissue, nerves, and blood vessels. A similarity between early limb development and regeneration is that neither process will occur in the absence of an ectodermal/epidermal covering. A significant difference is that limb development is initiated in an aneural environment, whereas the initiation of regeneration is dependent on the presence of nerves. If nerves are absent in adults, dedifferentiation may be initiated, but it does not progress, whereas in larvae, dedifferentiation proceeds unchecked (rev. Carlson, 1974c). A blastema does not form in either case.

From molecular studies, Gardiner and colleagues (1999, 2002) have subdivided this early period into two phases: wound healing and dedifferentiation (Fig. 9-1, A). Simply creating a skin wound, whether over the end of an amputated limb or by simply removing skin from the lateral surface, elicits the rapid (within 1–2 hours) expression of genes, such as *Msx-2* and *MMP-9,* in the underlying tissues. Expression of these genes is, as is epidermal wound healing itself, independent of innervation, but it does not occur if the amputation surface is covered by full-thickness skin, which inhibits regen-

A. Limb

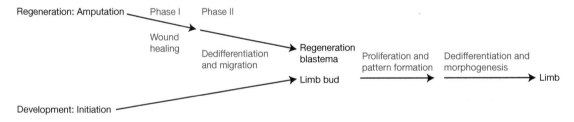

B. Urodele Lens

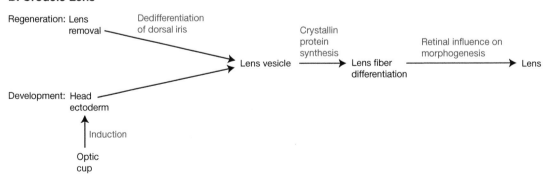

FIGURE 9-1   Diagrams showing converging pathways for development and regeneration. *(A)* Urodele limb. (Based on Gardiner and colleagues [2002].) *(B)* Urodele lens. The single arrows to the right indicate processes in which similar gene products guide both embryonic development and regeneration.

eration. This points out that the presence of epidermis, as well as deep tissue damage, is vital for the initiation of regeneration.

Innervation, as well as the lytic action of the matrix metalloproteinases, is required for transition into the second phase of regeneration, namely, dedifferentiation. The reexpression of several *Hox* genes is a prominent feature of this transition period. Two of these genes, *HoxD-8* and *HoxD-10,* are also expressed within the first day in non-regenerating lateral wounds, but two others, *HoxA-9* and *HoxA-13,* appear to mark the meaningful transition from the nonregenerating to the regenerating state (Gardiner and Bryant, 1996). The latter two genes, which are both spatially and temporally expressed colinearly in the embryonic limb bud, are expressed initially at the same time and place in the amputated limb—just as dedifferentiation is beginning. Not until the blastema is formed and begins to grow out does an embryo-like colinear pattern of expression of these genes take shape. Gardiner and Bryant (1996) speculate that the early expression pattern of these two genes may create an early center of distal morphogenetic information (through the HoxA-13 protein), which then serves as a morphogenetic reference point that allows positional values of distal stump tissues to become reassorted (see Fig. 7-14) so that intercalation can occur.

All of these changes during regeneration are occurring on a background of tissues in which the initial morphogenetic specification had already occurred in the early embryo. For example, the specification of forelimb versus hind limb does not need to be repeated in the regenerate. In fact, one of the anomalous molecular patterns seen in regeneration is the expression of *HoxC-10* in the regenerating forelimb. In the embryo, expression of this gene is limited to the developing hind limb and tail (Carlson et al., 2001). In addition to its functions in limb development, *HoxC-10* appears to play a unique role in regeneration, possibly involving the proliferation of dedifferentiating cells. *Tbx5* and *Prx1* are also expressed early after both skin wounding and amputation, and they are independent of innervation (Suzuki et al., 2005). These transcription factors may also serve as dedifferentiation markers.

The phase of dedifferentiation occurs under the continuing influence of the thickened apical epidermal cap and innervation. In the early 1960s, Rose (1962) proposed a "tissue-arc" sequence of interactions by which nerves maintain the integrity of the apical epidermal cap, which, in turn, promotes blastema formation. It is now known that the apical cap produces fibroblast growth factors (FGFs; e.g., FGF-1, -2, and -8), and that in denervated limbs, which are characterized by an attenuation of the apical cap, FGF-2, delivered by implanted beads, will rescue regeneration (Mullen et al., 1996). Neurotrophic factors, such as transferrin (Mescher, 1996), promote the continuing proliferation of the early blastemal cells.

At a point during the period of late dedifferentiation and early blastema formation, the regenerative process makes the transition into a phase in which both the morphology and underlying controls of the regenerative process begin to closely resemble those of the developing embryonic limb. The early blastema is a homogeneous-looking mass of cells beneath an overlying epidermal covering that can self-differentiate, as can the embryonic limb bud. As differentiation begins to occur within the outgrowing blastema, the manifestation of specific patterns of a differentiated tissue structure follows a course that is remarkably similar to that which occurs in embryonic development (Grim and Carlson, 1974). In parallel with the embryonic limb, the outgrowing regenerate becomes, for the most part, independent of innervation for its basic morphogenesis.

In the premolecular era, transplantation experiments suggested that the controls driving embryonic and regenerating limbs are similar. In small axolotls, Muneoka and Bryant (1982) grafted the tips of embryonic limb buds onto amputated limbs and regeneration blastemas onto amputated limb buds, all with the anteroposterior axis of the graft reversed in relation to that of the host (Fig. 9-2). Muneoka and Bryant found that supernumerary limbs of the expected type and location formed at the graft–host junction in the majority of cases after each type of graft, and they concluded that the pattern-forming mechanisms must be the same between embryonic and regenerating limbs.

Molecular studies have, for the most part, shown considerable similarities in patterns of expression of genes or gene products between developing and regenerating limbs. Mentioned earlier in this chapter is the establishment of colinearity between *HoxA-9* through *-13* once the regenerating limb begins to grow out. Similarly, sonic hedgehog (shh) is expressed in embryonic and regenerating limbs (Imokawa and Yoshizato,

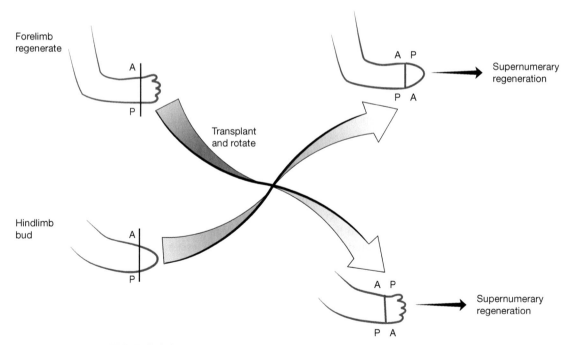

FIGURE 9-2 Exchange transplantation and 180-degree anteroposterior (A-P) rotation of embryonic limb buds onto limb stumps or regeneration blastemas onto amputated embryonic limb buds in the axolotl. In each case, supernumerary limbs regenerate from areas of A-P discontinuity. (Based on Bryant and colleagues' data [1987].)

1997), but it appears later in the regenerative process than it does in the embryonic limb bud. Another hedgehog molecule, newt-banded hedgehog (N-bhh), is a homolog of Indian hedgehog (Ihh), but it is expressed throughout the early blastema instead of being restricted to differentiating cartilage (Stark et al., 1998). *Msx-1* is prominently expressed in areas of labile limb mesenchyme, and it has been detected in embryonic and regenerating limbs. The T-box gene, *NvTbox1,* is expressed only in the forelimb during both embryogenesis and regeneration (Simon et al., 1997). A role for heat shock proteins (HSPs) in regeneration is only recently becoming recognized (Makino et al., 2005), but the expression patterns of *Hsp-70* in developing and regenerating limbs are remarkably similar (Lévesque et al., 2005). Retinoic acid is certainly a major player in development and regeneration (rev. Stocum, 2005), but as in so many systems, the exact role of the retinoids in normal processes remains tantalizingly obscure.

Although much more molecular mapping remains to be done, the overall scheme of expression of pattern-forming genes in regeneration appears to roughly recapitulate the spatiotemporal sequences seen in the embryo (Géraudie and Ferretti, 1998). Relatively little research has been conducted on the expression of tissue-specific genes in amphibian limb regeneration, but from the small amount of available data, one would expect parallels between the embryonic and regenerative differentiation of specific tissue types in the limbs. (This topic is examined in greater detail later this chapter with the discussion of mammalian tissue regeneration.)

AMPHIBIAN TAIL

In many respects, the amputated amphibian tail follows a course of regeneration similar to that of the limb. Like the limb, its regeneration can be divided into two broad phases: an early phase dominated by wound healing, dedifferentiation, and the formation of a blastema; and a later phase consisting of the outgrowth of a new tail from the blastema. The second, later phase is the one that invites comparison with embryonic development of the tail.

An early study (Holtzer et al., 1955) on tail regeneration in *Ambystoma* larvae showed some differences in pattern and process from the embryonic development of a tail. A significant difference was that the notochord, present in the embryonic and larval tail, is not present in the regenerate, although it does form in the regenerated tails of anuran larvae (Atkinson et al., 1976). Whereas in the embryo, the sequence of tissue formation is neural tube → muscle → neural arches → centra of vertebrae, in regenerating larval tails, outgrowth of the spinal cord is followed by the differentiation of vertebral centra → muscle → neural arches. In regenerating tails of adult urodeles (Mufti, 1973; Mufti and Simpson, 1972), as well as larvae (Holtzer et al., 1955), the regenerating axial skeleton forms a continuous cartilaginous rod that only partially undergoes segmentation during the later stages of regeneration. The regenerating spinal ganglia form from ependymal cells rather than the neural crest. Similarly, the regenerated spinal cord itself is derived from the ependymal tube (Egar and Singer, 1972). A later study by Echeverri and Tanaka (2002a) showed that during tail regeneration in the axolotl, labeled ependymal cells even create cartilage and muscle cells. Such meta-plasia, however, does not appear to occur in regenerating tails of *Xenopus* (Slack et al., 2004).

The overall differentiation of individual tissues in the regenerating tail follows a morphologic sequence quite similar to that in the embryo. In regenerating muscle, both the pattern of dystrophin expression and the sequence of expression of actin genes resemble those seen earlier in normal development (Arsanto et al., 1992; Khrestchatisky et al., 1988). Within the spinal cord in *Pleurodeles,* the homeobox genes *PwNkx-3,2* and *PwNkx-3,3,* normally expressed in embryos, continue to be expressed in the adult and are even more strongly expressed in the spinal cord during tail regen-eration (Nicolas et al., 1999). Nicolas et al. conclude that expression of these pattern-forming genes is not completely turned off even in the stable adult stage. A similar situation is seen in the axolotl, where shh is expressed in the floor plate of both the mature and regenerating spinal cord (Schnapp et al., 2005). As in embryonic develop-ment, shh restricts the ventral expression of dorsal genes, such as *Pax-6, Pax-7,* and *Msx-1,* and it stimulates the expression of *Sox-9* in ventral precartilaginous mesen-chyme. Much more work remains to be done before a coherent scheme of expression of pattern-forming genes during tail regeneration is established.

The tail of *Xenopus* tadpoles regenerates well, except for a short refractory period between stages 45 and 47. In experiments that stimulated regeneration of the tail during these stages, Beck and colleagues (2003) found the key to be reactivation of the bone morphogenetic protein (BMP) and Notch signaling pathways that are used during embryonic development of the tail. By knocking them out, the researchers were able to inhibit normal regeneration, and by activating either of the pathways separately, they could stimulate regeneration of the tail during the refractory period. BMP signaling was required for the formation of all tail tissues, whereas Notch alone, which is down-stream of BMP, resulted in formation of the spinal cord but was not sufficient to guide the formation of muscle in the regenerating tail.

AMPHIBIAN LENS

Stimulated by an inductive influence from the optic cup, the overlying ectoderm of the embryo invaginates and undergoes a transformation into a lens (rev. Spemann, 1938). The completely removed lens does not regenerate in higher vertebrates, but in amphib-ians and some fish, excellent regeneration occurs after lentectomy (rev. del Rio-Tsonis and Eguchi, 2004; Reyer, 1977; Yamada, 1977). However, in most urodeles, the cellular origin is different. More than a century ago, Colucci (1891) presented the first histologic evidence that the regenerating lens in newts arises directly from the dorsal iris, an adult derivative of the optic cup. A few anurans also regenerate a lens, but in *Xenopus,* for instance, the cornea, rather than the dorsal iris, creates the regenerating lens (Freeman, 1963).

As is the case in limb regeneration, regeneration of a lens from the dorsal iris involves a prominent phase of dedifferentiation, during which the cells at the margin of the iris shed their pigment granules and other differentiated characteristics to produce a population of embryonic-appearing cells. These cells then go on to evolve into a lens by following pathways of differentiation similar to those that take place in the embryonic lens. Thus, as in the amputated limb, an early phase in which regeneration follows a unique course is followed by a later phase in which embryonic patterns are closely recapitulated (see Fig. 9-1, B). This generalization is based on three historical phases of research. The first involves principally morphologic studies; the second involves immunocytochemical analysis, mainly of lens crystallin proteins; and the third involves molecular analysis.

Early histologic studies (rev. Reyer, 1962) clearly showed that shortly after lentectomy, the pigmented epithelial cells at the rim of the middorsal iris become depigmented and begin to swell (see Fig. 2-2). This is accompanied by the reentry of these cells into the mitotic cycle and a general increase in RNA synthesis (Yamada, 1977). These cells then form a lens vesicle. The cells comprising the posterior wall of the lens vesicle thicken and withdraw from the mitotic cycle before they differentiate into clear lens fibers, as they do in the embryo, whereas the cells of the anterior wall remain cuboidal in shape as the lens epithelium, which serves as the precursor for the formation of new lens fibers.

With the advent of electrophoresis and immunocytochemistry, the types and sequence of synthesis of the various lens crystallin proteins were followed during embryogenesis and regeneration (del Rio-Tsonis and Eguchi, 2004; rev. Yamada, 1977). The urodele lens contains α-, β-, and γ-crystallin proteins. The earliest differentiating lens fibers contain β- and γ-crystallins, and at a later time α-crystallin can also be detected. This sequence recapitulates quite closely that which occurs in the developing lens. In lens regeneration from the cornea in *Xenopus*, there is a slight divergence between the embryonic and regeneration pathways (Mizuno et al., 1999a). In the embryo, all three crystallins appear simultaneously at the lens placode stage, but during regeneration, the presence of α- and β-crystallins precedes lens fiber differentiation. γ-Crystallin does not appear until differentiation of lens fibers occurs.

More contemporary molecular studies have clarified even further the relation between the development and regeneration of the lens. These have been assisted by the availability of *in vitro* models of transdifferentiation, which are based on chick, as well as amphibian, cells. An interesting parallel between regeneration of the lens and the limb is the role of thrombin and phosphorylated retinoblastoma protein in dedifferentiation and allowing the entry of differentiated cells into the mitotic cycle (Imokawa et al., 2004; Tanaka et al., 1997; Thitoff et al., 2003). During the early differentiation stage, molecules other than the crystallins show strong parallels in expression between embryonic development and regeneration. *Pax6* and *SOX2,* which are critical for many aspects of embryonic eye development, are expressed throughout the iris shortly after lentectomy, but they soon become restricted to the region of dedifferentiation in the dorsal iris (Hayashi et al., 2004; Mizuno et al., 1999b). Another homeodomain family protein, Prox 1, which may regulate crystalline gene expression in the embryonic lens, is expressed at high levels during much of the differentiative phase of lens regeneration (del Rio-Tsonis et al., 1999).

FGF is thought to be one of the major mediators of the influence of the neural retina on the lens. In normal ontogenesis, it is involved in stages from induction of the lens to later orientation of the lens fibers. Hayashi and colleagues (2004) have shown that injections of FGF-2 near the dorsal iris stimulate the development of a second lens, and that ultimately the original lens regresses in response to the presence of the new lens. Of a group of injected growth factors including FGF-1, -2, -4, -7, -8, -10, epidermal growth factor, insulin-like growth factor (IGF), and vascular endothelial growth factor, only FGF-2 was effective in inducing the formation of a new lens. After lentectomy, FGF-2 levels in the dorsal iris also increased, leading Hayashi and colleagues to conclude that FGF-2 is the molecular stimulus for lens regeneration. Another set of signaling molecules, shh and Ihh, is expressed during development and regeneration of the lens, although they are not present in the intact lens (Tsonis et al., 2004). These authors also showed that inhibition of the hedgehog pathway interfered with lens regeneration. The exact role of these molecules and their receptors, ptc-1 and ptc-2, remains to be determined, but they appear to regulate aspects of proliferation and differentiation.

From the preceding data, one can construct a general scheme similar to that which outlines the parallels, or lack thereof, between embryonic development and regeneration of the limb (see Fig. 9-1). In both limb and lens, the preparatory phase leading to and culminating with the onset of dedifferentiation is unique and does not share many common properties with the inductive phase of embryologic development, but as differentiation and morphogenesis begin to get under way, the embryonic and regenerative processes converge to a phase in which both the morphology and molecular control follow a common pathway. It is interesting that these processes converge, even though there are substantial differences in what triggers regeneration in these two systems. Limb regeneration follows amputation, with an epidermal covering and the presence of nerves being necessary for dedifferentiation to occur, whereas the stimulus for lens regeneration is the relatively atraumatic removal of the lens without the need for direct tissue interaction. Notable, however, is the parallel between the important roles of FGF-2 in the early stages of both processes.

MAMMALIAN SKELETAL MUSCLE

Both the embryonic development and regeneration of mammalian skeletal muscle have received considerable attention, especially with respect to molecular controls. Therefore, muscle is used here as the model for comparing tissue regeneration with embryonic development. In making this comparison, this section examines regeneration of a generic limb muscle at four levels: gross morphogenesis, tissue structure, cellular level, and molecular level.

Gross Morphogenesis

At the level of a muscle as an organ, virtually no similarity exists between the processes that cause a muscle to take shape in the embryo and that which occurs during tissue regeneration. In the embryonic limb bud, premyoblasts migrate into the limb bud from the ventrolateral region of the somites. These cells, which are mobilized under the influence of hepatic growth factor (HGF/scatter factor) and which express Pax-3 while they are migrating, differ from the premyoblasts of the dorsomedial myotome by not expressing myogenic regulatory factors while they are migrating. Once in the limb bud, they settle down as dorsal and ventral muscle masses (Čihák, 1972). The formation of dorsal and ventral muscle masses is followed by a second phase in which the common muscle masses begin to split up and primordia of individual muscles take shape. This phase of morphogenesis depends principally on information contained within the associated connective tissue, rather than the muscle cells themselves, because myogenic cells derived from grafted somites that originated at different axial levels can partici- pate in normal muscle formation in the limbs (Chevallier et al., 1977). Muscle mor- phogenesis at this level is independent of innervation. The tendons and connective tissue sheathing (e.g., epimysium, perimysium, and endomysium) of the muscles are derived from the lateral mesoderm of the limb buds, not the somites.

Few experimental models or clinical situations result in the regeneration of an entire muscle. The most instructive model is minced muscle regeneration, in which a regener- ated muscle must take shape from a completely formless mass of implanted muscle fragments (Carlson, 1972b; Studitsky, 1959). A muscle regenerating from a mince takes the shape of a generic muscle, with proximal and distal tendon connections and often a major site where nerves and blood vessels enter the regenerate, but a minced muscle regenerate almost never grows back to the form of the original muscle (Fig. 9-3, A). The origin of the mince does not affect the final morphology of the regenerate. The surgical sponge material Gelfoam (Pharmacia & Upjohn, Bridgewater, NJ), was soaked in a noncellular muscle extract and implanted in place of the removed gastrocnemius muscle in rats. Within days after implantation, tendon connections had been established with the Gelfoam mass (Carlson, 1972b). The form of the implanted Gelfoam was identical to that of an early minced muscle regenerate even though it contained no muscle cells (see Fig. 9-3, B), showing that physical factors in the limb could account for all aspects of gross morphogenesis of a regenerating mince. This is in contrast with embryologic development, where overall morphogenesis is virtually independent of physical factors. What does account for gross morphogenesis of an embryonic muscle remains, for the most part, a mystery.

Tissue Structure

The internal tissue structure of an embryonic muscle takes shape as its overall gross morphogenesis proceeds. The orientation of the muscle fibers assumes its definitive form as soon as the muscle fibers begin to take shape. Nowhere is this more evident than in the human tongue, where the three-dimensional orthogonal arrangement of muscle fibers is present as soon as the myotubes are recognizable, well before the onset of contractile function. In a limb muscle, the myotubes begin to form basal laminae and become associated with the fibroblasts that will produce the various ensheathments early in the development of the muscle. The first neural connections are also made at

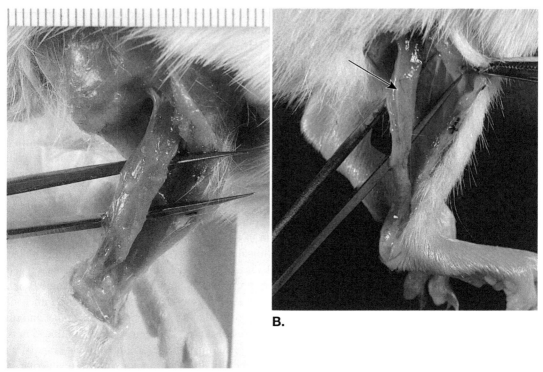

FIGURE 9-3 Minced muscle regenerates in the rat. *(A)* Typical regenerate of the minced gastrocnemius muscle. Although the regenerated muscle has proximal and distal tendon connections and is vascularized and innervated, its gross shape is that of a generic muscle and does not duplicate that of the gastrocnemius. *(B)* A 13-day Gelfoam (Pharmacia & Upjohn, Bridgewater, NJ) "regenerate." Here, the gastrocnemius muscle was removed and pieces of the surgical sponge material, Gelfoam, were implanted instead of the usual mince. The Gelfoam had been molded into the shape of a generic muscle, like that in *A*, and had proximal and distal tendon connections. It was vascularized, and the sural nerve (arrow) can be seen running over its surface. Despite its gross similarity to a minced muscle regenerate, this Gelfoam "regenerate" contains no muscle fibers. (Reprinted from Carlson, B.M. 1972. *The regeneration of minced muscles.* Basel, Switzerland: S. Karger, by permission.)

about the same time. Muscle spindles are induced by sensory nerve fibers during the period when the developing muscle fibers are in the myotube stage (Zelená, 1964). As the nuclear bag and chain fibers differentiate, these fibers begin to be surrounded by a specialized layer of connective tissue cells and fibers, the spindle sheath.

Often in the tissue regeneration of a mammalian muscle, the internal tissue structure is largely determined by that of the original muscle, especially in cases in which there is no substantial mechanical damage to the muscle. In such cases, for example, after exposure to various local anesthetics or toxins, the muscle fibers break down within the existing basal laminae and connective tissue sheaths, and regeneration occurs within a preformed matrix. A slightly different model is whole muscle-free transplantation, in which, after grafting, most of the muscle fibers break down because of ischemia. If the muscle is autotransplanted in its normal orientation and with tendons sutured, the mechanical conditions during early regeneration do not differ greatly from those of the intact muscle, and little internal reorganization of the regenerating muscle occurs. In a more extreme situation, Carlson (unpublished data) grafted the musculature of a rat tongue into the leg. Although muscle fiber breakdown and regeneration occurred according to schedule, the orientation of the regenerated fibers was still the three-dimensional orthogonal arrangement characteristic of the tongue, rather than the nearly parallel arrangement of fibers seen in the limb muscle into whose bed the graft was placed. In the minced muscle model, the internal architecture of the original mince is totally chaotic. When regenerating myotubes first begin to form, their orientation corresponds to that of the implanted minced fragments, but as the mince develops tendon connections and mechanical forces are imposed on the mince, the myotubes straighten out, and their orientation becomes parallel to the lines of tension within the regenerate.

In the embryo, muscle spindles are induced by sensory nerve fibers (Zelená, 1957). Mammalian muscle spindles regenerate (Milburn, 1976; Rogers, 1982), but only if the original spindle capsule remains intact. In a minced muscle regenerate, spindles are not found (Zelená and Sobotková, 1971), presumably because the capsules are destroyed by the mincing process. Spindles can form in an aneural regenerating muscle as long as the capsule is present (Rogers and Carlson, 1981), but little evidence exists that sensory nerve fibers can induce the formation of new spindles in an adult. In this case, it appears that the control of intrafusal fiber formation and differentiation has been ceded from the nerves to the capsule.

Cellular Level

In the main stages of muscle fiber formation and differentiation, there is a strong similarity between embryonic development and regeneration. In both processes, myogenic precursor cells withdraw from the cell cycle to produce spindle-shaped mononuclear myoblasts, which fuse to form multinucleated myotubes that have chains of nuclei running down the center (Fig. 9-4). The synthetically active nuclei of the myotubes guide the formation of massive amounts of contractile proteins, which begin to assemble into well-organized sarcomeric units. As the myotubes fill up with contractile material, the nuclei move out to the periphery, at which time the myotube can be properly called a *muscle fiber.*

Despite these similarities, a number of unique aspects characterize both development and regeneration. The principal differences are the origins of the myoblastic cells and the environment in which early regeneration begins. The myogenic cells initially migrating into the limb bud are clearly of somitic origin, and they are kept in a nondifferentiating state by *Pax-3* until they have settled down in the limb bud. Once in the common muscle masses, they begin to synthesize myogenic regulatory factors that stimulate their transition into committed myoblasts and the later stages of muscle fiber

A. Embryonic development

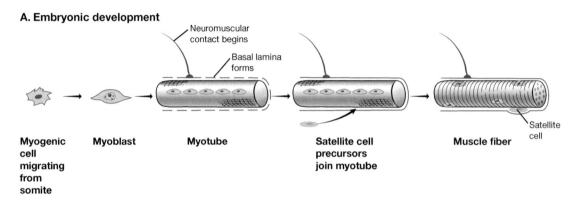

B. Regeneration

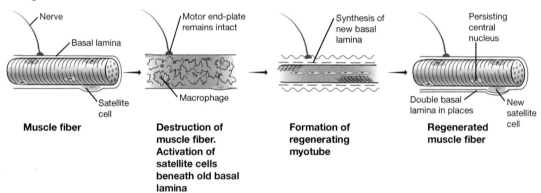

FIGURE 9-4   Comparison between the development and regeneration of a mammalian skeletal muscle fiber.

differentiation. These early myogenic cells are the ones that differentiate into primary and secondary myotubes in the embryo (Carlson, 2004). At a later time, they are joined by cells from an apparently different lineage, which will become situated next to the early myotubes as satellite cells (Feldman and Stockdale, 1992; Hartley et al., 1992). Although some controversy exists regarding the lineage relations between primary and secondary myoblasts and satellite cells, it is noteworthy that *Pax-7* mutants in mice form embryonic muscle fibers but lack satellite cells (Seale et al., 2000).

Most of the cells that take part in mammalian muscle regeneration are derived from satellite cells (Snow, 1977). Satellite cells are normally quiescent and are located in a unique environment between the muscle fiber and its basal lamina. After muscle damage, satellite cells are exposed to a complex environment consisting of breakdown products of muscle fibers, hypoxia, and growth factors secreted by macrophages and released from the damaged muscle. In contrast to embryonic myogenesis, the early stages of muscle regeneration all occur in a highly structured environment that was set in place by the original muscle.

With minor exceptions, the morphologic stages of muscle fiber differentiation are quite parallel between developing and regenerating muscles. In both cases, further growth of the multinucleated muscle fiber depends on the incorporation of satellite cells into the muscle fiber. One difference is the relation between the muscle fiber and its nerve supply. Whereas an embryonic muscle fiber must create the neuromuscular junction *de novo* in conjunction with a motor nerve terminal, the formation of the specialized postsynaptic region of a muscle fiber occurs completely within the old basal lamina. The presence of the motor nerve is not even necessary, because the postsynaptic region will differentiate in a denervated muscle (Burden et al., 1979; Hansen-Smith, 1986). Another difference is the relation between nerves and muscle fibers with respect to functional type, for example, fast versus slow. According to the most recent evidence, the earliest outgrowing nerve fibers, whether fast or slow, do not determine whether embryonic muscle fibers become fast or slow. Rather, the assignment of basic muscle fiber type is determined intrinsically, and the outgrowing motor nerve fibers match up with the corresponding types of muscle fibers (Thompson et al., 1990). In contrast, the functional type of regenerating muscle fibers appears largely to be determined by its motor innervation, because cross innervation of a regenerating fast muscle by a slow nerve or vice versa results in transformation of its functional type to one corresponding to that of the nerve (Gutmann and Carlson, 1975).

A final difference between embryogenesis and regeneration is the location of the myonuclei in a terminally differentiated muscle fiber. Whereas the nuclei are always located peripherally in normally developing muscle fibers, regenerating fibers are characterized by the persistence of substantial numbers of central nuclei.

Molecular Aspects

The molecular underpinnings of embryonic and regenerative muscle development can be subdivided into several major categories, including "positional signals" (Zhao and Hoffman, 2004), myogenic regulatory factors, growth factors, and contractile proteins. The complement of contractile proteins and associated enzymes is largely responsible for determining the contractile properties of the muscle.

In the early development of the limb musculature, several molecular pathways are active in the initial specification of myogenic cells and their migration into the limb buds. These include shh signaling from the notochord and ventral neural tube to regions of the somite and members of the Wnt pathway. The ventrolateral cells in the somite are exposed to HGF/scatter factor through their elaboration of the c-met receptor, and through the mediation of BMP-2, -4, and -7, they express Pax-3 while they are migrating to the limb bud (rev. Carlson, 2004). These early positioning pathways are not active in muscle regeneration (Zhao and Hoffman, 2004), presumably because the precursor cells in muscle regeneration are already specified.

The differentiation of embryonic muscle is regulated to a considerable extent by a group of four myogenic regulatory factors: Myf5 and MyoD, which act during early stages of development, and myogenin and MRF, which act on successively later stages of differentiation (Fig. 9-5). In normal young adult muscle, these regulatory factors are not expressed in either the muscle fibers or the quiescent satellite cells, but on injury, the activated satellite cells and regenerating myotubes reexpress these molecules in roughly the same sequence that occurs during embryonic myogenesis (rev. Chargé and Rudnicki, 2004).

Both developing and regenerating muscle are responsive to a wide variety of growth factors, but because of the complexity of the adult body and the postinjury environment, satellite cells and their progeny are exposed to a considerably wider array of growth factors and cytokines than are embryonic myoblasts. Nevertheless, certain generalizations are possible. Some growth factors exert a pronounced effect on the proliferative activities of myogenic cells in the embryo and satellite cells in postnatal muscle. These include FGFs, HGF, IGF-1, and IGF-2 (Chargé and Rudnicki, 2004; rev. Hawke and Garry, 2001). Members of the transforming growth factor (TGF)-β family typically play a more negative regulatory role in early muscle differentiation. Of particular importance is TGF-β family member myostatin, which acts as a negative regulator of overall muscle growth during normal development (McPherron et al., 1997). Myostatin also plays an important role in muscle regeneration (Kirk et al., 2000; Yamanouochi et al., 2000). In addition to its later effects on regeneration, myostatin released from degenerating muscle fibers may play a role in initially holding proliferation of the satellite cells in check. HGF and activation of the Notch pathway by Delta, its ligand, appear to have a particularly important role in activating quiescent satellite cells in the immediate postinjury period (Tatsumi et al., 1998; Wagers and Conboy, 2005). Currently,

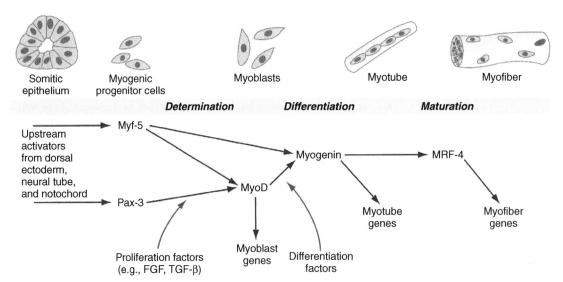

FIGURE 9-5 Myogenic regulatory factors in the embryonic development of muscle. (Reprinted from Carlson, B.M. 2004. *Human embryology and developmental biology*. 3rd ed. Philadelphia: Mosby. 178–180, by permission.)

IGF-1 appears to be one of the most important growth factors acting on the differentiation phase of muscle development and regeneration.

The induced differentiation of skeletal muscle from bone marrow stromal cells follows a course quite similar to that seen in normally differentiating skeletal muscle (Dezawa et al., 2005). Following specific stages of induced differentiation, the initial stem cells down-regulate genes, such as *Pax 3,* and sequentially up-regulate later genes characteristic of developing muscle, such as *Pax 7,* followed by *MyoD* and *myogenin* and, finally, *Myf6/MRF4.* Whether the methods of inducing myogenic differentiation of the stem cells recapitulate the embryonic pattern as closely is currently unclear.

The main contractile proteins of a muscle fiber consist of actin and myosin, together with regulatory proteins, such as troponin and tropomyosin. Of these, studies on the synthesis of the myosin proteins have been the most carefully conducted and have proved to be the most instructive. Myosin is a complex molecule, consisting of two helical heavy chains associated with four light chains. There are both fast and slow myosins, corresponding to the basic fast and slow muscle fiber types. During embryonic development, both fast and slow myosin heavy and light chains go through a series of isoform replacements, which for myosin heavy chain includes embryonic, neonatal, and adult isoforms (Fig. 9-6). The same fundamental sequence of isoform replacements occurs in regenerating muscle (Maréchal et al., 1984; Plaghki, 1985). However, the synthesis of fast myosins is regulated by thyroid hormone, whereas that of the slow myosins appears to escape the influence of this hormone (d'Albis et al., 1989). Patterns of synthesis of other proteins related to muscle contraction follow the same general course as the myosins, with an overall similarity between embryonic development and

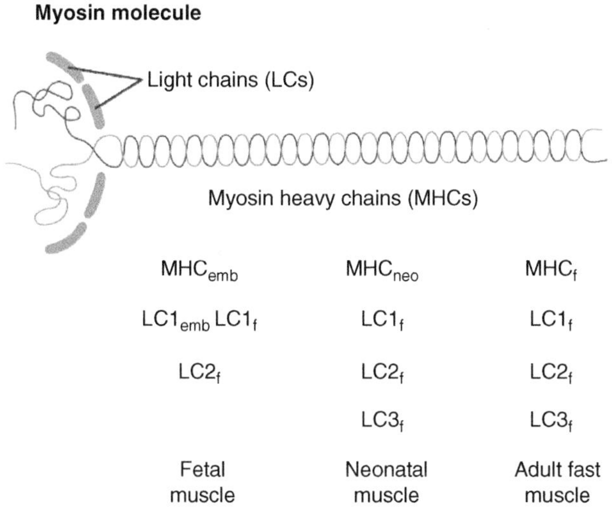

Myosin molecule

FIGURE 9-6   The myosin molecule and its isoforms. (Reprinted from Carlson, B.M. 2004. *Human embryology and developmental biology.* 3rd ed. Philadelphia: Mosby. 178–180, by permission.)

regeneration. Even the development of contractile properties during regeneration is modeled after that of fetal and neonatal muscles, with a slowly contracting initial phase, followed by a rapid speeding up to attain normal fast speeds of contraction for regenerated fast muscles (Carlson and Gutmann, 1972).

Regenerating cardiac muscle provides an interesting contrast to skeletal muscle with respect to the recapitulation of embryonic stages of cell differentiation. Rather than arising from a population of undifferentiated satellite cells, new heart muscle in lower vertebrates arises from the proliferation of existing differentiated cardiomyocytes. In comparing molecular aspects of heart development versus regeneration in the zebrafish, Raya and colleagues (2004) have found that none of a group of heart development genes (*nkx2.5, tbx5,* and *CARP*) was up-regulated during regeneration, and none of the genes up-regulated during regeneration (*msxB, msxC, notch 1b,* and *deltaC*) appears to be active during embryonic development. One would expect that there would be greater correspondence in the appearance of actin and myosin proteins.

SUMMARY

As an overall summary, one can generalize that whenever regeneration proceeds by the epimorphic mode and is accompanied by dedifferentiation and blastema formation, the overall process of regeneration closely recapitulates that of embryonic development once the blastema phase is reached. However, the process of generating the cells that constitute the blastema is significantly different from that by which the corresponding embryonic primordium is established. Tissue regeneration does not appear to use any of the embryonic mechanisms in setting up its gross or architectural organization, but at the level of differentiation of individual cells, regenerating tissues usually coopt the main pathways leading to the appropriate cellular differentiation.

CHAPTER 10

Regeneration and Aging

Maintenance and repair mechanisms are expensive, and the degree to which they are developed should depend upon how much they are likely to prolong the lives of individuals in a population. Where the probability of death from extrinsic causes is low, maintenance and repair may prolong the lives of many that survive to old age and may therefore be strongly selected.

—Robert Ricklefs and Caleb Finch (1995)

Everyone who is older than 50 years is keenly aware that skin wounds do not heal as quickly as they did when they were younger. Some changes begin to occur *in utero,* where the healing of skin wounds is essentially scarless. For the skin, the qualitative transition from essentially true regeneration to wound healing characterized by scarring has already taken place shortly before birth. Other structures or organs show qualitative or major quantitative differences in their regenerative capacity at different times throughout the life cycle. In some cases, damage during the early postnatal period is restored by a continuation of what would best be called *embryonic mechanisms.* In other cases, the adult pattern of repair is already fixed in place by the time of birth. In yet other cases, for instance, the peripheral nervous system, damage shortly after birth has much more serious effects than it does later in life.

Overall, however, the pattern is an inexorable slowing of regenerative processes with increasing age and a lesser degree of likelihood that regeneration will be quantitatively as successful as it is earlier in life. This is true not only of vertebrates. Zamaraev (1973) has reviewed the effects of aging on a large number of invertebrate systems. Although in most systems there is an age-related decline in regeneration, regenerative capacity actually appears to increase with age in some invertebrate groups, such as ascidians, starfish, and phoronids.

In examining the effects of age on regeneration, several questions recur regardless of the system. Are there common features of aging effects on regeneration that cut across all systems? Does aging have a fundamental effect on the progenitor cells on which regeneration of new tissue depends? What is the role of the overall aging environment on regeneration? Can the effects of aging on regeneration be reversed? This chapter begins with a summary of what is known about aging effects on regeneration of a variety of structures and continues with a discussion of general principles that can be derived from the individual data sets.

AGING AND REGENERATION IN INDIVIDUAL ORGAN SYSTEMS

Studies on the influence of aging on regeneration have been heavily skewed toward analysis of physiological regeneration or structures that regenerate by hypertrophy. Some systems have been investigated extensively, but for others the effect of aging on regeneration has received only passing attention.

Skin

In the chick embryo, epidermal (ectodermal) wounds do not heal at all until the 10th day of development (see p. 63). At that point, the epidermis develops the *in vivo* capability of covering a defect. During the fetal period, wounds of mammalian skin heal by scarless regeneration (Longaker and Adzick, 1991). In rats, a transition between scarless regeneration and typical wound healing occurs between the 16th and 18th days of embryonic development (Ihara et al., 1990), and in an opossum, the transition from the fetal to the adult pattern of healing occurs in the pouch young at day 9 (Armstrong and Ferguson, 1995). Human fetuses show a similar pattern of fetal healing, and the tendency toward scarless healing has been a stimulus for attempting a variety of *in utero* surgical procedures. The basis for the transition from the fetal to the adult healing pattern remains not well understood, although most emphasis has been placed on the *in vivo* growth factor environment.

During the aging process, which is commonly defined as the period from mid-adulthood to the end of life, the literature on epidermal aging and healing is much less uniform than would be expected (Ashcroft et al., 1995). Although there is a general consensus that overall cutaneous healing becomes slower and less effective with increasing age (Fig. 10-1) (Goodson and Hunt, 1979; Swift et al., 1999), analysis of individual components of a healing skin wound do not always show a consistent age-related decline. Many studies, especially on humans, have not been well controlled for comorbidity effects, such as nutritional status or the presence of other pathologic agents, among others. With respect to keratinocyte turnover (physiological regeneration), there is an increase in epidermal turnover time that is underlain by a decrease in epidermal labeling index (Grove and Kligman, 1983). In addition, aging epidermal cells show a decreased mitogenic response to growth factors (Rattan and Derventzi, 1991). Swift and colleagues (1999) have focused on angiogenesis during wound repair. They found that delayed angiogenesis in old mice was associated with a decrease in angiogenic mediators (fibroblast growth factor-2 and vascular endothelial growth factor), as well as a decrease in endothelial responsiveness to angiogenic factors. Overall, the diversity of research results on wound healing and aging led Ashcroft and investigators (1995) to wonder whether many of the age-related alterations in skin healing and regeneration may originate in the cellular microenvironment, rather than from intrinsic changes.

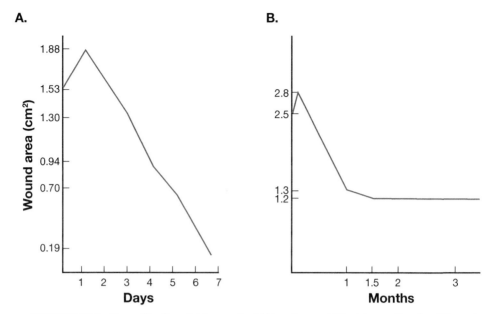

FIGURE 10-1   Rates of healing of skin wounds of *(A)* newborn and *(B)* adult rats. (Based on Sidorova's experimental data [1976].)

Blood

The hematopoietic system is another that must support physiological regeneration throughout the life span of the individual. Some studies on mice, including serial transplantation experiments in which marrow-derived stem cells were transplanted into several successive generations of recipients, have indicated only a small decline in hematogenic capacity with old age (rev. Hornsby, 2001). In contrast, Geiger and van Zant (2002) have suggested a greater decline, although the number of stem cells is still sufficient to maintain a normal blood cell count. The major effect of aging is that there is a smaller reserve in the event of a massive need for new blood cells. A difference between aging humans and mice is that, in humans, there is a much smaller reserve hematopoietic capacity in the marrow.

Liver

During adult life, the liver is normally mitotically quiescent, but after hepatectomy, most of the cells of the liver enter the mitotic cycle while the original mass of the liver is rapidly restored. After hepatectomy, hepatocytes, the normal parenchymal component of the liver, become mitotically active, but as a backup, oval cells, derived from the bile duct system, can act as a source of new cells. Within a day after partial hepatectomy there is already a large peak of DNA synthesis and mitosis among the remaining cells in the liver (Fig. 10-2). The character of the regenerated functional units (lobules) varies, depending on the age of the rat at hepatectomy. In 7-day-old rats (at the time of surgery), the regenerated lobules are of normal size, whereas in 24-day-old rats, the lobules are 50% larger than those of control rats (Liozner, 1974). This result points to the formation of new lobules in very young rats and the hypertrophy of existing lobules in older ones.

Liver regeneration in old animals is characterized by a reduction in rate (Bucher and Glinos, 1950; Sidorova, 1976), a delay in the onset of DNA synthesis (see Fig. 10-2) (Bucher et al., 1964), a reduction in overall proliferative response of the hepatocytes (Iakova et al., 2003), and a reduction in the number of hepatocytes that undergo DNA synthesis (Stocker and Heine, 1971). As was the case with marrow-derived stem cells, serial transplantation of hepatocytes into younger hosts showed almost no diminution in their repopulating capacity over time (Overturf et al., 1997).

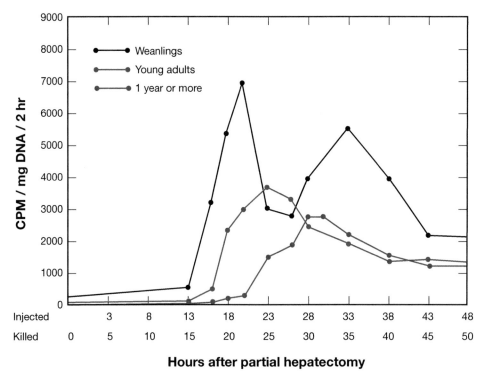

FIGURE 10-2 Effect of age on DNA synthesis in the regenerating rat liver. With increasing age, the peak period of DNA synthesis occurs later. (Based on Bucher and Malt's experimental data [1971].)

Kidney

After unilateral nephrectomy in very young rats (5 days old), the remaining kidney hypertrophies to 90% of the weight of two corresponding control kidneys. The degree of hypertrophy declines progressively with increasing age at nephrectomy, until by 2 years of age, the hypertrophic kidney attains only 60% to 70% of the combined mass of control kidneys (MacKay et al., 1932). This decline corresponds to an age-related decline in mitotic index of the remaining kidney. According to Sidorova (1978), 3 days after unilateral nephrectomy in 4-month-old rats, there were 13.8 mitoses/2000 cells, whereas in 3-year-old rats, activity had declined to 2.7 mitoses/2000 cells. During the early weeks after birth, new nephrons still form, and after unilateral nephrectomy, nephrons still continue to form for a while (Fig. 10-3). Shortly thereafter, new nephrons do not form in the remaining kidney. Instead, existing nephrons enlarge. In younger animals, proliferation accounts for a significant proportion of the enlargement, but as

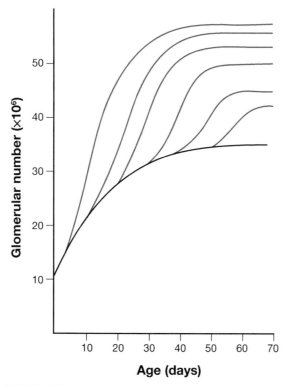

FIGURE 10-3   The influence of age on glomerular numbers after unilateral nephrectomy in immature rats. The black curve represents the normal postnatal increase of glomerular number in one kidney. The red lines represent numbers of additional glomeruli that form at different ages after removal of the contralateral kidney. (Adapted from Goss, R.J. 1978. *The physiology of growth.* New York: Academic Press, by permission.)

the animals age, cellular hypertrophy takes over as the dominant mechanism (Bucher and Malt, 1971; Goss, 1964b).

Lungs

Like the kidneys, the lungs continue to produce functional units (alveoli) during early postnatal ontogenesis (perhaps as late as 18 months in humans [Zeltner et al., 1987]); also like the kidneys, the remaining lung undergoes compensatory hypertrophy after unilateral pneumonectomy. The similarities between these two organs continue. If one lung is removed from rats during the early days after birth, hypertrophy proceeds by greatly increased proliferation and the formation of new alveoli. Later in life, the response of the remaining lung to unilateral pneumonectomy is gross hypertrophy, but internally, the gross hypertrophy is based on the enlargement of existing alveoli, rather than the formation of new ones (Romanova, 1971). Only recently (Maden and Hind, 2003, 2004; Massaro and Massaro, 1997) has it been shown that retinoic acid, which is necessary for the initial development of alveoli, is also able to induce the regeneration of new alveoli in adults. Maden and Hind (2004) speculate that retinoic acid in the adult may reawaken molecular pathways that were used during ontogenetic alveologenesis.

Sidorova (1978) has summarized a large number of experiments on organs that regenerate by hypertrophic mechanisms by noting the following characteristics:

1. Although mass may be recovered to a considerable extent, normal gross morphogenesis is not restored.
2. In young individuals, restoration is accomplished largely by cell proliferation. In older individuals, cellular hypertrophy is the predominant mechanism.
3. In most internal organs, consisting of multiple functional units (e.g., liver lobules, nephrons, alveoli), the formation of such units after loss of tissue occurs only in late fetal life or early after birth, and then weakly. After that, gross hypertrophy is based on increasing the size of existing functional units.

Skeletal Muscle

Because of its importance with respect to practical issues of mobility in old age, the effects of aging on skeletal muscle regeneration have received considerable attention. Muscles are more susceptible to injury in older than younger individuals, especially the injury caused by lengthening contractions (Devor and Faulkner, 1999; Faulkner et al., 1995).

Old muscle regenerates, but not as well as young muscle. This is reflected in its histologic structure (Sadeh, 1988), mass, and contractile properties (Carlson and Faulkner, 1989; Carlson et al., 2001; Marsh et al., 1998), and the generally deficient regeneration is a reflection of the gradual deterioration of muscles with increasing age (Brooks and Faulkner, 1994; Doherty, 2003).

Many factors appear to contribute to the declining efficacy of muscle regeneration with increasing age (Carlson, 1995). There are notable changes in the progenitor cells, the satellite cells, of muscle. In rats, both the absolute numbers and percentages of satellite cells decline by a factor of between 2 and 3 from 1 to 24 months of age (Gibson and Schultz, 1983). In humans, as well, the percentage of satellite cells decreases by roughly 50% between young adults and individuals in their mid-70s (Kadi et al., 2004; Renault et al., 2002). Not only the number of satellite cells but also their proliferative potential decreases with age, although in both rats and humans, the major decrease in proliferative potential occurs between birth and young adulthood (Table 10-1; Fig. 10-4) (Renault et al., 2000; Schultz and Lipton, 1982). Conboy and colleagues (2003) have attributed the reduction in proliferative potential to insufficient up-regulation of the Notch ligand, Delta, and the consequent diminished activation of Notch in old rats. A consistent observation *in vivo* is that there is a delay in the mitotic response of satellite cells after injury in old compared with young animals (McGeachie and Grounds, 1995). Once cell division gets under way, there is a tendency for the formation of more interstitial connective tissue among the regenerating muscle fibers in older individuals.

One of the significant factors affecting the quality of both normal and regenerating muscle in older individuals is its innervation (Larsson and Ansved, 1995). Overall, the number of nerve fibers supplying a given muscle declines dramatically with age, in some cases by as much as 90% in very old humans (Brooks and Faulkner, 1994). The general tendency is for axons that supply the largest and fastest muscle fibers to die. Then the denervated muscle fibers of that motor unit are either secondarily reinnervated by sprouts arising from other motor axons (usually slow), or they undergo denervation atrophy and ultimately degenerate. If the nerves that supply an old muscle are preserved, whereas the muscle itself is caused to degenerate by transplantation, the number of motor units is not decreased over that of a normal muscle; but in 2-year-old rats, the total number of motor units in a muscle is 2/3 the number in 4-month-old rats (Cederna et al., 2001). When the nerves to a damaged old muscle are also damaged, the recovery of motor units is substantially less than the number in control muscles.

TABLE 10-1

Number of Divisions in Culture of Satellite Cells from Humans of Different Ages

Age	Number of Divisions
5 days	55–65
5 months	46
9 years	30
15 years	28
26 years	19
28 years	21
31 years	17
52 years	20
79 years	17
81 years	15
86 years	15

Based on Renault and colleagues' data (2000).

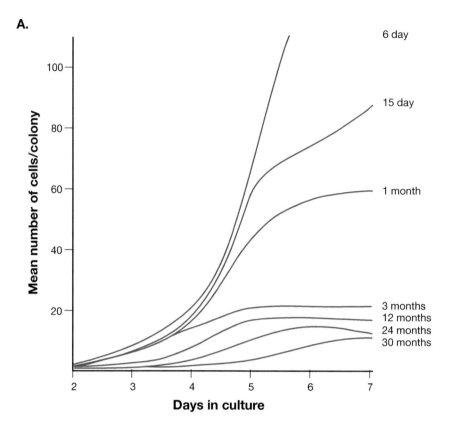

A.

B.

FIGURE 10-4 Changes in the proliferative potential of muscle satellite cells with age. *(A)* Mean numbers of satellite cells per colony over days of culture derived from single satellite cells taken from rats of different ages. (Based on Schultz and Lipton's data [1982].) *(B)* Numbers of divisions of cultured satellite cells over time from satellite cells taken from human donors of different ages. (Based on Renault and colleagues' data [2000].)

Peripheral Nerve

Peripheral nerves show both qualitative and quantitative changes in their regenerative capacity throughout the life cycle. Probably the least expected reaction to nerve transection occurs in fetal or newborn animals, in which transection of an axon typically leads to death of the neuron rather than to regeneration (Romanes, 1946; Schmalbruch, 1984). In mice, the period of most profound susceptibility of transected neurons is during the first week after birth (Fig. 10-5) (Pollin et al., 1991).

This seemingly paradoxical reaction may well be related to the normal phenomenon of pruning back on the excess numbers of nerves that form in the embryo. When there are more motor axons going to a muscle than there are muscle fibers, a normal phase of neuronal cell death prunes the excess numbers of motor neurons so that only those neurons that are connected to an end organ will survive (Buss and Oppenheim, 2004). If a nerve is transected during this pruning stage, it is likely that the nerves with amputated axons have descended below a minimum threshold of trophic support, and thus cannot survive on their own.

This situation soon changes, and shortly after the natural pruning period, transected peripheral nerves in young individuals regenerate well (see Fig. 10-5). With adulthood and advancing old age, peripheral nerve regeneration becomes progressively less vigorous (Drahota and Gutmann, 1961). In an early study, Gutmann and coworkers (1942) noted a marked delay in outgrowth of axons from the proximal stump of a transected nerve, but once the regenerating axons entered the distal stump, the rate of regeneration appeared to be the same in both young and old individuals. Clinical studies (Sunderland, 1978) have shown a decline in both the rate and degree of recovery with advancing age. Moyer and colleagues (1960) have found almost identical rates of func-

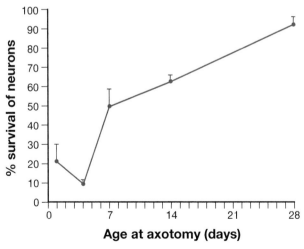

FIGURE 10-5 Percentage of motoneurons surviving 4 weeks after axotomy in rats of varying ages. Postaxotomy survival is low in rats 1 and 4 days old. (Based on Pollin and colleagues' data [1991].)

tional recovery (~2.6 mm/day) of spinal nerves in kittens younger than 1 week and cats older than 10 years. Subsequent research has confirmed that in older animals, there is a longer latency period at the start of the regenerative process (Tanaka and Webster, 1991; Vaughan, 1992; Verdú et al., 1995). At a given period after the nerves have been crushed, there is a marked reduction in the numbers of axons distal to the crush site in old animals (Fig. 10-6).

Considerable uncertainty continues concerning the reasons underlying the poorer regeneration of old axons, but a number of differences between axonal regeneration in old as compared with young animals have been identified. As part of the long latency period, the chromatolytic response in the neuronal cell bodies is delayed, and there is a slower removal of distal myelin from the peripheral segment of the nerve. In a regenerating nerve, transected axons typically produce multiple sprouts early in the regenerative process (see Fig. 4-3). Normally, large-diameter axons produce more sprouts than do small ones. Vaughan (1992) has found fewer axonal sprouts in regenerating old nerves. One option is that there is decreased trophic support for axonal sprouting in old animals. Another is that, in old animals, large-diameter axons are among the first to degenerate; thus, the decreased sprouting could also be a reflection of the character of the remaining population of axons.

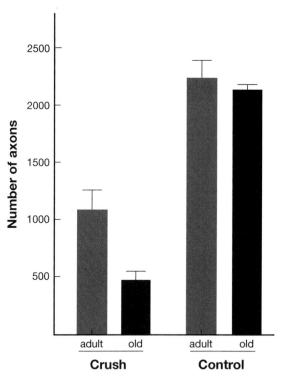

FIGURE 10-6   Numbers of myelinated axons in fascicles of posterior tibial nerves of young and old mice 2 weeks after crush injury. (Based on Tanaka and Webster's experimental data [1991].)

Corresponding to the early reports that once begun, axonal regeneration in the distal nerve segment is not retarded in old animals, is the finding by Willcox and Scott (2004) that aging sensory neurons appear as capable as young ones of inducing several growth-associated molecules (α_1-tubulin, neurofilament light subunit, and GAP-43) during the regenerative phase. Nevertheless, regenerated old axons have a great increase in microtubular density and a 50% reduction in neurofilament density (Vaughan, 1992).

Bone

It is common clinical knowledge that the younger the patient, the better fractures heal. A fractured femur in a newborn infant becomes firmly united within a month, whereas a comparable fracture in a person 50 years old is likely to take 4 months to reach the same stage. In children, certain long bones can regenerate from the periosteum, but for most bones, this capacity is lost after adolescence (Liozner, 1974). Despite the importance of fractures and their subsequent healing in the elderly, there have been surprisingly few studies on the course and mechanisms of fracture healing in old age.

A different type of regenerative response is seen in skull bones. Defects created in skull bones of young puppies and kittens heal completely with well-formed bone, but by 3 months in dogs, only isolated islands of regenerated bone are seen (Polezhaev, 1972a,b). Even in humans some regeneration of defects in the parietal bones exists up to 3 years of age.

In laboratory mammals, many mechanical properties of healed long-bone fractures are inferior to those seen in young animals (Bak and Andreassen, 1989). The less successful fracture healing in old animals is associated with a reduced population of precursor cells in the periosteum (Tonna, 1978) and a diminished inductive potential in the matrix of old bone (Syftestad and Urist, 1982). At a cellular level, there is an age-related decline in the osteogenic capacity of adipose tissue–derived stem cells. Fracture healing in older women is further complicated by the common tendency for osteoporosis after menopause.

EPIMORPHIC SYSTEMS

Because historically many experiments on epimorphic regeneration have been conducted on wild-caught animals that could not be readily aged, relatively little emphasis has been made on the effects of age on epimorphic regenerative processes. In amphibian limb regeneration, a greater correlation often appears between size (e.g., cross-sectional area of the limb) and the speed or success of regeneration than with age. The most profound aging effects on limb regeneration in amphibians occur in relation to metamorphosis in anurans. Most tadpoles regenerate limbs quite well, but as metamorphosis gets under way, regenerative capacity is lost in a proximodistal direction. Typically, postmetamorphic frogs and toads completely lose their ability to regenerate limbs, but *Xenopus* is able to produce spike regenerates after metamorphosis (Dent,

1962). A notable exception to this rule is the Kenyan reed frog *(Hyperolius),* which is able to regenerate complete digits with several joints throughout adult life (Richards et al., 1975). These frogs have highly specialized sucking pads at the tips of their digits. After amputation of the forearm, only a nonjointed spike forms, but interestingly, sometimes a patch of the characteristic cells of the sucking pads appears at the end of the spike regenerate (Richards et al., 1977).

Cultured blastemal cells appear to defy aging *in vitro.* Whereas most types of cells undergo replicative senescence after a defined number of divisions, blastemal cells from lizard tails have been cultured for as long as 8 months (Simpson and Cox, 1967), and limb blastema cells from newts have been shown to exhibit no senescence changes after more than 200 doublings in culture (Ferretti and Brockes, 1988). The basis for this phenomenon is unknown, but it appears to be a function of some change(s) that occurs during the process of dedifferentiation.

GENERAL CHARACTERISTICS OF REGENERATION IN OLD AGE

Despite the wide variety of organs and structures that can regenerate in old age, one can make a few generalizations (Table 10-2). Most investigators have observed a slower rate of regeneration with increasing age, regardless of the system, but when this has been dissected further, the most frequently noted common element is that there is an increased latency period before the full-blown regenerative response gets under way. The basis for the delay in the latency period (and the latency period itself, regardless of age) is poorly understood in any system, but it is likely to be related, in part, to the time needed to move the progenitor cells into the mitotic cycle. The fact that in most systems there is a smaller number of progenitor cells may compound the problems of initiating the regenerative process. In other cases, a weaker initial inflammatory process and a slower removal of tissue debris left behind from the injury could delay the initiation of regeneration.

Once the progressive phase of regeneration has begun, there appear to be differences among systems in the rate of regeneration. In some cases, it is retarded, but in others (e.g., axonal regeneration), there may be little difference between young and old. Where cell proliferation is involved, a reduced rate of proliferation likely plays an important role. A general characteristic of regeneration in older individuals, as in their normal tissue structure, is a greater proportion of connective tissue and a relatively lesser

TABLE 10-2

General Characteristics of Regenerating Systems in Old Compared with Young Age

1. Slower overall rate of regeneration	5. Less cellular proliferation
2. Longer latency period	6. Increase in amount of connective tissue in regenerate
3. Smaller number of progenitor cells	
4. Weaker inflammatory and phagocytic responses	7. Final regenerate less complete

amount of parenchyma. This may be greatly influenced by the hormonal status and overall humoral environment of the body at a given age. Regardless of the cause, the final regenerate in an old individual is often less complete than a comparable regenerate in a young one. This may be reflected in mass, mechanical properties, or other quantitative characteristics (e.g., number of axons in a regenerating nerve).

MAJOR ISSUES IN REGENERATION AND AGING

The collective literature on regeneration and aging points to a general retardation in rate and a reduction in overall success of regeneration with increasing age. Considerably less information exists on the reason for this. One factor could be a general age-related decline in the number and proliferation of stem cells and progenitor cells in a variety of organ systems (Geiger and van Zant, 2002; Kuhn et al., 1996; Limke and Rao, 2003). Caplan (2006) has proposed six different periods of mesenchymal stem cell titers from the embryonic period until late life, and that over an individual's life span, titers of mesenchymal stem cells decrease by one or two degrees of magnitude. And van Zant and Liang (2003) point out the tendency of stem cells, at least those in the hematopoietic system, to lose the breadth of their developmental potential during aging. For example, stem cell differentiation in transplants of old marrow is skewed toward producing myeloid cells at the expense of the T- and B-lymphocyte lineage. A fundamental question is whether the diminished regeneration in old age is due to a reduction in the regenerative capacity of the organ or structure or whether the regenerative capacity remains but is impeded by an unfavorable environment.

The question of environmental influences has been addressed by several cross-age transplantation experiments in which tissues from old animals were transplanted into young animals and vice versa. In an early experiment, Horton (1967) examined aging effects on the cycle of hair growth in mice, which can be delayed as long as 4 weeks in old animals. He transplanted skin of 30- to 33-month-old mice to 3-month-old recipients and found that in the old-to-young grafts, the rate of hair growth was accelerated so that it corresponded to that in the young host mice. As the host mice aged, the hair in the skin grafts, which were then 54 to 57 months old, grew correspondingly more slowly to match the reduction in rate of the host hair.

A similar approach was used in reciprocal grafts of mammary glands between young and old mouse hosts. The old mammary glands grew well in the young hosts, but young glands transplanted into old hosts did no better than old mammary gland grafts (Young et al., 1971). Serial transplants of mammary glands (Daniel et al., 1968) and muscles (Carlson, unpublished data) have shown that it is possible to keep a structure alive far longer than the normal life span of the animal.

Cross-age grafting has also been used in studies of muscle regeneration (Carlson and Faulkner, 1989; Carlson et al., 2001). Grafts of the extensor digitorum longus muscle regenerate two to three times more poorly in 2-year-old than in 4-month-old rats. Old muscles were grafted into one leg of young hosts, and in the other leg, the young muscle was autografted (Fig. 10-7). Conversely, a muscle from the young host

A.

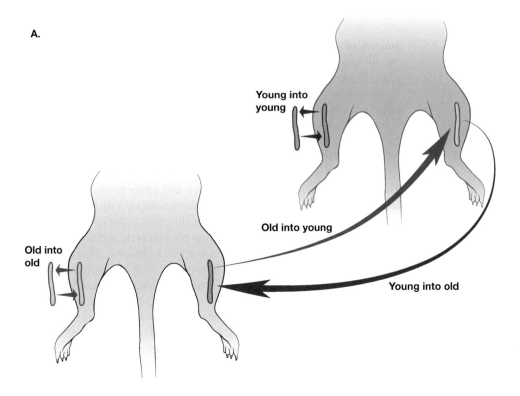

B. Comparative results of cross-age grafting of the rat EDL muscle

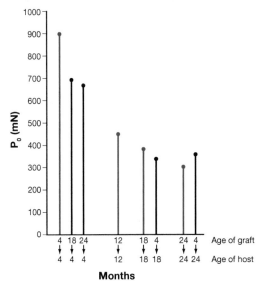

FIGURE 10-7   *(A)* Scheme of muscle cross transplantation between young and old rats. *(B)* Comparative functional results of cross-age grafts (black) of the rat extensor digitorum longus muscle versus same-age grafts (red). The maximum tetanic force (P_o) of same-age grafts steadily declines with age, whereas the P_o of the cross-age grafts is closer to that of the same-age grafts of the host versus the donor. (Based on Carlson and Faulkner's data [1989].)

was grafted in place of the corresponding muscle of the old donor. In these experiments, the degree of regeneration of the cross-age graft consistently matched up with that of the host autograft, in both young and old animals.

All of these cross-age grafting experiments have shown that the old structure in question has the capacity to regenerate as well *in vivo* as the corresponding young structure, and that the environment of the old animal inhibits the regeneration of even young tissue. What the experiments do not show is whether there may still be a difference in the maximal regenerative capacity between the young and old tissue.

Within the central nervous system, the age-related decrease in stem cell–based neurogenesis can be reversed by modifying environmental experience, diet, exercise intensity, or levels of glucocorticoid hormones (Bernal and Peterson, 2004; Brazel and Rao, 2004). In other systems, such as skeletal muscle, enhancement of the growth factor environment, such as insulin-like growth factor I (IGF-I), can also increase regeneration over normal levels for a given age.

Attempts to identify the basis underlying the old–young differences in regeneration have been conducted mainly on regenerating muscle. Carlson and Faulkner (1989) attribute a significant component of the regenerative deficit in old rats to inadequate reinnervation of the muscle grafts. Barton-Davis and coworkers (1998) obtained some improvement in the regeneration of old muscle in mice by inducing overexpression of IGF-I in the muscles.

Conboy and colleagues (2003) found diminished activation of Notch caused by insufficient up-regulation of its ligand, Delta, in old muscle and related that to the poor proliferative potential of old satellite cells in muscle. They then constructed old/young parabiotic mice and found improved regenerative potential in the satellite cells of the old muscles of the pairs together with a marked up-regulation of Delta (Conboy et al., 2005). It was not possible to determine whether the effect was due to bloodborne stimulators from the young member of the pair or the dilution of inhibitors coming from the old member. They found a similar enhancement of proliferation of the hepatocytes in the old parabionts.

One of the persistent questions in studies of aging and regeneration is the extent to which telomere shortening with age adversely affects regenerative processes. In humans, the telomere caps on the ends of chromosomes shorten with each cell division, and unless there is telomerase-mediated restoration of telomere length, a line of cells reaches a point of mitotic stagnation when telomere length is reduced by a certain extent. In postembryonic humans, telomerase activity is limited to germ cells, stem cells, progenitor cells, and activated lymphocytes (Djojosubrato et al., 2003). Mice, in contrast, have a much more active telomerase system during adult life. If they are engineered for a deficiency in one of the telomerase subunits (mTERC$^{-/-}$), they show defective regeneration of organs such as the liver and a decreased life span (Rudolph et al., 2000). Mean telomere length in human muscle (biceps and masseter) does not change significantly past adult life (Renault et al., 2002), and in this study, the decrease in regenerative capacity with age was attributed to a reduced number of satellite cells.

Telomerase activity is not present in most cells in the human brain, but it is present in the neural stem cells of the subventricular zone (Caporaso et al., 2003). If this is true for most other stem or progenitor cell populations, it represents an adaptation for maintaining the capacity to produce new cells throughout the life of the individual.

SUMMARY

There is little doubt that, for most organs, the success of regeneration diminishes with increasing age. The most pronounced changes occur early in life and toward the end of the normal life span. Common to most systems is a decrease in the rate of regeneration, which is often due to an increased latency period at the beginning of the process. One of the most important issues regarding the effect of aging on regeneration is whether the decline in regeneration is due to a decrease in the intrinsic capacity of the organ to undergo regeneration, or whether it is due to an inhibitory effect of the overall environment of the aging individual. The sum of a number of experiments suggests that, although there is likely a decrease in the regenerative capacity of most organs in old compared with young individuals, there remains sufficient reserve capacity to regenerate completely under ideal environmental circumstances. The extent to which telomere shortening exerts a significant effect on regeneration with increased age remains an open one, and it may be related to the composition and state of the stem-cell population of the organ in question.

CHAPTER 11

Influence of the Environment on Mammalian Regeneration

Recently the Soviet scientist Studitsky showed that it is possible to regenerate an entire muscle from minced muscle tissue, which is deemed to be living substance in a non-cellular state. (Translated from Czech.)

—Milan Hašek and Vera Hašková (1953, p. 105)

One of the fortuitous side effects of the tragic period of Lysenkoism (Medvedev, 1969; Soyfer, 1994) in Soviet science was a considerable interest in the phenomenon of regeneration. At the time when the power of the Lysenko doctrine was at its peak (Anonymous, 1948), regeneration research was focused principally on the amphibian limb and invertebrate systems. Lysenkoism essentially denied the role of genetics and stressed the importance of the environment in the control of biologic processes. An extreme subdoctrine (Lepeshinskaya, 1945, 1951, 1952) even claimed that, in many developmental and regenerative processes, cells did not arise from preexisting cells; rather, that they took shape from a proteinaceous soup, called "living substance" *(zhivoe v'eshchestvo)*. This theory led Studitsky (rev. 1959) to devise the minced muscle regeneration model in mammals. He attributed the successful results of these experiments to the formation of the newly regenerating muscle fibers from Lepeshinskaya's "living substance" (Studitsky, 1953).

Shortly after the 1948 Congress that elevated Lysenkoism to official doctrine, a new Soviet initiative to study the regenerative powers of mammals was formulated (Vorontsova, 1949, 1953). This resulted in a major Soviet program to investigate the regenerative powers of mammals (Khrushchov, 1954), which was summarized in one of the first monographs devoted to regeneration in mammals (Liosner, 1960). Together with purely descriptive work, considerable emphasis was placed on how the normal or experimentally altered environmental conditions *(uslovie)* affected the course of regeneration of a variety of mammalian tissues and organs (Liosner, 1972). Research on mammalian regeneration has blossomed all over the world in recent decades, and one of the consistent contemporary themes in mammalian regeneration studies is the extent of the influence of environmental conditions on regenerative processes. This chapter examines examples of environmental influences on regeneration.

WHAT IS MEANT BY ENVIRONMENT?

The Russian word *uslovie* is difficult to translate into a single word or phrase that conveys its meaning in the context of regeneration biology, but in effect, it refers to almost any outside influence on the cells that take part in the regenerative process. One of the most important influences is the microenvironment, which includes the various components of the extracellular matrix, ranging from basal laminae to the entire non-cellular substrate of a connective tissue. Not only are the physical features of the substrate important for functions such as cell attachment or migration, but the substrate is now recognized as a biologic sponge that absorbs and binds bioactive molecules, such as growth factors. When released, these growth factors exert powerful influences with respect to both stimulating local cells to enter the mitotic cycle and guiding their differentiation as regeneration proceeds.

The cellular and tissue environment can also play a vital role in mammalian regenerative processes. This can range from general interactions between the microvasculature and precursor cells of regeneration, to influences of local inflammatory cells, to the specific trophic interactions between axonal endings and regenerating muscle fibers. Local or systemic pathology can either influence or be influenced by regenerative processes.

Especially in tissue regeneration, the local mechanical environment is a major determinant of not only form and architecture, but also the pace and extent of tissue differentiation. Other physical features of the environment include bioelectrical potentials or currents and general physiologic conditions, such as oxygen saturation and pH. More broadly based environmental features include the hormonal and nutritional status of the individual and the immunologic environment, especially in the case of regeneration occurring in transplanted cells or tissues. As discussed in Chapter 10, the age of the individual is a powerful influence on the relative success of a local regenerative process. Finally, external influences, such as x-radiation or applied chemical or pharmaceutical products, can have pronounced effects on regeneration.

THE INFLUENCE OF SUBSTRATE

All cells interact with some form of substrate, which in the broadest context includes not only extracellular matrix, but other cells as well. For certain types of regeneration, such as nerve and muscle fibers, the critical substrate consists largely of existing extracellular matrix belonging to the original damaged tissue. In skeletal tissues, a major aspect of regeneration consists of producing new bone or cartilage matrix materials. The composition of the extracellular matrix in bone marrow plays a vital role in maintaining "stemness" of the progenitor cells that constitute the hematopoietic system. In a clinical context, implantation of an acellular, biologically derived tissue matrix provides the basis for tissue regeneration based on cellular infiltration.

Two of the best examples of the importance of residual substrate involve the basal laminae that surround muscle fibers and peripheral nerve fibers. In each case, regeneration occurs within the old basal laminae, which provide a permissive microenvironment for the proliferation and fusion of myoblastic cells, in the case of skeletal muscle, and axonal outgrowth, in the case of peripheral nerve. The binding and release of growth factors, such as fibroblast growth factor (FGF), by the basal lamina is important, as is the function of the basal lamina as a semipermeable cellular barrier that keeps unwanted cells and their products out of the area where regeneration is occurring. In addition, these basal laminae provide good adhesive sites for basic cellular activities—proliferation, movement, and extension.

Probably no better examples of the importance of substrate can be found than those associated with axonal regeneration. In the case of peripheral nerves, the many attempts to provide an artificial substrate that can fill in a gap in a transected and retracted peripheral nerve demonstrate the importance of a microenvironment that will allow neurite extension across the gap (rev. Fine et al., 2000; Yannas, 2001). It now appears that a substrate that is not conducive to neurite extension constitutes a major reason for the poor regeneration in the brain and spinal cord. This has been clearly shown in the experiments of Aguayo and colleagues (Bray et al., 1987), who obtained impressive regeneration of a wide variety of central neurons (Table 11-1) when their processes were supplied with segments of peripheral nerves in which to grow (see Fig. 4-4). On the other side of the coin, the nearly impenetrable nature of glial scars in the damaged central nervous system shows that an inappropriate substrate can impede regeneration as clearly as a favorable substrate can facilitate the process.

Tissue engineering approaches have shown that supplying foreign substrates from both different organs and different species can be a highly effective means of guiding the regeneration of structures as complex as the esophageal wall. Badylak (2002, 2004) emphasizes the importance of native as opposed to chemically modified matrices as effective substrates for tissue replacement. Such native matrices promote cellular infiltration, deposition of new matrix material by the host, degradation of the implanted matrix, and the remodeling of tissue without scar tissue formation.

Various components within the tissue matrices promote tissue-specific properties among different types of regenerating structures. For example, type I collagen is impor-

TABLE 11-1

Types of Central Neurons Shown to Regenerate into Peripheral Nerve Grafts

Cerebral cortex	Basal ganglia
Somatosensory	Thalamus, hypothalamus
Motor	Hippocampus, amygdala
Visual	Deep nuclei of cerebellum
Retinal ganglion cells	Brainstem nuclei
Olfactory bulb	Spinal cord

Adapted from Bray, G.M., et al. 1987. Regeneration of axons from the central nervous system of adult rats. *Prog Brain Res* 71:373–379, by permission.

tant where strength is needed, whereas type III collagen is associated with flexibility and extensibility of structures, such as the regenerating urinary bladder. Type IV collagen, however, is important for its ability to bind with specific types of cells, such as endothelial cells. Other components of the substrate, such as fibronectin, laminin, and various growth factors, promote cellular attachment and/or differentiation, whereas hyaluronic acid, in particular, greatly facilitates cell migration. Some of the common types of growth factors bound to extracellular matrices are listed in Table 4-2. Because natural tissue matrices are donor cell free and their components are not strongly immunogenic, they elicit a weak host immune response.

CELLULAR ENVIRONMENT

In a number of regenerating systems, cells themselves constitute a highly influential part of the local environment. This is well illustrated in the regeneration of human fingertips, amphibian limbs and tails, and ear-hole punches in rabbits and mice. In all of these cases, the presence of a naked wound epidermis abutting the damaged underlying tissues is required for the initiation of the regenerative process. Although the exact mechanism of the epidermal influence is not known in any of these systems, the secretion of various forms of FGFs and matrix metalloproteinases by the epidermis is likely to be a significant component of the cellular environment.

In an unusual experimental model, pieces of 15- and 19-week human fetal skin were grafted onto athymic nude mice in two locations—as a cutaneous patch or subcutaneously. A week later, an incisional wound was made in the pieces of grafted skin. The cutaneous fetal skin grafts healed by scarring, whereas those implanted beneath the skin regenerated the wounds without forming scars (Lorenz et al., 1992). The mechanism underlying this differential response could not be determined, but a combination of interactions with both host cells (including endothelial cells) and fluids was viewed as delaying maturation of the skin and determining the nature of its subsequent healing response in the subcutaneous grafts.

Early in tissue regeneration, interactions with inflammatory cells are important determinants of the overall success of regeneration. In the presence of large numbers of acute inflammatory cells, mammalian skeletal muscle regeneration is effectively shut down. This may be due to the release of massive amounts of reactive oxygen species, which inactivate the satellite cells. In contrast, the presence of large numbers of macrophages is of vital importance to most mammalian regenerative processes. Not only do they phagocytize tissue debris resulting from the traumatic event, but their secretion of a wide variety of cytokines stimulates the proliferation of tissue-specific progenitor cells.

An instructive example of the role of highly specific cell–cell interactions is that between Schwann cells and axons in peripheral nerve regeneration (Fawcett and Keynes, 1990). Although basal laminae and perineurial fibroblasts play a role in axonal regeneration, Schwann cells are vital in this process. Axonal progression into a variety of types of nerve grafts is weak unless Schwann cells from the proximal nerve stump accompany the outgrowing axons. The interaction between Schwann cells and axons is mediated to a large extent through cell-surface adhesion molecules, namely, L1/ NgCAM, N-cadherin, and integrins. If these molecules are bound by antibodies, almost all axonal growth on Schwann cells *in vitro* is inhibited. Later in the regenerative process, the Schwann cells recapitulate embryonic development by wrapping themselves around the regenerating axons to remyelinate them.

Schwann cells are not the only influential cellular players in the microenvironment of a regenerating peripheral nerve (Zochodne, 2000). The unmyelinated nerve fibers in the nerve sheath (nervi nervorum) release perivascular peptides that initiate a neurogenic inflammation. Their effect on blood flow, via factors such as nitric oxide (NO), in the microvasculature of the nerve sheaths strongly influences the overall physiologic environment of the section of traumatized nerve.

The importance of the cellular environment is well illustrated in a model of spinal cord repair. Within the olfactory system, olfactory ensheathing cells support axonal extension from newly forming olfactory neurons into the central nervous system by surrounding the new neurites with a variety of myelin that has characteristics of peripheral myelin. These cells allow good penetration and integration of olfactory neurites with targets in the brain. In rat models, when olfactory ensheathing cells are grafted into a traumatic lesion of the spinal cord, they support the rapid regeneration of neurites across the gap that they fill and appear to allow functional connections with host nerves (Raisman, 2004). A challenge in the use of endogenously derived olfactory ensheathing cells for the therapy of human spinal cord lesions is the small number of these cells *in vivo* and the difficulty of expanding this population *in vitro* before transplantation.

LOCAL GROWTH FACTORS

We now recognize that almost any area in which healing or regeneration is occurring is awash in growth factors and cytokines. These can be secreted by invading inflammatory cells, released from the local extracellular matrix, or synthesized by the regenerating cells themselves. All are part of the local environment in which regeneration occurs, and the vigor of the regenerative process depends on their presence and amount. This recognition was not always the case. As a student in the early 1960s, I distinctly remember hearing a well-known scientist comment on the prescient work of Valy Menkin (1956), who reviewed mechanisms underlying inflammatory processes. He scoffed, "That Menkin, he sees growth factors every time he turns around."

Shortly after wounding, platelets and blood-derived inflammatory cells initiate a variety of growth factor cascades (Singer and Clark, 1999). Platelets from the blood secrete platelet-derived growth factor, a chemoattractant for macrophages, as well as a stimulator of fibroblast proliferation. The macrophages secrete a host of cytokines and growth factors, including members of the epidermal and transforming growth factor families, which stimulate cell motility and proliferation, and FGFs and vascular endothelial growth factor (VEGF), which increase vascular permeability and stimulate angiogenesis.

Once a suitable environment for early regeneration has been established, the local release of a number of growth factors, such as FGFs and hepatocyte growth factor, stimulates the proliferation of tissue-specific progenitor cells (Black et al., 2004; Chargé and Rudnicki, 2004). After their initial stimulation by external sources, the progenitor cells themselves often become the source of growth factors that modulate the rate of proliferation and the transition from the proliferative to the differentiative phase of regeneration. Considerable variation exists among regenerating systems in the control of differentiation by local growth factors. In some, such as skeletal muscle, it involves cellular fusion, the activation of myogenic regulatory factors, and the rates of synthesis of the various contractile proteins.

Later local growth factor influences can affect a variety of functions. For example, local production of nerve growth factor within a regenerating sensory nerve has a strong effect on the rate of axonal elongation (Fawcett and Keynes, 1990), and insulin-like growth factor-1 modulates several stages of muscle regeneration (Shavlakadze et al., 2005). A common pattern in most regenerating systems is for the production of growth factors to decrease as regeneration approaches completion and the regenerated tissues become stabilized.

VASCULAR ENVIRONMENT

With the exception of a few tissues, such as the lens, the local microvascular environment exerts a major influence on mammalian tissue regeneration. For example, in bone regeneration, the presence or absence of a local blood supply determines whether the cells within the fracture callus will initially differentiate into cartilage (a result of poor vascularization) or bone (good vascularization).

In almost all cases after an injury, parts of the traumatized area are ischemic because of the interruption of the local vascular supply. For regeneration to occur, it is necessary that these regions become revascularized. By definition, the region of ischemia is hypoxic. Recent research has uncovered a remarkable mechanism by which hypoxic tissues initiate an angiogenic process that will lead to their revascularization (Maxwell and Ratcliffe, 2002). This process is initiated by the production of hypoxia-inducible factor-1 (HIF-1) by the ischemic tissues. As described on page 185, secreted HIF-1 then goes on to induce the increased transcription of VEGF, which stimulates angiogenesis. The ingrowth of blood vessels into the ischemic area of damage is accompanied by the local release of macrophages, which themselves release many of the growth factors that stimulate the proliferation of tissue progenitor cells.

Although the local vasculature obviously exerts a vitally important influence on regeneration, the local environment also exerts a powerful reciprocal influence on the regenerating blood vessels (LeCouter and Ferrara, 2002). It is now well established that the capillary endothelia of different organs have tissue-specific characteristics (Auerbach et al., 1985). The endothelial cells of many secretory organs, such as the liver, are fenestrated, whereas those in organs with a blood–tissue barrier are designed to keep bloodborne proteins out of the tissues with adaptations such as occluding junctions.

A variety of transplantation experiments have shown that invading capillaries assume the characteristics of the tissues in which they reside. For example, blood vessels growing into mesodermal tissues transplanted into the brain do not develop the characteristics of neural capillaries with barrier functions; rather, they take on the general appearance of connective tissue capillaries. In contrast, capillaries that invade neural tissue transplants develop the barrier-like phenotypes of brain capillaries (Stewart and Wiley, 1981). Similarly, when endothelial cells from mouse yolk sac were cultured with embryonic brain, they took on the characteristics of brain endothelial cells (Yu and Auerbach, 1999). In skeletal muscle regeneration, not only do the regenerating capillaries differentiate into arterial and venous segments (Grim et al., 1986), but the capillary/muscle fiber ratio develops according to whether the muscle fibers are fast (few capillaries) or slow (rich capillary supply).

SYSTEMIC HORMONAL INFLUENCES

The systemic hormonal environment influences a host of mammalian regenerative processes. One of the systems most profoundly affected by hormones is the female reproductive system—in particular, physiological regeneration of the uterine endometrium. In the last half of the menstrual cycle, the endometrium is built up largely through the action of progesterone so that it is capable of receiving and nourishing an implanting embryo. If implantation does not occur, the corpus luteum of the ovary regresses because of a lack of hormonal support and, consequently, produces less progesterone. The sharp reduction in progesterone results in a loss of interstitial fluid and an infiltration of the endometrial stroma with leukocytes, as well as spasmodic contractions of the spiral arteries within the endometrium. This results in patches of local ischemia and the ultimate shedding of the affected regions of endometrium. The massive surface wound created by menstruation is then healed, without scarring, by an estrogen-mediated migration of epithelial cells derived from remnants of the uterine glands and a build-up of the endometrial stroma. Further estrogen priming leads to the expression of progesterone receptors on the sensitive endometrial cells. This allows them to receive the full benefit of the later increased levels of progesterone, which propels endometrial regeneration, including development of the vasculature, to its peak.

The ovaries of fish are also profoundly influenced by hormonal conditions. Many fish lay up to several hundred thousand eggs per year, and it is not practical for the ovaries to store enough eggs to serve the lifetime reproductive needs of the fish. In such species, actual regeneration of the gametes occurs.

Another system that is highly dependent on hormonal influences is antler regeneration in deer (Goss, 1969). In most species of deer, development of the pedicle, which is the anatomic base for antler formation, is dependent on testosterone. Castrated male fawns do not develop pedicles and, consequently, never form antlers. Female fawns treated with testosterone develop small pedicles, but they do not produce antlers. That these treated female deer, as well as estrogen-treated male deer, do not develop antlers leads to the suspicion that estrogenic influences may have an inhibitory effect on actual antler formation.

Under normal conditions, the increased testosterone secretion leading into the breeding season results in full ossification of the antlers and the shedding of the velvet. When a buck with growing antlers is castrated, the antlers remain permanently encased in velvet and are not shed. If the vascularized antlers do not freeze during the winter, new antler material is added the next spring, and over several years, bizarre antler growth develops on heir heads (Fig. 11-1).

The hormonal cycles of deer, like those of many animals, are highly light dependent. Artificial prolongation of the length of daylight shortly after the antlers are shed results in a rapid acceleration of the regrowth cycle, and bucks exposed to these conditions produce two sets of antlers in a single year. Conversely, artificially shortening of daylight during late spring results in premature shedding of antlers by summertime.

An example of organ specificity to the systemic hormonal environment is a testosterone-sensitive muscle. Several muscles in male vertebrates, which are special-

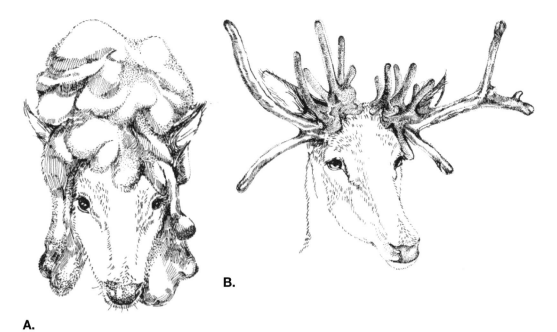

B.

A.

FIGURE 11-1 Varieties of antler formation in castrated deer. *(A)* Peruke formation in a roebuck. This wiglike formation remains viable and in a permanent velvet condition. *(B)* Excessive antler formation in a whitetail deer. If the antlers are kept from exposure to freezing weather, they keep forming new branches each year without losing the previous year's growth. (Reprinted from Goss, R.J. 1969. *Principles of regeneration.* New York: Academic Press, by permission.)

ized for reproductive functions, are highly sensitive to the effects of testosterone. These include the thumb musculature of frogs, the temporalis muscle of guinea pigs, and the levator ani muscle in rats. Of these, the levator ani has been most extensively investigated. This muscle, which is involved in ejaculation, is found in both sexes, but in the absence of testosterone in females, it undergoes profound atrophy. In male rats, it also undergoes atrophy after castration, but it can be restored with the addition of testosterone to the system. The levator ani regenerates after crushing *in situ,* but the presence of testosterone is required. When this muscle was transplanted to the site of a nontestosterone-sensitive limb muscle and was innervated by the nerve to that muscle, it retained its sensitivity to testosterone, showing that its intrinsic hormonal sensitivity could not be modified by its new mechanical and neural environment (Carlson et al., 1979b).

ROLE OF THE OVERALL BODY ENVIRONMENT

In addition to circulating hormones, the overall environment of the body can play a significant role in the regeneration of many structures. The influence of the aging body on the success of regeneration is discussed at length in Chapter 10, and the significant improvement of regeneration when a structure is cross-age transplanted into a younger host is also outlined (see p. 218). Similar transplantation approaches have been used in studies of diseased organs, which regenerate better when placed into normal disease-free hosts (Gulati and Swamy, 1991).

The nutritional status of the body is an important variable. Although epimorphic limb regeneration in newts and morphallaxis in planarians proceed remarkably well in starving animals, most regenerative processes in mammals are adversely affected by starvation. Starvation reduces cell division in tissues as diverse as epidermis, marrow, intestinal epithelium, liver, and kidney (Goss, 1978). Not unexpectedly, starvation also retards the rate of regeneration or hypertrophy of many internal organs.

Strain differences among inbred animals can also affect the course of regeneration. In SJL/J mice, several aspects of muscle regeneration (amount of leukocytic infiltration, rate of myoblast recruitment, and myotube formation) are accelerated or more prominent than they are in BALB/c mice. When muscles from BALB/c mice are grafted into SJL/J hosts, the pattern of regeneration in the grafted muscles corresponds to that of the host (Roberts et al., 1997), demonstrating, as in cross-age grafting, the importance of the overall environment of the host body in regeneration. Much of the host influence was attributed to the speed or efficiency of the leukocytes of the host in bringing about breakdown of damaged muscle fibers or in stimulating proliferation of muscle precursor cells.

IMMUNOLOGIC ENVIRONMENT

One of the most poorly understood aspects of regeneration is the relation between the immunologic environment and restorative processes. Historically, many ideas have been put forth, ranging from immune cells serving as precursor cells for limb regeneration blastemas at one extreme to no relation between immune function and regeneration at the other. Most authors and researchers who have dealt with this topic have postulated a modulating influence of the immune system on regeneration.

Historically, there have been scattered attempts to determine whether immunologic influences on amphibian limb regeneration exist (rev. Sicard, 1985). Overall, the results have been inconclusive, with only suggestions of a possible effect. An early hypothesis was based on a comparison between regeneration and tumor formation in amphibians (Prehn, 1970). The essence of the hypothesis was that mild immunostimulation may enhance limb regeneration, whereas excessive immunostimulation impairs the process. Mescher and Neff (2005) have reviewed the contemporary literature on immunity in amphibians and have pointed out the complexity of the cutaneous immune system. In anurans, a correlation exists between the loss of limb regenerative capacity during metamorphosis and increasing immune competence. These authors also point out that the period of scarless healing of skin wounds in the mammalian fetus precedes the development of many components of the cellular immune system. Interestingly, regeneration of the liver appears to be reduced in germ-free mammals.

An incontrovertible fact is that, in some mammalian systems, regeneration can occur in the face of massive immune rejection. In the minced muscle model involving the implantation of a mince from a different strain of rat, regenerated cross-striated muscle fibers can be seen forming among dense masses of small lymphocytes (Fig. 11-2). Despite the robust early regeneration, the regenerated tissue derived from the foreign mince is ultimately rejected (Carlson, 1970b).

Babaeva (1972, 1985, 1989; Babaeva and Zotnikov, 1987) has conducted an extensive series of investigations involving the effect of organ damage on immune functions and the effects of immune cells on cell proliferation and regeneration of internal organs.

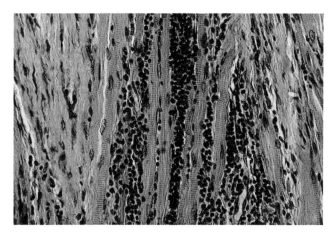

FIGURE 11-2 Photomicrograph of minced rat skeletal muscle fibers regenerating to the cross-striated stage in the face of a strong cellular immune response. The donor muscle came from a strain of rat different from that of the recipient. Ultimately, the regenerated muscle fibers are destroyed.

As an example of the experimental paradigm, donor mice were partially hepatecto-mized (70% resection of the liver). Then spleen cells were removed from the donor at defined intervals after hepatectomy and were injected into syngeneic recipients. The mitotic index of the partially resected donor liver peaked at 44 hours (Fig. 11-3). If splenocytes (mainly T lymphocytes) were removed from the donor mice before the mitotic index in the donor increased (2 and 17 hours) and then injected into normal recipient mice, the mitotic index of the hepatocytes of the normal host liver increased substantially. In contrast, if the donor splenocytes were removed from the donors at 48 hours after hepatectomy and then injected into normal recipients, the stimulatory effect on mitosis of the recipients' hepatocytes was completely abolished. In fact, if the liver of the host had been partially resected, addition of hepatocytes taken 48 hours after liver resection in the donors suppressed proliferation of the hepatocytes of the regen-erating host liver. This effect was reported to be reasonably organ specific, although some increase in mitosis of reticuloendothelial cells was also noted. Similar results were reported on proliferation of renal tubule epithelial cells after unilateral nephrec-tomy of the donor mice.

The overall interpretation of these results was that somehow lymphocytes are involved in the regulation of organ stability in the normal body and in the return to equilibrium after damage. This follows the hypothesis of Burwell (1963), who postu-lates a role of lymphoid tissue in morphostasis, involving the modulation by lymphoid

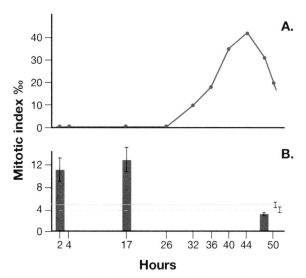

FIGURE 11-3 The effect of injecting splenocytes from a donor with a regenerating liver on hepatocyte proliferation of the host. *(A)* In the donor mouse, the mitotic index peaks 44 hours after partial hepatectomy. *(B)* When splenocytes were removed from the donor 2 or 17 hours after partial hepatectomy, the mitotic index of the hepatocytes of the recipient was increased over normal. In contrast, if the splenocytes were removed from the donor after the mitotic peak of the donor liver, there was no proliferative effect on the liver of the host. Solid blue line indicates mitotic index of hepatocytes in nonoperated mice; dashed blue line indicates mitotic index in control recipients. (Based on Babaeva's experimental data [1989].)

cells of chemical regulators of organ growth. This was the era in which chalones (Bullough, 1962) and other negative regulators of organ growth (Weiss and Kavanau, 1957) were being proposed to control growth equilibrium in internal organs. Although some of these concepts are no longer in vogue, the experimental results are striking and deserve follow-up. We still do not really understand why a liver stops growing at its normal size or why the kidney remaining after unilateral nephrectomy undergoes compensatory hypertrophy. More importantly, the role of the lymphoid system as it relates to any regenerative process continues to be incompletely understood.

A significant role of complement in the initiation of regeneration is beginning to emerge. It has long been known that C3b and C5a are associated with muscle fiber necrosis and the ingression of macrophages into areas of muscle fiber damage (Engel and Biesecker, 1982). More recently, C3a and C5a have been shown to be necessary for initiating liver regeneration as well (Strey et al., 2003). In amphibians, C3 and C5 are associated with the regeneration of both limb and lens (del Rio-Tsonis et al., 1998b; Kimura et al., 2003). The role of the related molecule, CD59, as a mediator of proxi-modistal identity in the regenerating amphibian limb is especially intriguing (Morais da Silva et al., 2002).

A relatively recent model of mammalian regeneration is that of holes punched into the ears of mice. Wild-type mice do not heal such holes, but two strains of mice, the MRL (Clark et al., 1998) and Athymic Nude-*nu* (Gawronska-Kozak, 2004), heal ear holes well by a tissue mechanism that involves the formation of a blastema-like structure. Both strains of mice have some immune deficiencies, with those of the Athymic Nude-*nu* being quite severe. Gawronska-Kozak (2004) postulates that thymic hormones enhance suppressor T-cell activation after injury, and that the absence of the thymus in the nude mice creates a general T-cell insufficiency that may allow regeneration to proceed. Considerably more research must be done to verify this hypothesis.

MECHANICAL ENVIRONMENT

In many systems, the mechanical environment plays a vital role in determining the success of a regenerative process. The importance of the local mechanical environment in the differentiative and morphogenetic phases of both bone and skeletal muscle regeneration is discussed in some detail in Chapter 8. In essence, both the gross form and internal architecture of a bone or muscle regenerating as an isolated tissue are shaped largely by the tissue's response to the forces of tension or pressure, or both, acting on the regenerating tissues. Similarly, tension, in particular, can strongly affect the amount of tissue that regenerates. This is seen in the Ilizarov technique that is used to lengthen regenerating bone (see p. 173) and in the considerable increase in length of an amputated muscle that is subjected to tension through connections made by a regenerating tendon.

Another system illustrating a role of the mechanical environment as a modulator of regeneration is cardiac muscle. The ventricles of newts, as well as those of newborn rats, are capable of considerable regeneration based on cell proliferation (Oberpriller and Oberpriller, 1991; Rumyantsev, 1991). In newts at all stages of their life cycle and in rodents during their first days of postnatal life, most of the cardiomyoytes are diploid. Within a week, the rodent cardiomyocytes undergo a high degree of polyploidization (Brodsky, 1991). This is correlated with a strong reduction in their proliferative ability, whereas the diploid hearts of newts remain capable of regeneration throughout life (Oberpriller et al., 1988). It is almost as if when the cardiomyocytes become polyploid, they become strangulated in their own excess DNA and are no longer capable of regeneration. Small pieces of ventricle from newborn rats have been transplanted into skeletal muscle and beneath the kidney capsule of adult hosts, where they can grow and be maintained in a beating state for months (Brodsky et al., 1988; Jockusch et al., 1983). In cells transplanted under the kidney capsule, the program of polyploidization of the transplanted cardiomyocytes was incomplete, and about 40% to 60% of these cells remained diploid (Brodsky et al., 1988). This suggests that removal of the ventricular tissue from its normal functional environment can affect a process as fundamental as cytokinesis.

In skeletal muscle, the mechanical environment plays a role in the initial activation of satellite cells (Anderson, 2000; Anderson and Wozniak, 2004). Shear stresses caused by components of muscle shifting past one another during contraction result in the release of NO, which, in turn, stimulates hepatic growth factor release, a potent satellite cell activator. The relation between NO release and satellite cell activation, however, is a complex one, with low pulses of NO seemingly inducing quiescence and larger releases caused by major muscle contractions or injury stimulating activation.

The mechanical environment also plays a role in the regeneration of tissues within implanted biologic matrices. These tissue matrix implants are acted on by the local tissue environment in which they are placed, and they develop lines of stress. After the wave of inflammatory cells has subsided, the parenchymal cells that form on the implanted matrix orient themselves along lines of stress (Badylak, 2002).

One of the tissues most responsive to its mechanical environment is the vasculature, both during normal life and in regeneration. Individual cytoskeletal components (actin stress fibers, microtubules, and intermediate filaments) of endothelial cells respond to shear stresses by internally reorienting themselves, resulting in the elongation of the cells in the direction of shear stress (McCue et al., 2004). This even carries through to mitosis, where the daughter cells align themselves along lines of shear stress. Stretch not only affects structural components of endothelial cells, but it also causes upregulation of membrane receptors and growth factors in these cells (Zheng et al., 2004). These effects even indirectly affect the recruitment of circulating endothelial cell progenitors (Wang et al., 2004).

BIOELECTRIC ENVIRONMENT

One of the most enigmatic elements of the environment surrounding regenerating structures is the bioelectric field, which is measured as either electrical potential or currents. Although the existence of bioelectric fields has been recognized for more than two centuries and clear electrical effects have been demonstrated on a number of regenerative processes, it has proved to be difficult to integrate this knowledge with other elements that affect the regenerating tissues and organs (Borgens et al., 1989; Levin, 2003; Marino, 1988).

All living beings are associated with some sorts of electrical fields and phenomena (Lund, 1947). These range from the resting membrane potential present across the membranes of cells to the highly dynamic action potentials that sweep across the surface of activated nerve and muscle cells. In addition, there are the fields created by the piezoelectric properties of bones and collagen and the electrical potentials that traverse epithelial sheets. In aquatic animals, the skin acts like a battery, where a 40- to 80-mV potential (positive on the inside) is driven by the active pumping of Na^+ from the outside to the inside of the integument (Borgens, 1989a).

When a structure is damaged or removed, the electrical properties in that area are changed. The normal limb of the newt is characterized by inflowing electrical current (Fig. 11-4, A). When a limb is amputated, a strong outflow of current, leading out from the amputation surface, can be measured (see Fig. 11-4, B). This change is also reflected in a difference in electrical potential between the amputation surface and the base of the limb (Becker, 1961; Monroy, 1941). As the limb regenerates, the electrical properties of the limb gradually return to normal.

Similar studies have been conducted on invertebrates. For example, the earthworm, *Eisenia foetida,* has approximately 90 segments, and regardless of the level of amputation, regeneration proceeds until that number of segments is restored. Moment (1949) measured the potential difference (approximately 15 mV) between the anterior and posterior cut ends of the worms before amputation. He then amputated the worms through the 50th segment and found that the potential dropped to −12 mV at the cut surface immediately after amputation, and then slowly returned to normal as the worm regenerated its missing parts by 60 days after amputation.

A general rule of thumb is that regenerating or outgrowing structures (e.g., neurites, keratinocytes, osteoblasts, pollen tubes, and neural crest cells in the embryo) are attracted to a region of negative charge or the cathode in *in vitro* experiments (Robinson and Messerli, 2003). Corneal epithelial cells not only migrate toward the cathode, but their plane of division is oriented perpendicular to the vector of the electrical field (Zhao et al., 1999). Conversely, electropositive conditions are associated with tissue destruction or resorption, a good example being osteoclastic resorption of bone. Based on studies of regeneration and bioelectric potentials in regenerating planaria and coelenterates, Rose (1970) has postulated that a longitudinal electrical field with differences in potential acts like a biologic electrophoresis apparatus, transporting morphogenetically active charged molecules to critical locations in the regenerating system.

Healing and regenerating bone provides a good example of how stressors in the mechanical environment are translated into electrical charges, which are then reflected in charge-specific cellular activity. The piezoelectric properties of bone were first described in 1957 by Fukada and Yasuda. According to Wolff's law, the deposition of new bone is associated with compressive forces applied to a region of bent long bone, whereas the resorption of bone is associated with tensile forces. Electrical measure-

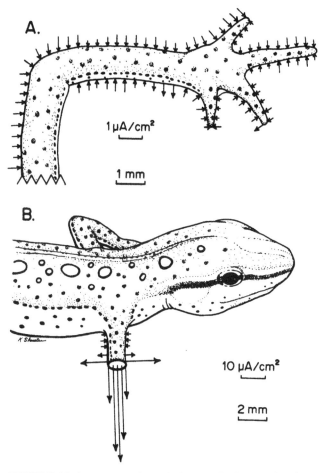

A.

1 μA/cm²

1 mm

B.

10 μA/cm²

2 mm

FIGURE 11-4 The flow of electrical current in the normal and amputated forelimb of the newt. The normal limb *(A)* is characterized by small currents flowing into (arrows) most areas of the limb. *(B)* The period after amputation is characterized by the strong outflow of current from the amputation surface and slightly proximal to it. (Courtesy R. Borgens)

ments (rev. Bassett, 1972) have shown that a net negative electrical charge is associated with the deposition of bone, and a positive charge is associated with bone resorption (see Fig. 8-5). The mechanism by which electrical activity is translated into cellular behavior in bone and other tissues remains poorly understood.

One way of testing the importance of electrical fields in regeneration is by changing the environment that produces the fields. By blocking sodium channels with substances such as amiloride, both the healing of corneal epithelial wounds (Sta Iglesia and Vanable, 1998) and amphibian limb regeneration (Borgens et al., 1979) can be inhibited. Similarly, the placement of aquatic animals in a sodium-deficient medium results in the retardation or inhibition of regeneration of urodele tails and limbs (Borgens

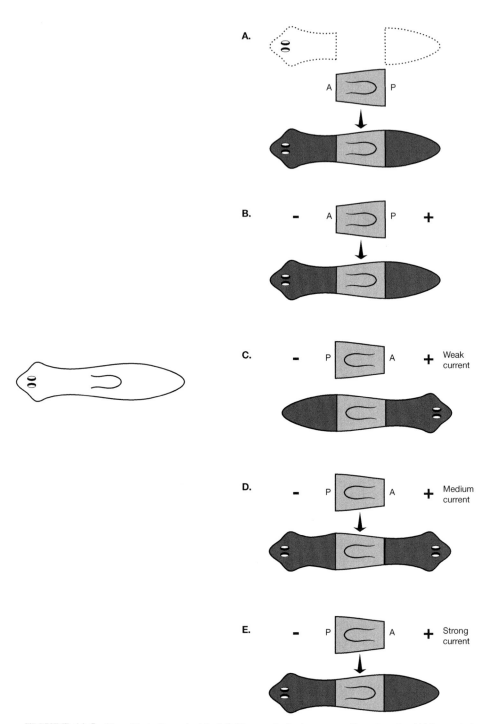

FIGURE 11-5 The effect of an electrical field on polarity in regenerating planaria. *(A)* In control experiments, a head and tail end *(red)* regenerate from a midbody segment in normal polar orientation. *(B)* If the anterior end of a midpiece is oriented toward the cathode, regeneration of head and tail is always of normal polarity regardless of current strength. *(C–E)* If the anterior end of a midpiece faces the anode, the regenerates are of normal polarity in weak current strength *(C)*, two-headed when exposed to moderate current strength *(D)*, and of completely reversed polarity in the face of strong current strength *(E)*. (Based on Marsh and Beams's experimental data [1952].)

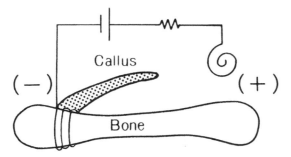

FIGURE 11-6   Formation of a bone spur *(black dots)* at the cathode when a bone is electrically stimulated. (Based on Yasuda's report [1974].)

et al., 1979; Nawata, 2001). Marsh and Beams (1952) subjected pieces of planaria to
artificially reversed electric fields stronger than the natural ones and were able to
produce heads from caudal amputation surfaces and tails from cephalic surfaces (Fig.
11-5).

A number of experiments have shown a stimulation of regeneration by applied
electrical fields. The electrically induced enhancement of axonal regeneration has been
reported for both peripheral and central neurons in vertebrate systems ranging from
lampreys to mammals (Borgens, 1989b; Sisken, 1988). Common to most of these was
an attraction of the outgrowing neurites to the cathode. The application of artificial
currents to amputated limbs of normally nonregenerating frogs also produced a significant level of relatively unorganized tissue outgrowth, together with a significant
increase in regeneration of nerves to the area (Borgens et al., 1977a). This was accomplished by making the amputation surface electronegative to the base of the limb. When
the current was reversed (i.e., tip of limb stump electropositive), tissue destruction
occurred. Similar attempts to induce limb regeneration in mammals through electrical
stimulation have produced disappointing results.

A major clinical application of electrical fields in healing is in the treatment of
fractures, especially those that produce nonhealing unions (McGinnis, 1989). A stimulus for this was Yasuda's (1974, 1977) demonstration that bony growth could be induced
in normal bone at the cathodal end of an applied electrical current (Fig. 11-6). A wide
variety of strategies has been used to produce such healing. Some of them involve the
application of constant direct currents (DC), whereas others apply employ pulsed DC
or alternating current (AC). Over recent decades, U. S. Food and Drug Administration
(FDA)—approved electrical healing devices have been used to treat thousands of
fracture patients (Bassett et al., 1982; Brighton et al., 1977; Lavine et al., 1977).

SUMMARY

With the increasing numbers of studies on the regeneration of tissues and organs in
mammals, the importance of the environment in which regeneration is occurring is

becoming increasingly apparent. The environment consists of many components and features, ranging from the physical and chemical nature of the substrate on which regeneration is occurring to the overall electrical or hormonal environment of the body. As experimentation becomes more sophisticated, the microenvironment and even nanoenvironment is increasingly being recognized as playing a significant role in determining the type and intensity of the response of cells to injury. An area that has not yet been addressed in depth is the hierarchy of influence of different, and sometimes competing, environmental influences on specific regenerative processes. Epimorphic regeneration traditionally has been considered to occur independently of environmental influences, but as research focuses on events at a lower level, this process is also recognized to be highly dependent on small-scale internal environmental influences. The extinguishing of limb regenerative capacity during metamorphosis in anurans is a classic example of a change in the macroenvironment that has a profound effect on epimorphic regeneration.

CHAPTER 12

Stem Cells, Plasticity, and Regeneration

The history of science makes it certain that the knowledge derived from research on stem cells will eventually lead to enormous benefits for human health, even if they are unpredictable.

—Bruce Alberts (2006)

Since the late 1990s, a sea change has been seen in the attitude of the research community toward cellular plasticity, stem cells, and transdifferentiation. Earlier, the possibility of dedifferentiation was denied by most researchers, and transdifferentiation was generally regarded as a biologic curiosity that occurred only in odd systems, such as urodele lens or retinal regeneration. If a role of hematogenous stem cells in regeneration was even mentioned in the literature, it was most often brought up to be refuted by existing lines of evidence. Yet, within just a few years after stem cells burst onto the scene, they have often been viewed as a panacea that can cure an amazing variety of ills. In this early stage in the history of stem cell biology, there is a need to examine the relative roles of traditional regenerative and healing mechanisms and stem cells as they apply to both natural and engineered regenerative phenomena.

Limb regeneration was one of the first fields in which the origin of regenerating cells was discussed and experimentally tested. In response to early discussions of a hematogenous origin of blastema cells, Butler (1935) conducted experiments involving x-radiation showing that a blastema does not form if the distal part of a limb stump is irradiated, but that a blastema does form if that part of the limb stump is protected while the rest of the body is irradiated (see Fig. 2-1). This experiment clearly showed the local origin of the regeneration blastema, but it did not exclude the possibility that cells beyond the distal limb stump could, to some extent, participate in the regenerative process. This latter possibility was never seriously tested.

In mammalian tissues, two main models of regeneration took shape. In one model, well exemplified by the liver, parenchymal cells constitute the principal source of regenerating cells, but if there is a deficiency of hepatocytes, then oval cells from the ducts could become activated and compensate for the deficiency in the normal primary cellular source.

The other main model of tissue regeneration requires the presence of a population of tissue-specific progenitor cells that could be called into play after damage to that tissue. Examples of this are the periosteum and endosteum of bone and the satellite cells of skeletal muscle. In these tissues, it was assumed that the primary tissue-specific cells (osteocytes of bone and muscle fibers in muscle) do not survive the injury, and that the progenitor cells have the capacity to restore any tissue deficit. Some of the classic models of pancreatic regeneration involve almost complete destruction of the parenchyma and subsequent regeneration from ductal cells. The origin of these progenitor cells has not been easy to pin down. Whether the progenitor cells come from embryonic rests that did not differentiate together with the rest of the tissue or whether they arise from separate lineages continues to be discussed.

How do stem cells fit into these schemes? A principle articulated decades ago in the limb regeneration literature can serve as a framework. This is the principle of *double assurance,* another term for compensation. Experiments first conducted a century ago involved removing the bones from a limb stump; researchers observed that the regenerated limb contained a completely normal skeleton despite the absence of bones in the stump. With these and other experiments that were conducted at the tissue level, it became apparent that when the first-line regenerative response was denied, a secondary mechanism was called into effect to produce the missing tissue. In a more contemporary context, many of the early single-gene knockout experiments produced disappointingly minor developmental disturbances. It was not until double knockout experiments were tried and resulted in major developmental disturbances that the principle of compensation by a closely related molecule also became apparent at the genetic level. The level of genetic redundancy is such that, for many important genes, a molecular backup system is available to compensate for a single mutation. A similar situation also appears to apply at the cellular level. Many damaged tissues can normally regenerate adequately with local cellular resources, but if for some reason these are deficient, a cellular backup system can be brought into play. In the case of the liver, for example, the backup system can act at several levels. The first level is the parenchyma. The second level is the oval cells, which represent a local progenitor population. A third level of compensation may be stem cells originating from outside the organ itself.

WHAT IS A STEM CELL?

Much of our knowledge of stem cells comes from the numerous studies on the hematopoietic system, which serves as a model, although not necessarily the only one, for understanding the properties and capabilities of stem cells (Körbling and Estrov, 2003; Weissman, 2000). Even in this well-studied system, many major questions remain. A generic definition for stem cells is "self-renewing populations of cells that undergo symmetric and asymmetric divisions to self-renew or differentiate into multiple kinds of differentiated progeny" (Cai et al., 2004, p. 585).

Many different types of stem cells currently have been identified. The first dichotomy in a classification of stem cells is between embryonic (ES) and adult stem cells. ES cells are derived from the inner cell mass of blastocysts. These have been described as totipotent because a single cell from the inner cell mass can, in theory, go on to form any of the differentiated cell types in the body. Permanent cell lines derived from single ES cells can be perpetuated *in vitro*. From such founder lines, subsets of cells can be exposed to defined culture conditions and steered to differentiate along certain specific pathways before being implanted into damaged or deficient tissues of organs. This seemingly straightforward approach shows great promise, but it is fraught with technical difficulties. In addition to understanding how to direct the ES cells toward differentiation in a desired direction, other important issues are the potential developmental lability of these cells and their antigenicity. Ensuring that implanted cells do not form teratomas is a significant challenge. Because descendants of ES cells are likely to be immunologically different from the host in which they are implanted, options are to match donor and host, reduce the antigenicity of the donor cells, or suppress the immune response of the host. Superimposed on all of these scientific problems are political and ethical issues involving obtainment of human ES cells.

Adult stem cells are cells in the postnatal body that retain stem-cell properties throughout life. Although the existence of hematopoietic stem cells in the bone marrow has been both discussed and demonstrated for many decades, the recognition of the presence of stem cells in other parts of the body is much more recent. Several levels of "stemness" are possible. True stem cells must meet the criteria of the definition given at the beginning of this section. Although much of the time stem cells are mitotically quiescent, when a stem cell does divide, several outcomes are possible (Fig. 12-1). The division could be a symmetric one in which both daughter cells remain stem cells. Another option is for the division to be an asymmetric one, in which one daughter cell remains a stem cell, but the other one becomes a more restricted intermediate type of cell. Yet, a third option is for each daughter cell in a symmetric division to become an intermediate type of cell.

If one looks at developmental potential, the primary stem cell, sometimes called a *long-term stem cell,* has the greatest ability to create a wide variety of differentiated cell types (Fig. 12-2). Some of their descendants, which remain mitotically active, become more restricted in their developmental fates so that they can no longer produce as many cell types as could the primary stem cell. These have been called *short-term stem cells.* Further down the line are transit-amplifying cells. These cells may be restricted to producing only one type of progeny, and they produce the bulk of the cells needed to repair a damaged organ. Finally, their progeny leave the mitotic cycle and differentiate into mature parenchymal cells of the organ in question.

To complicate matters further, a given tissue or organ may house more than one type of stem cell. For example, adult bone marrow is reported to contain at least four types of stem cells: hematopoietic stem cells, mesenchymal stem cells, multipotent adult progenitor cells, and endothelial precursor cells (Körbling and Estrov, 2003). These stem cells have different antigenic properties and have the potential to produce different sets of progeny (Fig. 12-3). At one time, it was thought that adult stem cells could be distinguished from tissue-specific progenitor cells, such as satellite cells of muscle or basal cells of epidermis, that were thought to be limited to producing only one differentiated type of progeny, but it now appears that even these progenitor cells have sufficient inherent plasticity to create other cell types when exposed to different environmental conditions.

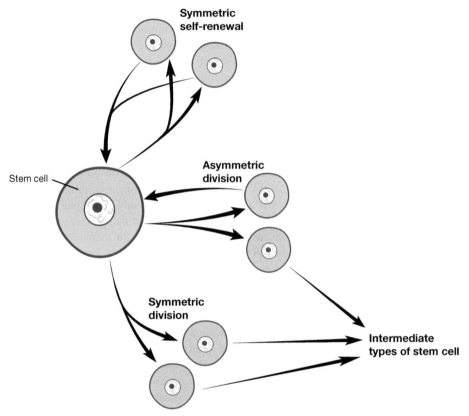

FIGURE 12-1   Types of cell divisions possible for stem cells.

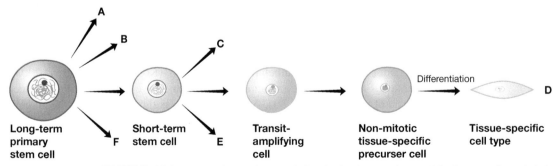

FIGURE 12-2   Scheme illustrating restriction in developmental potential of stem cells and their progeny.

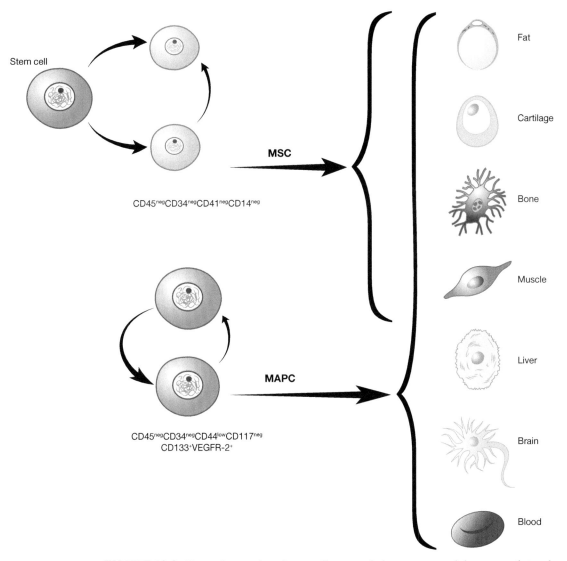

CD45^{neg}CD34^{neg}CD41^{neg}CD14^{neg}

CD45^{neg}CD34^{neg}CD44^{low}CD117^{neg}
CD133⁺VEGFR-2⁺

FIGURE 12-3 Types of mesenchymal stem cells present in bone marrow and the progeny that each can produce. MAPC, multipotential adult progenitor cell; MSC, mesenchymal stem cell. (Based on Bunting and Hawley's experiments [2003].)

WHERE ARE ADULT STEM CELLS FOUND?

At one time, only bone marrow was thought to contain adult stem cells, but since the 1990s, adult stem cells have been identified in a wide variety of tissues (Table 12-1). Stem cells are extremely difficult to identify by morphology alone. Marrow-derived stem cells can be identified by unique combinations of antigenic markers, but even with these it is easiest to identify them through fluorescence-activated cell sorting rather than in histologic preparations. As a result, it is not easy to localize stem cells in many tissues. There are exceptions, however. In the central nervous system (CNS), stem cells have been identified in regions such as the subventricular zone through their mitotic activity (Gage, 1998). At an even more specific level, stem cells associated with hairs are located in the small epithelial bulge region alongside the hair follicle (Sieber-Blum et al., 2004).

The situation is much less clear in other tissues, such as skeletal muscle. In muscle, tissue-specific satellite cells are located alongside individual muscle fibers, but there appears to be another population of stem cells in muscles that have been defined principally as coming from the "side population" capable of Hoechst 33342 efflux as identified by cell sorting (Deasy et al., 2004). To further complicate the situation, Dreyfus and colleagues (2004) have localized marrow-derived stem cells in muscle in both the connective tissue and beneath the basal laminae of the muscle fibers, the latter being a location thought to be reserved exclusively for conventional satellite cells.

TABLE 12-1

Types of Adult Stem Cells and Locations Where They Can Be Found

Ectodermal	Endodermal
Epidermis	Intestinal crypt
Hair follicle	Gastric gland
Neural	Pancreas
Neural crest	Liver
Retina	
Dental pulp	**Germ Cell**
Mesodermal	Primordial germ cells
Cardiac muscle	
Skeletal muscle	
Hematopoietic	
Mesenchyme	
Multipotent adult	
Endothelial	
Umbilical cord blood	
Umbilical cord matrix	

Adapted from Cai, J., et al. 2004. In search of "stemness." *Exp Hematol* 32:585–598, by permission.

An emerging concept is that of specific microenvironmental niches for the production and maintenance of stem cells (Moore and Lemischka, 2006). In certain situations, such as the intestinal crypt and the bulge region in hair follicles, the topography of the stem cell niche is reasonably well defined, but in others, it is much less clear. Interactions between stem cells and surrounding mesenchymal cells, involving both Wnt and bone morphogenetic protein (BMP) pathways and working through nuclear β-catenins in the stem cells, appear to be a common element in maintaining "stemness" in the niches and also in stimulating their production of daughter cells.

PROPERTIES OF ADULT STEM CELLS

In most cases, adult stem cells are best defined by the cell types that they can produce, rather than by intrinsic properties of the cells themselves. Currently, there is intense activity to identify antigens, cellular properties, or gene products that are common to all varieties of stem cells (Cai et al., 2004), but to date, only the most general of characterizations can be made. One important property of stem cells is the absence of specific lineage markers that characterize individual types of differentiated cells. Equally important in the identification of the differentiated progeny of stem cells is the expression of lineage markers, such as type II collagen for chondrocytes and alkaline phosphatase for osteocytes.

Adult stem cells, although quiescent for long periods, must be able to enter the cell cycle when stimulated. Expansion of a stem-cell population is triggered by external events, which are often topographically removed from the stem cells themselves. This means that the initiating event must be communicated to the stem cells by some means, probably chemical in most cases. It is likely that the initiating signal is different for most types of stem cells, although the existence of some common element has not been ruled out. Then the stem cell must have the proper receptors to receive the signal and to activate the mitotic apparatus.

Stem cells are often long lived, and they must be protected from DNA damage, which means that they must possess a functional system for DNA repair. Because they may need to divide many times over the course of a lifetime, human stem cells must also have higher telomerase activity than most somatic cells to maintain telomere length toward the end of life. Given these properties for cellular maintenance, stem cells must be able to respond to an activating signal by entering the mitotic cycle. Entry into mitosis by typical somatic cells is regulated at the G1/S checkpoint, but stem cells spend more time in the S-phase, and the checkpoint for them is at the S/G2 junction (rev. Cai et al., 2004). One of the least understood aspects of stem cell division is what determines whether the daughter cells will be channeled into a self-replication cycle or will enter an asymmetric division, with one remaining a stem cell and the other becoming a more restricted cell heading toward a specific cell lineage. Wnt signaling maintains stem cells in a condition of "stemness" and prevents them from differentiating (Kléber and Sommer, 2004).

Recent research has clarified considerably the molecular basis for maintaining the state of stemness (Bernstein et al., 2006; Lee et al., 2006). This research has focused on chromatin (histone)-based repression of gene activity. Members of the polycomb group proteins catalyze the methylation of histone H3 at specific lysine loci, and this complex silences the regulatory genes, whose proteins turn on more than 200 key developmental genes that are involved in cellular differentiation. The juxtaposition of a silencing site with one that has a stimulating function in what is called a *bivalent domain* allows for greater flexibility and control of gene expression in both stem cells and early ES cells. Like undifferentiated cells in the cleaving embryo, true stem cells express the transcription factor *Oct-4*.

Another important but poorly understood property is translocation. The final progeny of most stem cells mature and settle down in a location other than that of the original stem cell, but what causes them to translocate and how they do it are, for the most part, unknown. The pathway of translocation is known in at least one system. In the CNS, the stem cells in the subventricular zone send their progeny via a rostral migratory stream to the olfactory cortex (Fig. 12-4, A), where they differentiate into neurons or glial cells (Lois and Alvarez-Buylla, 1994). Ciliary currents created by the ependymal cells set up a chemorepulsive gradient (Slit-2), which guides neuroblast migration into the olfactory area (Sawamoto et al., 2006).

Once stem cell progeny reach the area in which they will engraft, they must deal with the issue of integration with the other cells of that tissue. This is a tissue-specific phenomenon that is likely dealt with in a specific way by each tissue in the body. Definitive differentiation of stem-cell progeny may well involve interactions with other cells or the matrix of that tissue. A final critical property of stem-cell progeny is avoidance of neoplastic transformation and overall maintenance of genomic stability.

STEM CELLS IN REGENERATION AND TISSUE RECONSTRUCTION

Stem-cell progeny are known to populate a wide variety of human tissues and organs. This has been demonstrated in cases of sex-mismatched bone marrow or organ transplants by the use of Y-chromosomal markers. The analyzable combinations include female individuals who have had bone-marrow transplants from male donors or male individuals who have received organ transplants from female donors. In the latter case, the presence of cells containing the Y chromosome in the transplant shows that host cells have populated the transplant. In the former case, the presence of cells bearing the Y chromosome in any nonmarrow tissue indicates a cellular mosaic between host tissue and grafted marrow cells. According to one report, after a single male bone-marrow stem cell was injected intravenously into lethally irradiated mice, progeny of that cell were found in the skin, the kidneys, the liver, and the epithelial lining of lung and small intestine (Krause et al., 2001).

A.

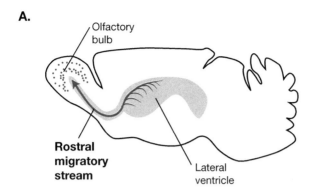

B.

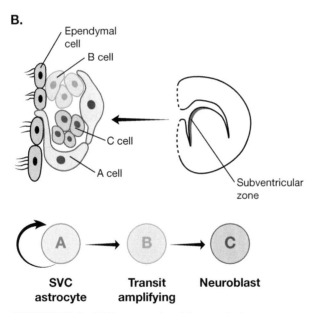

FIGURE 12-4   *(A)* Representation of the rostral migratory stream, which carries cells generated in the subventricular zone along the lateral ventricle into the region of the olfactory bulb in the rat. *(B)* Organizational details of cell arrangements in the subventricular zone as seen in cross section. The sequence below shows the pathway of cell transformation. A cells represent subventricular zone (SVC) astrocytes; B cells represent rapidly dividing transit-amplifying cells; C cells represent neuroblasts. (Adapted from Bottai et al. [2003]; Alvarez-Buylla and Lim [2004].)

Careful analysis in these studies is essential because the Y-chromosome–bearing cells could be macrophages or some other bloodborne cell, rather than a parenchymal cell or a permanent component of the stroma. In transplant situations, male-derived cells have been found in the liver, heart, kidney, endothelium of blood vessels, bone, epithelia of gut, buccal cavity, and skin (rev. Körbling and Estrov, 2003). The reported percentages of immigrant cells vary considerably, from less than 1% to more than 20%. One important observation from the clinical studies is that the percentage of immigrant cells in an organ is typically greater if rejection or some pathologic process is occurring in that organ. This confirms laboratory data suggesting that stem cells are more likely to be called into play when an organ is damaged.

A significant issue in the interpretation of experiments or natural phenomena involving stem-cell populations of tissues is the demonstration that under some circumstances stem cells can fuse with other cells (Terada et al., 2002; Ying et al., 2002). The original demonstrations of stem cell fusion were made *in vitro*. It has been more difficult to verify the presence or absence of fusion *in vivo*. Especially in cases in which analytic capabilities are somewhat limited, as in the human situations described earlier, claims of engraftment and conversion of stem cells to organ-specific parenchyma may not be completely substantiated if fusion or the definitive identification of the phenotype of the stem-cell derivative is in question.

Because of the difficulty of analysis in human cases, most research on stem cells in regeneration has been conducted on experimental animals. One of the most important reasons for using laboratory animals is the ability to track the fate of implanted stem cells. Three commonly used methods for marking stem cells involve: (1) the use of green fluorescent protein (GFP)–marking of cells, (2) the detection of Y chromosomes by *in situ* hybridization, and (3) the detection of β-galactosidase in transgenic cells. Such tracking, combined with tissue-specific differentiation markers, is one of the most effective ways of determining where stem cells are going and what they turn into. At this stage in the development of stem-cell biology, much of the research effort is concentrated on what amounts to lineage tracing, but as stem-cell pathways become clearer, effort is turning toward understanding what makes a stem cell a stem cell and how to maximize its participation in regenerative or restorative processes.

EXAMPLES OF STEM-CELL PARTICIPATION IN REGENERATION

The promise of stem-cell therapy for the treatment of many currently intractable pathologic conditions has led to considerable research on both the presence of stem cells in organs and the ability of stem cells to populate a variety of organs. Organ-specific studies have shown a number of common issues that are generic to stem-cell therapy, as well as those that are unique to the organ in question. This section summarizes what has been learned and what remains to be sorted out in three intensively investigated model systems: heart, CNS, and skeletal muscle.

Heart and Cardiac Muscle

One of the real surprises to those who pioneered the isolation and maintenance of ES cells was how often the resulting embryoid bodies contained groups of spontaneously beating cardiac muscle cells. Subsequent studies have shown that the cardiomyocytes differentiating from stem cells go through an orderly sequence of developmental gene expression that closely mirrors that which occurs during *in vivo* cardiac embryogenesis, and that these cells develop the physiologic and pharmacologic properties of terminally differentiated cardiomyocytes (Bohler et al., 2002; Sauer et al., 2004).

A major possibility of using stem-cell therapy in the regenerative reconstruction of the heart is based on the pathology of a myocardial infarct. In a typical infarct, a segment of the ventricular wall is deprived of oxygen because of blockage of a coronary vessel, and the cardiomyocytes in that area die. This necrotic area becomes invaded by macrophages and fibroblasts, which ultimately transform into myofibroblasts and produce scar tissue. Between the necrotic area and the relatively undisturbed heart wall is a perinecrotic zone, in which some cellular reparative activity occurs. Over time, the scar tissue of the necrotic zone expands (ventricular remodeling) because of the continuous intraventricular blood pressure, and the pathology of congestive heart failure develops.

A number of strategies have been designed to deal with this problem. One of the more common ones is transplantation of a flap of skeletal muscle (commonly latissimus dorsi) over the region of the infarct. Another has been to inject skeletal muscle myoblasts into the area of the infarct or into vessels that lead into the infarcted region (Horackova et al., 2004). Little evidence exists of a transformation of skeletal muscle into cardiac muscle, and one of the significant issues is whether any skeletal muscle that grows or persists on the heart can adapt itself to the continuous beating function required of cardiac muscle. Implanted skeletal myoblasts do not become coupled electrically with neighboring cardiomyocytes (Leobon et al., 2003). They contract independently of the myocardium, which could lead to the development of cardiac arrhythmias. An issue with the injection of myoblasts is the disappearance of more than 90% of the myoblasts from the site of injection within a few hours. The same phenomenon of disappearance of the cellular injectant occurs when myoblasts are injected into dystrophic muscles. One explanation is the removal of the foreign cells by natural killer cells.

The demonstration of the presence of host cells marked by Y chromosomes within transplanted hearts derived from female individuals (Quaini et al., 2002) has stimulated considerable research and even clinical trials (Schächinger et al., 2004; Strauer et al., 2002) on stem-cell therapy for the treatment of certain types of heart disease. A strategy that has been touted as a successor to the above treatment is the introduction of stem cells into areas of heart damage. This has been attempted on a number of occasions, especially with marrow-derived adult stem cells. The stem cells have been introduced by direct injection into the infarcted region or into nearby coronary vessels. According to a number of reports, significant numbers of the stem cells have become engrafted into the ventricular wall as new cardiomyocytes (Orlic et al., 2001a,b). In contradiction to these claims, two later reports (Balsam et al., 2004; Murry et al., 2004) have found no or little evidence of conversion of either directly injected hematopoietic stem cells or derivatives of transplanted bone marrow (in irradiated mice) into cardiac muscle cells. Other studies (e.g., Jackson et al. [2001]) have found a low percentage (e.g., 0.2%–0.3%) of donor-derived cardiomyocytes and a higher percentage (several percent) of donor-derived endothelial cells in the perinecrotic region.

These contradictory results or interpretations point out some important issues in research on any type of stem-cell–based regenerative therapy. First, the donor cells must be clearly labeled, so that later they or their progeny can be clearly identified (Thiele et al., 2004). In addition, it is important to know whether the donor population is homogeneous or whether it possibly includes more than one type of stem cell. The site of introduction is significant (e.g., into the organ itself or into a distant site, such as bone marrow). Once one gets past qualitative issues, such as whether conversion of the stem cells into the desired phenotype actually occurs, then quantitative issues (e.g., yield) become important. On the other side of the equation, it is critically important to precisely identify the phenotype of the stem-cell derivative once it is situated in the desired tissue. For research purposes, unequivocal markers (genetic or antigenic) of the phenotype of the regenerative product must exist. For example, in the case of cardiac muscle, the antigen must be unique to cardiac muscle cells, and the antibody demonstrating it should not cross-react with any other type of cell. In addition, the possibility of autofluorescence must be carefully controlled. Most careful laboratory studies use a definitive marker (e.g., Y chromosome, GFP, or an inserted gene) for the introduced stem cells and then one or more specific phenotypic markers for the tissue to be regenerated. Ideally, one should be able to demonstrate the lineage marker and the phenotypic marker on the same cell. Unfortunately, for clinical studies, this is often not always possible. A recent study, based on Cre/lox recombination between graft cells and host cardiomyocytes in mice, showed that fusion is the predominant mechanism of cardiomyocyte generation from bone-marrow–derived cellular grafts (Andrade et al., 2005).

The current status of the field of stem-cell therapy for damaged cardiac muscle is still very much fluid. The outstanding question remains whether stem cells from the bone marrow or some other location within or outside of the heart can populate a region of myocardial damage and undergo conversion into tissue-specific cells. In fact, it remains unknown whether some form of stem or progenitor cell exists in the normal adult heart. Parmacek and Epstein (2005) have reviewed several recent studies that suggest the presence of some form of stem cell in the heart. One of the more promising is a population of cells derived from the secondary (anterior) heart field in the embryo and that express the islet-1 antigen (Laugwitz et al., 2005). These cells have been isolated from very young animal and human hearts, but they are rare or, in humans, apparently nonexistent in adult hearts.

One group (rev. Sussman and Anversa, 2004) has reported the identification of a population of undifferentiated cells with stem-cell characteristics in the heart of the adult rat. These cells express the surface antigens, c-kit, MDR1, and Sca-1, which are found in hematogenous stem cells. It was suggested that groups of these cells are stored in niches, principally in the atria and apex, where they are exposed to low levels of mechanical stress. To be functional in repair, these cells would then have to be able to migrate toward areas of tissue pathology.

A significant consideration in the case of cardiac infarcts is what types of stem cells to introduce into the heart and whether they will actually engraft in the area around the infarct. One strategy is to introduce both endothelial stem cells, to provide a vascular base, and cells that would have the potential of differentiating into cardiomyocytes (Itescu, 2004). Another issue under active investigation is how best to deliver the stem cells to the heart. Tissue-engineering approaches combining cells and substrates (Zammaretti and Jaconi, 2004) are discussed in Chapter 13. Despite the uncertainty concerning what cells or structures the introduced stem cells produce, there still appears to be clinical evidence of improved cardiac function (Couzin and Vogel, 2004). Much of the research on stem-cell therapy for the diseased heart is still occupied with technical issues, such as proper identification of cells and delivery methods (Müller et al., 2005). These aside, the fundamental biologic questions remain those that confront all areas of stem cell biology (Table 12-2).

Central Nervous System

For most of the twentieth century, the CNS was considered to be a terminally differentiated tissue with no capacity to replace injured or destroyed cells. The series of elegant autoradiographic studies by Altman during the 1960s (Altman and Das, 1965, 1967) that showed areas of DNA-synthesizing cells in the mammalian brain was largely ignored or sidestepped by the neuroscience community because the results did not fit into the dominant paradigm of the times. Later, Goldman and Nottebohm (1983) showed that adult neurogenesis occurred in a nucleus that controls song in the brain of canaries. Widespread acknowledgment of adult neurogenesis in the brain did not occur until after Luskin (1993) and Lois and Alvarez-Buylla (1993) demonstrated that phenomenon in the subventricular zone of the mammalian brain.

The subventricular zone (see Fig. 12-4, B) is a region just beneath the ependyma that lines the lateral ventricle of the brain (Alvarez-Buylla and Lim, 2004; Bottai et al., 2003). Within a specialized extracellular matrix is a group of cells designed to provide replacements for the olfactory system. The subventricular stem cell is a modified astrocyte that can divide to produce more of its own type or generate an intermediate type of rapidly dividing transit-amplifying cell, which ultimately produces type A neuroblasts (rev. Doetsch, 2003). These neuroblasts migrate to the olfactory bulb in a defined rostral migratory stream, and within the olfactory bulb they differentiate into neurons. Another major zone of proliferative activity within the adult brain is the subependymal region associated with the dentate gyrus of the hippocampus. Adult neural stem cells *in situ* are located in clusters close to small blood vessels. Shen and colleagues (2004) have shown that this relation is a functional one, and that products from the endothelial cells stimulate the self-renewal, inhibit differentiation, and enhance neuron production in these stem cells.

Since the demonstration of the pockets of neural stem cells in the subventricular and hippocampal regions in the early 1990s, there has been a veritable explosion of experimental results showing the ability of a wide variety of stem-cell types to produce

TABLE 12-2
Important Biologic Questions on Stem-Cell–Based Tissue Regeneration

Stem cells themselves
What is the appropriate source (e.g., the organ itself, bone marrow, other organs)?
Can the stem cells used differentiate into the desired cell type?
How can one prevent stem cells from differentiating into undesirable cell types?
Into what other types of cells can they differentiate?
Are the stem cells immunologically compatible with the host?
Are the stem cells capable of expanding their numbers to meet the requirements of the host organ?
Will the stem cells divide symmetrically or asymmetrically?
How many stem cells need to be introduced into the host?
Is the age of the donor of the stem cells relevant?
Is the sex of the donor of the stem cells relevant?
Is there a role for the genetic modification of the stem cells?

Introduction into the organ and engraftment
What is the best route for introduction (e.g., direct injection, injection into the local vasculature, transplantation into the marrow)?
If introduction of stem cells is done outside the desired host organ, how can one ensure that they enter that organ and not other organs?
If stem cells are introduced directly into the organ, should they be placed in the region of pathology or near it?
When should stem cells be introduced in relation to specific acute pathologies (e.g., myocardial infarcts)?
Does a diseased organ give off stem-cell homing signals at specific stages of its pathology?
What is the basis for the massive loss of introduced cells within a few hours of injection?
Should introduction of stem cells be accompanied by introduction of an artificial or natural substrate or preceded by its introduction?

Differentiation and functional integration
What are the substrate requirements for the desired differentiation in the host organ?
At what stage of restriction or differentiation should the cells be when they are introduced into the host?
How can differentiation be channeled into the desired phenotype?
What is the role of natural or added growth factors and/or substrates in guiding differentiation of stem cells?
How can differentiation into undesirable phenotypes be avoided?
How can expansion of the stem-cell population be controlled once the cells are engrafted in the host organ?
How can the engrafted stem cells ultimately replace existing pathology (e.g., scar tissue)?
How can the stem cells form new histologic functional units (e.g., renal tubules or liver lobules) or integrate into existing ones?
Do introduced stem cells stimulate regeneration or induce cellular responses from the host organ?

neurons or neuron-like progeny. In addition to sources within the adult CNS, neurons have been produced from ES and fetal stem cells, and in the postnatal body, from sources as diverse as neural crest derivatives, bone marrow, umbilical cord blood, dermis, and hair follicles (Table 12-3). Although neurons or neuron-like cells have been produced from a variety of tissue sources, many produce only a low yield of neurons; therefore, their potential efficacy in stem-cell regenerative therapy is questionable.

Stem-cell therapy has been proposed for the treatment of a wide variety of pathologic conditions in the CNS, but at this point, a small number of possibilities have emerged as prime targets for potential trials. Parkinson's disease has received the most attention. Parkinson's disease involves the degeneration of a specific neuronal subpopulation, namely, the dopaminergic nigrostriatal fibers. There have already been numer-

TABLE 12-3

Sources of Stem Cells That Can Differentiate into Neurons or Neuron-like Cells

1. Embryonic stem cells
2. Fetal cells
3. Adult cells
   A. Nervous system
      1. Subventricular zone along lateral ventricle
      2. Hippocampus—dentate gyrus area
      3. Substantia nigra
      4. Cortex—temporal and frontal
      5. Septum
      6. Glia
      7. Optic nerve
      8. Spinal cord
   B. Neural crest derivatives
   C. Bone marrow
      1. Bone-marrow–derived stem cells
      2. Mesenchymal stem cells
   D. Umbilical cord blood
   E. Dermis
   F. Hair follicles

ous cases of treatment of this disease with transplants of dopaminergic neurons obtained from human fetuses (Lindvall et al., 1990). Neural stem-cell treatment has also been proposed for multiple sclerosis, strokes, and spinal cord transection lesions. In the case of multiple sclerosis, the object of stem-cell therapy would be to introduce cells that would home in on demyelinated axons and differentiate into myelin-producing cells, presumably oligodendrocytes. For strokes, the strategy would be similar to that outlined earlier for treatment of myocardial infarcts. In the case of spinal cord trauma, the transplanted stem cells would have to provide any of a variety of functions, from being a source of axons to providing a substrate suitable for axonal regeneration from residual spinal neurons.

For any stem-cell therapy in the nervous system, a variety of specific issues must be addressed. One of the first is the function to be restored and the desired differentiated cellular outcome, specifically whether the need is for neurons, glial cells, or both, and then what specific type of cell within the category. Once that is decided, the best source of stem cells for introduction needs to be determined. Then the issue becomes whether to try to obtain a certain degree of differentiation toward the desired cell type before transplantation or to let the environment surrounding the engrafted cells determine the differentiative outcome. One of the final issues is how to introduce the cells into the body. For specific localized lesions, direct introduction into or near the lesion may be best, whereas for diffuse lesions, such as in multiple sclerosis, injections into the bloodstream may be more effective. Intraventricular injections have also proved to be effective for the distribution of neural stem cells in a number of animal experiments.

At a purely biologic level, most of the issues confronting those working with neural stem cells are quite similar to those seen in other systems. Once introduced into the host, the neural stem cells must survive long enough to become engrafted. If introduced systemically, they must be able to find their way selectively to the appropriate region in the CNS with a minimal number of cells settling in other locations. In the brain and spinal cord, this also means penetrating the blood–brain barrier. Once inside the blood–brain barrier, neural stem cells appear to be attracted to degenerating neural tissue (Aboody et al., 2000). This mutual affinity is a convenient one for those interested in applying neural stem cells to regenerative situations. The control of proliferation *in situ* is a function of both the intrinsic properties of the implanted stem cells and the microenvironment surrounding them. The importance of growth factors is illustrated by the dramatic expansion of the stem-cell population in the subventricular zone after epidermal or fibroblast growth factors have been infused into the brain ventricles (Gage, 2000).

In addition to proliferating, the implanted stem cells must differentiate along appropriate lines. This, too, is largely a function of the local environment. When neural stem cells were introduced into neurogenic regions of the brain, they differentiated into cells appropriate for that environment (Suhonen et al., 1996). However, when neuronal stem cells were grafted into nonneurogenic parts of the brain, neuronal differentiation did not occur. The presence of injury appears to change the environment so that neural stem cells are able to differentiate in areas of the brain that would not normally support such differentiation (Magavi et al., 2000; Nakatomi et al., 2002). Certain molecular environments, such as BMPs together with leukemia inhibitory factor, suppress differentiation in stem cells while promoting proliferation. In the adult neural stem cell environment *in vivo,* important developmental signaling pathways, such as Notch, sonic hedgehog, BMPs, and Eph/ephedrins, remain active. Such pathways regulate the transition from proliferative to differentiative functions in the stem cells. Whittemore and colleagues (1999) exposed cultured neural stem cells to a wide variety of growth factor environments. They found that cells exposed to epidermal growth factor differentiated toward the astrocyte lineage, whereas those exposed to FGF-2 were more likely to differentiate into neurons. However, in this experiment, it was not possible to determine whether the growth factors selected for expansion-specific subpopulations with a bias for neuronal differentiation, or whether they altered the course of differentiation of individual cells. Conditions similar to this must be present at sites of implanted neural stem cells. Ultimately, the stem cells will have to differentiate into neurons or the appropriate glia. During development, this is accomplished by exposure of the cells to different combinations of growth factors. For example, stem cells in the subventricular zone preferentially differentiate into astrocytes rather than into oligodendrocytes or neurons if exposed to BMP signaling (Alvarez-Buylla and Lim, 2004).

Not only differentiation, but connections and homing are important components of a successful neural stem-cell strategy. Neurons derived from stem cells must produce dopamine to be effective in the treatment of Parkinson's disease, and neurons, in general, must make proper synaptic connections to be part of a functional pathway. Stem-cell–derived oligodendrocytes, on the other hand, must be able to home to areas of demyelination to be effective. Functional synaptic connections of newly generated neurons in both the olfactory bulb and the dentate gyrus have been demonstrated (Carleton et al., 2003; Kempermann et al., 2004). Much remains to be learned about the nature of "settling down" factors that determine the site of cellular landing in the nervous system.

Ultimately, complete integration with the host is the ideal outcome of neural stem-cell therapy. One of the real unknowns in neural stem-cell therapy is the extent to which implanted stem cells may themselves provide a new microenvironment that stimulates host cells to do things that they ordinarily would not do.

Skeletal Muscle

As has been the case with a number of other tissues and organs, research on skeletal muscle has gone through what now appears to be a typical sequence of events concerning the recognition of the potential for stem-cell participation in tissue regeneration. For many years after the discovery of satellite cells (Mauro, 1961), it was assumed that these local progenitor cells were the main cellular source of regenerating muscle, if not the only one. Two reports (Ferrari et al., 1998; Gussoni et al., 2002) challenged this assumption by demonstrating in laboratory animals and in humans with bone-marrow transplants that cells derived from the transplanted marrow could be found in regenerating muscle. These reports cast doubt on the long-standing axiom that satellite cells are the principal source of new material for regenerating skeletal muscle. They also served as the basis for a wave of enthusiasm concerning the possibility of stem-cell therapy for muscular dystrophy and some other difficult-to-treat muscle diseases (Deasy et al., 2004; Ferrari and Mavilio, 2002; Jankowski et al., 2002).

Further experimentation showed that adult stem cells derived from a variety of sources in addition to bone marrow have the intrinsic capacity to form skeletal muscle (rev. Goldring et al., 2002). These include neural stem cells and cells from the dermis, adipose tissue, and synovial membrane. Stem cells derived from these sources are also capable of differentiating into cells with the characteristics of bone, cartilage, adipose tissue, and blood. It was shown that nonsatellite cells derived from a population of muscle cells (the "side population" from fluorescence-activated cell sorting) can reconstitute the marrow in irradiated animals. These cells with stem-cell properties were shown to be of hematopoietic origin (McKinney-Freeman et al., 2002). The formation of ectopic skeletal tissue in muscle under experimental circumstances (Fig. 12-5) or in certain pathologic situations (e.g., fibrodysplasia ossificans progressiva) is evidence of the presence of multipotential stem cells in skeletal muscle.

Experiments involving the injection of cloned, single, marrow-derived stem cells into host mice showed that such cells can reconstitute the complete hematopoietic system, but that they can also fuse with skeletal muscle fibers (Camargo et al., 2003; Corbel et al., 2003). Some investigators postulated a natural progression from marrow-derived stem cells into muscle as tissue-specific progenitor cells (satellite cells), which could then be activated and participate in regeneration after some form of injury (Jankowski et al., 2002; LaBarge and Blau, 2002). Several theoretical possibilities exist

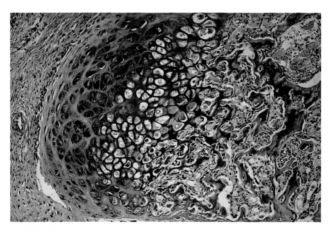

FIGURE 12-5 Nodule containing bone, cartilage, and bone marrow that formed in a rat minced gas-trocnemius muscle regenerate. In addition to these tissue types, the nodule even shows the major structural elements of an epiphyseal plate.

(Rudnicki, 2003). First, marrow-derived stem cells could migrate to muscle directly and settle down as satellite cells. Second, stem cells resident in muscle could pass through the basal laminae and become traditional satellite cells. Third, inflammatory cells arising in bone marrow could directly fuse with existing muscle fibers. One of the complicating factors in sorting through these options is that there now appears to be heterogeneity among the cells that occupy the traditional satellite cell niche alongside a muscle fiber (Sherwood et al., 2004).

Recent research involving the transplantation of normal muscles into immunosuppressed host LC3F nlacZ-E mice showed that less than 0.01% of the nuclei in the regenerated fibers of the muscle grafts were host derived (Washabaugh et al., 2004). Sherwood and colleagues (2004) carefully investigated various populations of fiber-associated and interstitial cells isolated from muscle by flow cytometry and determined that only a certain population of muscle-fiber–associated cells (characterized as follows: $CD45^-$, $Sca-1^-$, $Mac-1^-$, $CXCR4^+$, $\beta1$-integrin[+, $CD34^+$]) readily differentiated into muscle cells either *in vitro* or *in vivo*. These cells reside in the classic anatomic position of satellite cells, between the muscle fiber and the overlying basal lamina. Other cells in the same location and also marrow-derived interstitial cells did not typically form muscle derivatives unless cocultured with myogenic cells or associated with muscle damage *in vivo*. Much remains to be learned about the normal biologic functions of the marrow-derived stem cells that are found in normal muscle or that immigrate into an area of muscle damage. Their role as mediators of inflammation, immunity, or both needs to be separated from their role in reconstructing muscle fibers themselves.

Such research illustrates the complexity of populations of progenitor cells that previously were thought to be homogeneous. These studies also illustrate the limited ability of marrow-derived stem cells to differentiate into specific tissue types. Although they can do so, the percentage of conversion is extremely small. This represents a real challenge for those desiring to use adult stem cells in regenerative therapy.

INTEGRATING STEM-CELL BIOLOGY
WITH NATURAL REGENERATION

Although the field of stem-cell biology is still very young, a few major lessons have been learned. Probably most important is that there are significantly more stem cells than had been imagined previously. The two biggest questions are: (1) What do these cells normally do in the body? and (2) How does one get them to participate effectively in tissue reconstruction and regeneration? It is becoming clear that stem cells are part of a complex hierarchy of cells that are capable of participating in reparative processes throughout the body. They may, at the cellular level, represent part of the double assurance or backup mechanism that exists to compensate for defects at levels from proteins to entire structures.

Based on much recent research, several general principles are becoming clear. One is that bloodborne cells with remarkable powers of differentiation do enter tissues, especially after damage. What they do once they have entered the tissues remains largely a mystery. One option is differentiation into a phenotype appropriate for the parenchyma of that tissue. Another is fusion with cells of the tissue. The latter has been clearly demonstrated (Terada et al., 2002; Ying et al., 2002). A major open question is whether the fused stem cell takes on the phenotypic characteristics of the cell with which it fused or whether it fulfills some other function. Another important question is why stem cells fuse with parenchymal cells in some instances and not in others. Do different types of stem cells react differently when they enter a tissue?

One of the most intriguing issues in stem-cell biology is the proclivity of stem cells to home to damaged tissues in much greater numbers than to normal tissue. This has been demonstrated in a number of different systems. Many of these experimental systems have involved irradiation followed by marrow transplantation. These procedures themselves introduce a number of variables that need to be sorted out. However, some models that involve less trauma still indicate a tendency for selective homing to stressed or damaged tissues (Palermo et al., 2005). Because marrow-derived stem cells have been shown to home to a variety of different damaged tissues, one of the most intriguing questions is whether all damaged tissues emit a common type of signal that is received by bloodborne stem cells or whether tissue-specific signals are generated.

The factors that direct the differentiation of stem cells that have penetrated into damaged tissues are also poorly understood. Researchers studying virtually all systems that involve stem-cell participation in regeneration stress the importance of the microenvironment in influencing the differentiation of stem cells. One of the most clear-cut examples is the importance of stem cells' being associated with differentiating muscle cells in order to differentiate into muscle themselves. Similarly, in the case of bone, the inductive influence of the bone matrix is a critical factor in directing the differentiation of stem cells in this tissue. In addition, the role of the local vasculature and the resulting tissue oxygen concentration acts like a switch in determining whether stem cells will differentiate into osteocytes or chondrocytes.

A significant issue is the relation between aging and the function of stem cells in organ regeneration (van Zant and Liang, 2003). Although there is evidence from irradiation studies on dogs that hematopoietic stem cells in old individuals do not show an apparent decrement in function with age (Zaucha et al., 2001), nevertheless, serial bone-marrow transplantation studies have shown that ultimately the stem-cell population becomes depleted and fails to support the production of an adequate number of blood cells (Ogden and Mickelm, 1976). Investigators van Zant and Liang (2003) stress the importance of qualitative changes (e.g., restrictions in transdifferentiation potential and impaired homing) as opposed to purely quantitative changes. Old stem-cell populations tend to show increased apoptosis after stress, and a larger proportion of the remaining stem cells are engaged in proliferation.

SUMMARY

Stem cells have burst on the landscape of regeneration research. We now recognize that stem cells can be identified in a wide array of tissues and organs, and that many populations of stem cells have not only the capacity to replace tissue-specific defects of the organs in question, but to differentiate into other types of cells as well. Stem cells are resident in tissues and circulate in the blood, and the relative contribution of these populations to the repair of any tissues continues to be debated. Significant questions are: What characterizes "stemness"? How are stem-cell populations maintained throughout life? How are stem cells activated? What controls their differentiation?

At this stage in the development of stem-cell biology, there are far more questions than answers. Some of the questions have arisen because of overextrapolation of research results. Many others reflect important biologic issues. One of the major lessons learned to date in the stem-cell field is the importance of proper markers both for identifying the type and source of stem cells and for determining the cell types into which they transform. One of the fascinating, almost unexplored questions is the potential role of stem cells in facilitating or stimulating regenerative processes in a damaged tissue rather than becoming part of the regenerated tissue themselves.

CHAPTER 13

Tissue Engineering and Regeneration

The basic concept of TE [tissue engineering] includes a scaffold that provides an architecture on which seeded cells can organize and develop into the desired organ or tissue prior to implantation.

—Ulrich Stock and Joseph Vacanti (2001)

One of the most significant breakthroughs in the field of mammalian regeneration has been the application of principles of bioengineering to guiding tissue regeneration. The goal of tissue engineering is the replacement of damaged tissues and organs with new tissue that mimics as closely as possible the overall structure and function of the original tissue. This is a large step beyond tissue replacement with strictly artificial materials, which for decades has been successfully used clinically for the treatment of many conditions. Tissue engineering as a science is very young, only a few decades old, and it began with attempts to deal with some practical medical problems, such as replacing the skin of individuals who were so severely burned that ordinary skin grafts were impractical.

There are many approaches to tissue engineering, but all have to deal with the inherent complexity of the organization and function of tissues that is readily apparent on examination of any textbook of microscopic anatomy. Tissues are composed of cells that are embedded in an extracellular matrix, almost always of their own making. Most tissues are vascularized and innervated, and thus connected to the rest of the body. In addition to the cells that underwent primary development to form the tissue, most tissues also contain immigrant cells, many of which arrive from the blood or the immune system. It is now becoming apparent that many tissues contain cells with stemlike qualities that can be called into play when the need for repair arises. Coordinating the development, maintenance, and repair of tissues are the genetic instructions inherent in the resident cells. This is accomplished not only by the production of the molecules that comprise the cells themselves, but by the secretion of a wide variety of signaling molecules that influence both other cells and the extracellular matrix, resulting in the adaptation of that tissue for its particular physiologic and developmental circumstances.

Many approaches are possible for the regenerative engineering of any tissue, but they all share a number of common elements (Fig. 13-1). First, there must be a source of cells. One strategy is to raise a population of cells *in vitro* and to implant these cells, usually together with other components, into the tissue defect. Another broad strategy

Fundamental Components

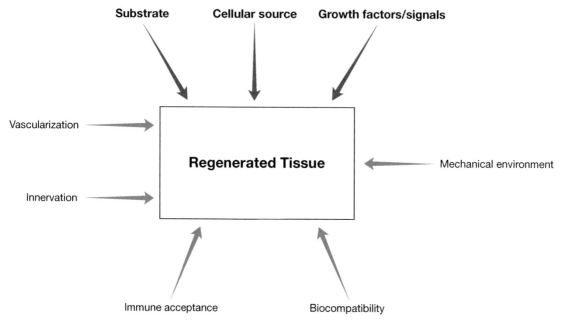

FIGURE 13-1 Diagram of elements that need to be taken into account when stimulating regeneration through tissue engineering.

is to implant a substrate into the defect and rely on cellular migration from the host into the implanted substrate as the source of cells.

A major focus of regenerative tissue engineering is devising substrates on which regeneration can occur. The choice of substrate is an important issue in contemporary tissue engineering (Lanza et al., 2000; Reis and Román, 2005; Yannas, 2001). Without getting into detail at this point, the major dichotomy is artificial versus naturally derived substrates, but substrates with a wide variety of properties can be prepared. Which is the best substrate often depends on the tissue and particular situation.

A critical element in constructing a tissue is the local signaling and growth factor environment. Although secreted factors are ultimately dependent on cells, many growth factors are bound to the extracellular matrix and play an important role in promoting the ingression or differentiation, or both, of cells in the regenerating system. Preservation of growth factor activity in natural matrices or addition of such activity to artificial matrices has received much attention.

With few exceptions, biologic tissues are richly vascularized, and any attempt to regenerate tissues, especially if cells are implanted, must take into account the following factors: (1) the ability of the injected cells to survive in an avascular environment until blood vessels grow in, or (2) the ability of the implant to attract appropriate cells once the implant becomes vascularized. Many promising tissue-engineering approaches are currently on the shelf because survival of the mass of implanted material is not possible in the absence of a blood supply.

Another cellular interaction that is important for many tissues is the reestablishment of proper innervation. Although reinnervation is absolutely essential for skeletal muscle, most organs are supplied with sensory or autonomic innervation, and such innervation is required for their final functional maturation and physiologic modulation.

For some tissues, especially those involved with locomotion or the circulatory system, where pulsatile flow is an important factor, the mechanical environment of the host is an important factor for the highest degree of success. Reintegration of the regenerating tissue into that environment is often required for full cellular differentiation, as well as for restoring overall function of the damaged component of the body.

Immune acceptance of implanted cells or matrices is another requirement of engineered tissue regeneration. Ideally, any implanted cells or substrates would not be immunogenic, but in many cases, it is necessary to inhibit the immune function of the host to the extent that the regenerated material remains intact. The greater the need for immunosuppression, the more severe the potential consequences are for the host.

Lastly, biocompatibility with anything that is implanted is an important consideration. This has been an issue for many decades in the replacement therapy of damaged tissues by nonbiologic materials, but it can also be important with natural tissue matrices, especially if they are derived from species foreign to the host. Biocompatibility issues tend to be chronic, rather than acute, as is the case with revascularization. Nevertheless, in the long run, they can negate the positive gains made by an otherwise successful system.

CELLULAR SOURCES

Any tissue-engineering approach to regeneration must involve a source of cells, which will ultimately fill the defect. One approach is to implant an artificial or an acellular natural matrix into the defect and allow it to become infiltrated by cells migrating into the implant from the host (Badylak, 2002, 2004). The other approach is to introduce cells into the defect, usually together with a matrix or substrate. The introduced cells can be derived from the host (autogenic), from other members of the same species (allogenic), or from a different species (xenogenic). Implantation of cells from each of these sources has its variants, with advantages and disadvantages.

Autogenic cells can be of several types. The first is differentiated cells derived from the host. The earliest example of this, although it is normally not considered in the category of tissue engineering, is transfused blood, which can be taken either from the host in advance of a projected need or from another donor. One of the most common clinical uses of autogenic cells is in the construction of artificial skin substitutes for individuals who have been badly burned (Bell et al., 1991; Parenteau et al., 2000; Yannas, 2001). The principal advantage of autogenic cells is the avoidance of immune rejection. A frequent disadvantage is that host cells are often in short supply. Those that are available typically have to be expanded in culture before reintroduction into the body. Under many circumstances, the delay during the period of expansion of the cell population is risky to the patient. For chronic conditions, this is normally not such a critical issue.

The other main source of autogenic cells for regeneration by tissue engineering is stem cells, derived either from the individual at the time of injury or from stem cells that might have been saved from umbilical cord blood shortly after birth and then banked. As with the use of remaining tissue cells, one of the main issues with autogenous stem cells is that of supply. Particularly when dealing with a tissue defect, such stem cells are already likely to be in short supply, and if the individual has a genetic disease, the stem cells would be expected to carry the same defect.

Allogenic cells, especially if obtained from cadaveric donors, are easier to obtain, but they have the same immunologic disadvantages that plague donated organs. Allogenic cells used in tissue engineering can be either differentiated cells or stem cells derived from embryonic or adult sources. Particularly in the treatment of pathologies resulting from genetic conditions, such as muscular dystrophy, allogenic cells have the advantage of introducing a normal genome to the tissue in question. Although the use of xenogenic cells has been considered for tissue-engineering purposes, the problems inherent in the use of foreign cells are magnified (George and Lechler, 2002). These include rejection (hyperacute rejection being a major issue) and safety issues (zoonotic infections), as well as ethical concerns.

Cells to be introduced into a defective area can be used in the same state as when they are removed from the donor—the classic example again being a blood transfusion—or they can be treated in some way. Sometimes all that is necessary is to increase their numbers—massively in the case of preparing artificial skin for burn patients. At this scale, the cells typically are grown in bioreactors, which maximize the growth potential of the cells to be implanted. Under other circumstances, the cells are engineered in some other manner, usually to produce large quantities of a product deemed to be essential for the treatment of the patient's condition. An example of this approach is the intrathecal implantation of encapsulated fibroblasts or myoblasts engineered to secrete increased amounts of ciliary neurotrophic factor in individuals with amyotrophic lateral sclerosis (Aebischer et al., 1996).

For successful regeneration to follow the introduction of exogenous cells, a number of conditions must be met. First, sufficient cells must be introduced to provide the basis for complete restitution of the organ. Typically, this involves expansion of the population before introduction and continued proliferation once the cells have been introduced into the defect. The latter requires a local environment compatible with maintaining the cells in the mitotic cycle. The introduced cells must not only proliferate, but they also must differentiate appropriately. Differentiation of virtually all cells is the result of their interactions with the local environment through the mediation of soluble molecules, such as growth factors, and the physical nature of the matrix in which the cells are situated. The closer the cell–matrix relations are to those found during normal development or regeneration, the more successful the level of differentiation is likely to be.

For proper function of the regenerated tissue, the cells must be appropriately positioned with respect to their surrounding matrix. In some cases (e.g., epithelia), the desired organization is a sheet or tube of cells resting on a basal lamina. In others, such as skeletal tissues, the regenerated cells and matrix must relate to each other in a way that supports weight and the mechanical demands of the structure. Similarly, regenerated muscle fibers should be oriented in a manner producing maximal contractile force of the muscle. For complex areas such as the digestive tract, the epithelial cells in any regenerated tissue would have to be able to participate in the required absorptive or secretory functions, whereas the cells within the wall should be able to participate in peristaltic movements and provide overall support.

In engineered tissue regeneration, implanted cells must be able to interact appropriately with cells growing in from the host. This includes nerves in the case of muscle or the microvasculature in the case of almost all tissues. Good structural and functional relations with the microvasculature are particularly important for cells that produce a secretory product and that may be sensitive to hormones or other biologically active molecules carried through the blood.

Finally, growth and proliferation of the regenerating cells must be limited to proportions normal for the tissue in question. In addition, the cells should not produce excessive amounts of biologically inappropriate molecules or tissue components, such as scar tissue.

MATRICES AND SUBSTRATES

One of the major challenges in tissue engineering is devising matrices that promote the regeneration of normally organized and fully functional tissues in the region of a defect. Implantable matrices typically fall into two general categories: biologically derived or artificial. Details on the construction and composition of specific matrices are more appropriately dealt with in books on tissue engineering (e.g., Lanza et al., 2000; Reis and Román, 2005), but some general principles apply to most or all of them.

Biologically derived matrices range from those built up from collagen fibers to decellularized whole-tissue matrices. In most cases, when a natural matrix is properly prepared, it does not provoke a serious immune rejection response. Natural matrices can be used as scaffolding for implanted cells or as substrates on which invading cells from the host can migrate and ultimately settle down and further differentiate.

Artificial matrices have been used for substrates, as well as for encapsulating implanted cells. The latter application will not be discussed further in this chapter other than to mention that prime requirements of encapsulating matrices are: (1) maintaining porosity so that secreted products of the encapsulated cells can escape, and (2) preventing their entombment by dense connective tissue deposits laid down by the host. Artificial matrices have been prepared from a wide variety of compounds. Because these are normally polymers, the variants of any particular type can be large. General categories of artificial matrix materials include polyglycolic or polylactic acids, polyhydroxybutyrate, and hydrogels. These have been fashioned into a wide variety of forms ranging from fibers to composite foams used in orthopedic repair applications. More advanced forms of such substrates may consist of composite materials that contain adsorbed hydroxyapatite or growth factors (e.g., bone morphogenetic protein [BMP]) in matrices designed for use in bone repair. A variety of injectable polymers can undergo a sol-gel transformation in response to factors as diverse as pH, temperature, ionic cross-linking, or solvent exchange (Gutowska et al., 2001).

Desirable Properties of Implanted Substrates

The list of desirable properties of an implanted substrate or matrix is intimidating, but straightforward (Table 13-1). A first requirement is that it be both biocompatible and nontoxic. The material must be sterile, and therefore able to be sterilized. Many biocompatibility issues have already been dealt with in earlier research on tissue replacement materials, but for every new material, one must be sure that it does not stimulate a scarring response or fail to become integrated with the surrounding tissues. Especially with artificial matrices, it is important that toxic organic solvents used during their fabrication be removed so that a toxic residue does not remain on the matrix. For many orthopedic applications, the implanted matrix material must also be able to bond with the neighboring skeletal tissues and have mechanical properties that are compatible with the needs for support until a natural matrix is produced.

Once implanted, an ideal matrix will provide an environment that allows maximal survival of accompanying implanted cells or that will encourage the rapid invasion of cells from the host. For survival of implanted cells, the immediate need is for permeability that allows good gas exchange and the passage of nutrients and waste materials. At this level, porosity is a major concern.

TABLE 13-1

Important Properties of Implanted Substrates or Matrices

 1. Biocompatible and nontoxic
 2. Nonimmunogenic
 3. Can be sterilized
 4. Preformed or injectable
 5. Ability to bond with host tissues (e.g., orthopedic applications)
 6. Large surface-to-volume ratio (promotion of cell seeding and matrix formation)
 7. Good permeability and porosity (for penetration of nutrients, fluids, and cells)
 8. Mechanical properties compatible with the needs of the tissue (e.g., in bone repair)
 9. Allows or encourages rapid vascularization
10. Encourages invasion and attachment of cells (through attached extracellular matrix molecules or growth factors)
11. Can ultimately be replaced by a natural matrix in an orderly fashion
12. Allows the organization of cells into normal tissue structure
13. Promotes cellular differentiation
14. Compatible with the ingrowth of nerves

Implantation of most matrices is followed by a uniform inflammatory response (Badylak, 2002). In the first day or two after implantation, the matrix contains a strong infiltrate consisting of equal portions of polymorphonuclear (PMN) and mononuclear cells. By the third day, the acute inflammatory response (PMN phase) has died down, and almost all of the cells in the infiltrate are mononuclear. Over the next 2 weeks, the number of mononuclear cells increases together with an increased vascular ingrowth. This is correlated with degradation of the implanted xenogenic matrix and its replacement by a host-derived neomatrix. By the third week, site-specific parenchymal and stromal cells appear in the matrix. These are oriented along lines of tension.

Many cellular functions are directly affected by their relation with the substrate. Among these are migration, adhesion and spreading, aggregation, and cell-specific functions. Migration, adhesion, and spreading all require a close relation between the cell and its substrate. Connections between cells and substrate molecules are accomplished largely through members of the integrin family located on the surfaces of the cells (Hynes, 1992; Ruoslahti, 1991). On the matrix side, a wide variety of molecules contain cell-binding domains (Table 13-2). Implants of natural matrices contain such molecules, but for artificial matrices, it is often important to treat them so that cell-binding peptides, such as RGD (arginine-glycine-aspartic acid), are adsorbed to their surfaces. This principle has been used effectively in the design of small-bore endothelialized vascular ingrowth matrices (Merzkirch et al., 2001).

It has been known for many years that the physical nature of the substrate also affects cell movements. For instance, migrating cells or outgrowing neurites follow microgrooves made in the substrate (Weiss, 1934, 1958). More recent studies on cultured cells have shown that the presence of micro peaks or valleys or the presence of small patches with more or less adhesive properties can profoundly affect cell migra-

TABLE 13-2

Extracellular Matrix Proteins and Their Cell-Binding Domains

Protein	Amino acid sequence	Role
Fibronectin	RGDS	Adhesion of most cells
	LDV	Adhesion
	REDV	Adhesion
Vitronectin	RGDV	Adhesion of most cells
Laminin A	LRGDN	Adhesion
	SIKVAV	Neurite extension
Laminin B1	YIGSR	Adhesion of many cells
	PDSGR	Adhesion
Laminin B2	RNIAEIIKDI	Neurite extension
Collagen I	RGDT	Adhesion of most cells
	DGEA	Adhesion of platelets, other cells
Thrombospondin	RGD	Adhesion of most cells
	VTXG	Adhesion of platelets

Adapted from Hubbell, J.A. 2000. Matrix effects. In: R.P. Lanza, et al., eds. *Principles of tissue engineering*. 2nd ed. San Diego: Academic Press, 237–250, by permission.

tions (Saltzman, 2000). In addition, it is now recognized that the nanoscale topography of the substrate can strongly influence cellular processes as diverse as adhesion, orientation, mitosis, surface antigen distribution, and differentiation (Stevens and George, 2005). Cells are even able to measure the stiffness of their substrate according to its resistance to intracellular contractions (Discher et al., 2005). These intracellular measurements determine many aspects of cytoskeletal organization, surface receptor expression, and overall cellular behavior. Mooney and colleagues (1992) demonstrated the influence of the substrate in the shifting of cell functions between differentiation and growth.

There is typically a close spatial and temporal relation between the invasion of cells into an avascular area and ingrowth of the microvasculature. For both cellular invasion and microvascular ingrowth, the physical and chemical nature of the substrate can facilitate or hinder these processes. For example, the RGD oligopeptide that is so important as a mediator of cell attachment can be made adherent to a number of artificial substrates. In addition, the presence of specific growth or angiogenic factors, attached to or associated with the implanted substrate, can exert a powerful stimulatory effect on cellular ingrowth and vascularization.

An important property of an implanted matrix is its ability to serve as a functional substrate during the early stages of regenerative tissue formation, but then to undergo controlled resorption as naturally produced matrix components begin to be laid down. Resorption that is too rapid can result in a structural or functional deficit, whereas delayed resorption of an artificial matrix delays the deposition of normal matrix components. For complex tissues, biologically derived matrices lead to a better reconstitution of a normal tissue organization than do artificial matrices (Badylak, 2002), but for homogeneous tissues, such as cartilage or bone, artificial scaffolding materials are often highly effective as a basis for tissue regeneration (Temenoff et al., 2004).

GROWTH FACTORS AND CYTOKINES IN TISSUE ENGINEERING

Successful approaches to tissue engineering involve more than just introducing cells and a matrix into a tissue defect. The effectiveness of these components is usually greatly enhanced by the addition of growth factors or cytokines to the mix. An immense amount of research conducted over recent decades has delineated the roles of growth factors in tissue formation and regeneration. A major challenge for tissue engineering is to devise practical methods for delivery of appropriate growth factors and cytokines in the proper locations and at the proper times to support the most vigorous regeneration and growth of tissues.

It is now apparent that from early embryogenesis to late in life, individual growth factors are reused at different times and in different parts of the body for specific purposes. Regeneration is no exception. Often during the course of regeneration of a structure, growth factors that were active during its embryogenesis or periods of active growth and differentiation are again called into play. Although the demarcation between growth factors and cytokines is not clearly defined, cytokines historically have been associated principally with inflammatory and immune functions. The foundation of the application of growth factors and cytokines to tissue engineering is understanding their role during normal ontogeny and tissue repair. This knowledge serves as the starting point for tissue-engineering applications, but it does not necessarily end there. For structures with poor natural regenerative capacity, a significant question is whether regeneration can be enhanced by the application of growth factors that may not normally be present or that are present in insufficient quantities.

Types of Growth Factors and Cytokines and Their Actions

Most growth factors are members of large molecular families, often with many varieties of a single named growth factor. For example, the transforming growth factor-β (TGF-β) superfamily contains several dozen individual growth factors, which play important roles ranging from the earliest embryonic inductions to bone formation in adults (Table 13-3). Another large growth factor family is the fibroblast growth factor (FGF) family, which contains at least 18 members, all of which are consecutively

TABLE 13-3

Transforming Growth Factor-β Superfamily of Growth Factors

Member	Representative Functions
TGF-β_1 to TGF-β_5	Inhibition (or stimulation) of cell proliferation
	Up-regulates synthesis of ECM molecules and receptors
	Stimulates or inhibits cell differentiation
Activin	Modulates the production of FSH from pituitary
	Mediates induction of mesoderm in embryo
Inhibin	Inhibits gonadotropin synthesis by pituitary
MIS	Regression of female duct precursors in male embryos
Decapentaplegic	Signaling molecule in limb development
	Embryonic patterning
Vg1	Early embryonic inductions
BMP-1 to BMP-13	Many embryonic inductive processes
	Induction of bone in embryos and adults
Nodal	Determination of left–right asymmetry in embryo
	Early embryonic inductions
GDNF	Early kidney induction
	Neural colonization of gut in embryo
	Survival factor for dopaminergic neurons in the midbrain
Lefty	Determination of left–right asymmetry in embryo
GDF-3/Vg2	Skeletal tissue ossification
	Development of lymphoid organs (thymus, spleen, marrow)
	Differentiation of adipose tissue

TGF, transforming growth factor; ECM, extracellular matrix; FSH, follicle-stimulating hormones; MIS, Müllerian inhibiting substance; BMP, bone morphogenetic protein; GDNF, glial cell line–derived neurotrophic factor; GDF, growth/differentiation factor.

numbered from FGF-1. Other growth factors, such as nerve growth factor and epidermal growth factor, are members of small families or are isolated molecules.

A single growth factor can exert multiple effects in the body. As an example, Table 13-4 illustrates the effects of TGF-β on components of the extracellular matrix. Growth factors not only influence the production of components of the extracellular matrix, but they are often bound to them in anticipation of future needs (see Table 4-2). The surfaces of the cells that are the targets of growth factors contain growth factor receptors, which can also consist of multiple isoforms. The binding of a growth factor to its receptor initiates a signaling pathway that ultimately leads to the nucleus of the cell and an effect on some aspect of gene expression.

Cytokines are small proteins or glycoproteins that were originally described as active agents in inflammatory and immune processes and in hematopoiesis (Mire-Sluis and Thorpe, 1998). Well more than 100 specific cytokines currently have been identified, with new ones being reported frequently. Like the growth factors, many cytokines are members of large molecular families. One of the largest, the interleukins, contains almost 20 members. Although most cytokines act in an autocrine or paracrine manner, some also exert more widespread endocrine effects. Most of these effects are involved in gearing up both the microvasculature and proinflammatory cells to respond to some form of tissue disturbance, whether caused by injury or pathogenic agents.

Growth factors and cytokines are powerful signaling molecules that typically act locally rather than systemically. Their localized production and distribution not only concentrates their influence, but also reduces the chances of their causing unwanted effects in ectopic locations. One of the major challenges in tissue engineering is to introduce bioactive molecules in the right place, at the right time, and in the correct amount to produce the desired effect. Two examples of the involvement of growth

TABLE 13-4

Effects of Transforming Growth Factor-β on the Synthesis of Components of the Extracellular Matrix

Stimulation	Inhibition
Fibronectin	Procollagenase
Fibronectin receptor	Plasminogen activator
Collagen	Osteocalcin
Osteonectin	
Osteopontin	
Integrins	
Proteoglycans	
Decorin	
Tissue inhibitor metalloproteinases	
Plasminogen activator inhibitor	
Progelatinase	
TGF-β	
Latent TGF-β binding protein	

TGF, transforming growth factor.

factors and cytokines in repair processes will illustrate the complexity facing the tissue engineer.

Wound Healing

The healing of a skin wound involves processes of coagulation, epithelialization, inflammation, and reconstruction of the dermis (see Chapters 3 and 4). Cytokines and growth factors play an important role in every phase of the process (Rohovsky and D'Amore, 1997; Singer and Clark, 1999). Starting with the blood clot, the aggregated platelets produce platelet-derived growth factor (PDGF), which attracts and activates macrophages and fibroblasts to the site of the wound. Cytokine signals generated by the coagulation cascade then attract neutrophils, the first responders, to the wound. Growth factors, such as TGF-β, released from the damaged extracellular matrix, together with monocyte chemoattractant protein 1, stimulate the infiltration of monocytes from the blood and their transformation into macrophages.

In addition to their phagocytic properties, macrophages can be described as cytokine factories, because they are known to produce several dozen cytokines and other biologically active molecules. Among the more potent products of macrophages is colony-stimulating factor, which promotes the survival and function of the macrophages themselves. The main products of macrophages are designed to facilitate the transition between the destructive and the repair phase of wound healing. These include TGF-α and -β, FGF-1 and -2 (formerly known as acidic and basic FGF), and PDGF, which attract fibroblasts to the wound site and stimulate their production of matrix material. Another set of factors, which are angiogenic in nature, includes interleukin-1, tumor necrosis factor-α, vascular endothelial growth factor (VEGF), PDGF, and the FGFs.

After the release of TGF-α and -β from the macrophages, fibroblasts and invading microvessels themselves produce factors, such as insulin-like growth factor (IGF) and keratinocyte growth factor, which facilitate epidermal migration and proliferation. The epithelial cells themselves produce TGF-α, which acts on itself in an autocrine fashion. Removal of the scab, which acts as a barrier to epidermal migration, is accomplished through the action of various metalloproteinases, which are produced by the fibroblasts and blood vessels.

Fracture Healing

Like most reparative processes, fracture healing occurs in phases. The initial phase consists of damage and the clearing of tissue debris. After this is a transition to a repair phase in which laying down new bone matrix is a major activity. This is followed by an extended phase of remodeling. Tissue engineering of bone involves providing immediate support through implanted matrices, discussed earlier in this chapter, and the stimulation of deposition of a natural matrix that will provide the basis for permanent fracture repair. As in the healing of a skin wound, growth factors and cytokines are involved in every step along the way to complete fracture repair.

Events during the early period after a fracture are similar to those that occur after a wound to the skin or most other structures, with clotting, inflammation, and angiogenesis being major components. The first major bone-specific process is formation of the soft fracture callus. This is initiated by the activation of osteogenic cells in the periosteum and endosteum and their migration over the fracture site. The osteogenic cells then lay down either a cartilaginous matrix or, when vascularization occurs, an osseous matrix. The initial matrix has a unique histologic character (woven bone); then it is resorbed and replaced with a mature bony matrix that undergoes continuous remodeling in keeping with its local mechanical environment. All of these processes are dependent on the actions of growth factors (Canalis, 2000).

Cytokine and growth factor involvement in the earliest stages after a bone fracture follows the general pattern seen in most wounds (see earlier), but superimposed on this background activity is the release from both the hematoma and the bone matrix of substances that directly affect the formation of new bone in the healing fracture. Prominent among them are the BMPs and TGF-β, which together with other growth factors, such as PDGF, FGF-1, and FGF-2, are stimulatory to the proliferation of chondrocytes and osteoblasts. BMPs and TGF-β stimulate the early deposition of bone matrix, but as new bone formation gets under way, TGF-β takes on a more regulatory role in relation to the stimulatory role of the BMPs (Roberts, 2000). Interestingly, most of the active growth factors also begin to be produced by the osteoblasts themselves, and thus act in an autocrine manner.

As the newly forming bone in the fracture callus matures, events in the remodeling stage are modulated by other combinations of growth factors. Osteoclast recruitment and the resorption of bone are mediated by IGF-1, IGF-2, and PDGF, with interleukin-6 and colony-stimulating factor acting more in the recruitment phase. Levels of IGF-2, PDGF, and TGF-β may respond to mechanical influences during the remodeling process.

The fields of orthopedic and dental surgery have been major beneficiaries of newer information on growth factor–induced bone formation. Many procedures have been devised to introduce osteoinductive materials into bone defects for a variety of clinical circumstances (Table 13-5). One of the major technical issues in BMP-mediated facilitation of bone repair is the choice of delivery systems (Mont et al., 2004; Rose et al., 2004; Seeherman and Wozney, 2005). An ideal system will deliver combinations of

TABLE 13-5

Clinical Situations in Which Osteoinductive Substances Have Been Introduced to Facilitate or Augment Bone Healing

Accelerating fracture repair	Distraction osteogenesis
Treating nonunions and pseudoarthroses	Bone grafting/spinal fusions
Filling in cranial defects	Repair of cleft palate and other facial clefts
Filling in mandibular defects	Augmentation of maxilla or alveolar ridge

the appropriate growth factors at the right time and place. It may also include a source of progenitor cells. The carrier system should have the suitable mechanical properties and have a pore size that allows rapid vascularization and the ultimate ingrowth of nerve fibers. Many of the most important growth factors have already been identified, and these small proteins are, in general, capable of maintaining their activity over extended periods when attached to an artificial matrix. Ideally, the growth factor cocktail would include angiogenic and neurotropic agents. Matrices include synthetic polymers of a number of types, collagen preparations with different patterns of cross-linking, inorganic bone matrix substitutes, and polysaccharide-based scaffolds. Growth factors adhere to all of these. Often after implantation, they are released in an initial burst and later at a more even rate. There have also been attempts to incorporate growth factors into microspheres or liposomes, which themselves may be attached to an artificial matrix. Good results have already been obtained, but there is room for tremendous improvement in the future.

VASCULARIZATION AND NEUROTIZATION

One of the greatest barriers to successful tissue engineering is a quantitative one. Many tissue constructs that can be assembled outside the body cannot be successfully implanted into the body because they are too large to become completely vascularized before the cells in the interior become necrotic. A good rule of thumb, based on experience with the free grafting of tissues, is that cells more than 1 to 2 mm from the surface of the graft are not likely to survive for more than a few hours after implantation into the body. In the case of free tissue transplants, the ischemic center of the graft often becomes revascularized, and parenchymal cells move in with the ingrowing blood vessels to repopulate the formerly necrotic center of the graft. For tissue grafting, the advent of microvascular surgery solved many problems, because connecting the nutrient vessels of the graft to vessels of the host provides an almost immediate functional blood supply. With vascularized grafts, tissue regeneration often does not occur because the initial necrotic event is avoided and most cells of the graft survive.

Innervation of an implanted tissue construct is a longer term affair. In the case of muscle, motor innervation is required for normal contractile function, whereas for many other tissues, sensory or autonomic innervation plays a modulatory role. A first-level strategy is to bring a nerve of the host as close as possible to the area of the implant, because axonal regeneration would start from that point. The time for complete innervation, as well as the extent of innervation of the regenerated tissue, depends on both the normal rate of regeneration of the nerve that supplies the axons and the nature of the terrain that the outgrowing axons encounter in the regenerating tissue.

Vascularization (Angiogenesis)

In a broad context of tissue engineering, angiogenesis can be viewed to encompass any attempt to stimulate blood vessel growth in the body, but this section covers only some principles and approaches directed toward increasing the vascularization of an implanted tissue or artificial matrix. The fundamental issue is how to stimulate the return of a functional circulation as early as possible so that cell survival will be promoted and necrotic scarring will be reduced. Successful stimulation of angiogenesis requires the presence of chemotactic factors, mitogenic factors, and a supportive extracellular matrix. *In vitro* studies (e.g., Ingber and Folkman, 1989b) have shown that whereas growth factors are the general stimulus for angiogenesis, the nature of the interaction between the outgrowing network of endothelial cells and the local underlying extracellular matrix determines specific cellular responses, which are translated into morphogenesis.

Currently, there are two main approaches for facilitating vascularization of an implanted tissue construct or matrix. One is to try to stimulate the ingrowth of new blood vessels into the construct as rapidly as possible. The other is to grow preformed vascular channels within the construct before it is implanted into the body and then get them connected with vessels of the host as soon as possible.

The basic strategy of simulating vascular ingrowth consists of introducing an angiogenic factor(s) into the area to be vascularized and letting the factor act on the nearby blood vessels of the host. Many options currently are available for the choice of an angiogenesis factor. Table 13-6 gives a partial listing of some of the many factors that are known to act either directly or indirectly on endothelial cells to stimulate their migration, proliferation, or both. Those factors that act indirectly modulate the activity of other angiogenesis factors, such as VEGF, which directly influence the migration and proliferation of endothelial cells. Once selected, the angiogenesis factor can be coated on an implanted matrix material, such as collagen or Teflon (DuPont) (Grand et al., 1992). Another option is to inject the growth factor into a matrix substance such as Matrigel (BD Biosciences) or incorporate it into a polymer pellet and allow it to diffuse into the tissues surrounding the matrix implant. With both of these methods, controlled delivery is a key issue. The angiogenic factor must get to the right place, at the right time, and in the right amount.

Controlling the concentration or binding state of angiogenesis factors can be critical, because some exert dramatically different effects at different concentrations. A good example is thrombospondin-1. A number of *in vitro* studies (rev. DiPietro, 2000) have shown that the presence of soluble thrombospondin-1 exerts an antiproliferative effect on cultured endothelial cells. Yet, when thrombospondin-1 is bound to the matrix, the substrate is permissive for endothelial cell proliferation.

The ability of endothelial cells to form tubes under certain conditions of *in vitro* culture (Folkman and Haudenschild, 1980) depends to a large extent on the nature of the substrate. When they are cultured on a substrate rich in laminin, endothelial cells quickly form tubular structures (Kleinman et al., 1982). Adding antibodies to the culture inhibits tubule formation. In contrast, a substrate rich in type I collagen is less successful in promoting rapid endothelial tube formation (Madri et al., 1988; Montesano et al., 1983). In an analysis of factors that promote the formation of capillary tubes *in vitro,* Ingber and Folkman (1989b) determined that within a given growth factor environment in the medium, it was the local physical nature of the interaction between

TABLE 13-6

Factors Known to Be Involved in the Stimulation of Angiogenesis

Angiogenin	Low O_2 tension
Angiopoietin-1	Neuropeptide Y
Angiotropin	Nitric oxide
Biogenic amines	PDGF
cCAF	PECAM-1
Dermagraft	Peptide KGHK
EGF	Peptide SIKVAV
Estrogen	Prostaglandin E_2
FGF-1 and -2	SPARC
Fibrin and fibrin degradation products	TGF-α and -β
Haptoglobin	Thrombin
HGF/scatter factor	Thrombospondin
Interleukin-8	Thymosin $\beta4$
Lactic acid	TNF-α
	VEGF

EGF, epidermal growth factor; FGF, fibroblast growth factor; HGF, hepatic growth factor; PDGF, platelet-derived growth factor; PECAM, platelet endothelial cell adhesion molecule; SPARC, secreted protein rich in cysteine; TGF, transforming growth factor; TNF, tumor necrosis factor; VEGF, vascular endothelial growth factor.

endothelial cell and substrate that determined whether endothelial tubes would form. A substrate with a high concentration of fibronectin, which promoted adhesivity with the endothelial cells, promoted spreading and growth, whereas one that was less adhesive or that was arranged three-dimensionally allowed the endothelial cells to retract and form tubes (see Fig. 13-3).

A significant step forward in the strategy of providing *in vitro*–cultured muscle with a preformed microvasculature consisted of culturing a mixture of myoblasts and endothelial cells on a three-dimensional biodegradable polymer scaffold. When embryonic fibroblasts were added to the mix, their secretion of VEGF stimulated the formation of a substantial microvascular network within the forming muscle (Levenberg et al., 2005). Over extended periods of culture, the implanted fibroblasts contributed to the formation of smooth-muscle–like sheaths around the endothelial tubes. On implantation into living hosts, there was evidence that vascular outgrowths from the host vasculature had anastomosed with the vascular channels within the muscle implants.

Many factors, such as the presence and amount of growth factors or the physical configuration of the matrix, exert a strong influence on the success of angiogenic processes *in vitro,* and presumably *in vivo* as well. For application of these principles *in vivo,* it will be important to ensure that the matrix to be implanted is in a form compatible with both endothelial cell migration and endothelial tube formation. An important issue for practical application is whether preformed endothelial tubes will survive implantation into a tissue defect and then be able to connect quickly with vascular sprouts emanating from the host.

Neurotization

Much less research has been done on the engineering of more successful neurotization of an implant than with vascularization. The precursor of such studies is the muscle transplant model. The main object here is to approximate a proximal nerve stump of the host with existing nerve channels in the graft. In the absence of such a connection, axonal regeneration into the graft is much less successful because the regenerating axons must make their way through the endomysial and perimysial connective tissue to reach muscle fibers. Neurotization of skeletal muscle, although necessary for a final functional result, is not as critical in the short term as is vascularization because the early steps in muscle regeneration are independent of innervation.

The tissue engineering of skeletal muscles has not advanced very far, largely because of vascularization issues; but once these issues have been dealt with, the next challenge will be to provide channels within the implant that are compatible with axonal extension. With this challenge, the experience gained with various bridges to facilitate the regeneration of axons across sites of transection will be useful.

Almost no attention has been paid to facilitating the autonomic innervation of engineered tissues or organs. Because autonomic fibers often do not run in discrete bundles by the time they reach an end organ, it is much more difficult to control the course of their regeneration.

In contrast to simple attempts to approximate cut ends of nerves with potential end organs, there had been an enormous amount of research designed to facilitate the repair of damaged nerves themselves (rev. Doolabh et al., 1996; Schmidt and Leach, 2003). The first level of approach has been to provide conduits with appropriate substrate characteristics for bridging gaps, mainly in peripheral nerves. The conduits have consisted of both natural (e.g., nerve or muscle segments) and a variety of artificial materials. The ideal conduit contains a parallel series of channels that incorporate characteristics such as porosity, support cells, substrate orientation, and controlled growth factor release (Fig. 13-2). Application of these principles and specifics to promoting axonal growth into end organs or within the CNS remains a major challenge of tissue engineering.

MECHANICAL ENVIRONMENT

The importance of the mechanical environment in healing and regenerative processes has been recognized since Wolff (1892) described the structural consequences of loading and unloading on the skeleton. Subsequent research has shown that the mechanical environment exerts effects at levels from gene expression to tissue differentiation to overall tissue architecture (Ingber and Folkman, 1989a). The mechanical environment itself can range from the connection between a small group of extracellular matrix molecules and a single cell to the gross mechanical forces that act across the hip joint on the head of the femur. For tissue-engineering applications, mechanical influences are increasingly being studied *in vitro,* with an eye on *in vivo* applications.

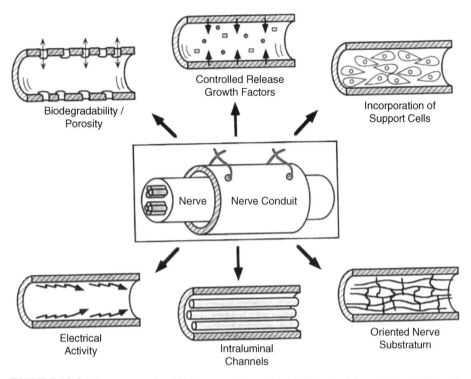

FIGURE 13-2 Components of an ideal nerve guidance channel. (Reprinted from Hudson, T.W., et al. 1999. Engineering strategies for peripheral nerve repair. *Clin Plast Surg* 26:617–628, by permission.)

A fundamental dichotomy in mechanical effects on cells is between an environment that promotes cellular growth versus one that promotes differentiation. This is well illustrated by Ingber and Folkman's (1989b) experiments on capillary endothelial cells *in vitro*. Endothelial cells in culture dishes were exposed to saturating amounts of the angiogenic factor, FGF. The substrate contained varying concentrations of fibronectin. When the density of fibronectin was high, the cells spread out in a monolayer and proliferated. When the density was very low, the cells could not spread and formed small clumps. However, with an intermediate density, the cells first spread, but then began to detach as their internal tensile forces (generated by the cytoskeleton) exceeded the attachment strength between cells and matrix. When this occurred, the endothelial cells began to form capillary-like tubules in the culture dish (Fig. 13-3). Substrate composition or structure that results in strong adhesion by cells promotes spreading and proliferation in other systems as well, whereas less adhesive substrates result in cell rounding and differentiation (Ben Ze'ev et al., 1988; Glowacki et al., 1983).

In vitro studies on skeletal muscle precursor cells have demonstrated the importance of mechanical stretch in early events of activation and differentiation of satellite cells—the precursors of regenerating skeletal muscle fibers. In normal muscle, satellite

A.

Strongly adhesive substrate
Cells form a monolayer

B.

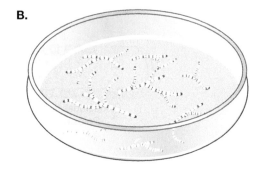

Less adhesive substrate
Cells retract and form tubules

FIGURE 13-3   The influence of substrate on the formation of endothelial tubes *in vitro. (A)* With a strongly adhesive substrate, cultured endothelial cells form a monolayer. *(B)* With a less adhesive substrate, the endothelial cells retract and form tubes.

cells are quiescent and have withdrawn from the mitotic cycle (G0 phase). Typical activation of quiescent satellite cells leading to reentry into the mitotic cycle involves a nitric oxide–mediated release of hepatocyte growth factor from muscle fibers and its binding to the c-met receptor on satellite cells (Wozniak et al., 2005). Mechanical stretch of cultured muscle leads to satellite cell activation within 2 hours through the mechanism outlined earlier in this chapter (Tatsumi et al., 2002).

At the level of cellular differentiation, mechanical stretch keeps the progeny of activated satellite cells in the myogenic pathway, rather than permitting them to enter the adipogenic pathway. This mechanism appears to involve Wnt signaling, because if the expression of Wnt is inhibited, spontaneous myoblast-to-adipocyte differentiation occurs *in vitro* (Akimoto et al., 2005). Such findings may be related to the appearance of fat cells in muscles of old or ill individuals who are relatively inactive.

The tissue organization of cultured muscle is strongly affected by its mechanical environment. The addition of a stretch component to the culture of skeletal muscle *in vitro* results in a strong parallel alignment of the developing muscle fibers (Vandenburgh, 1982), and the addition of fibroblasts to the cultures provides an important matrix that promotes the alignment of the muscle fibers (Kosnik et al., 2001). The combination of stretch and the formation of microvascular endothelial channels in muscle engineered *in vitro* has promise for providing an effective starting point for replacing damaged or missing muscles *in vivo*.

Similar principles are operative in the production of artificial ligaments designed for implantation in humans. A normal ligament is composed of a high content of parallel collagen fibers, with fibroblasts arrayed along the bundles. The intrinsic vasculature of a ligament is poor. A first step in the production of an artificial ligament is the formation of an aligned matrix of collagen fibers, which after implantation will ultimately be replaced by fresh fibers oriented along lines of natural mechanical tension in the body. Adding fibroblasts to the cultures and subjecting the tissue complex to cyclic bouts of traction results in a construct that is well designed for implantation and appropriate tissue replacement in the postimplantation period (Goulet et al., 2000). As with the implantation of other tissues in which extracellular matrix is a predominant component, an important tissue-engineering principle is that there should be an orderly transition between the removal of the artificial matrix and its replacement by a natural one.

IMMUNE ACCEPTANCE AND BIOCOMPATIBILITY ISSUES

For a regenerative tissue-engineering approach to be successful, it is necessary that an implanted construct not be viewed as foreign by the immune system of the host. In addition, the implanted materials should not stimulate local inflammatory responses at the interface between construct and host, which could lead to chronic inflammation, instability, walling-off by connective tissue, or even extrusion of the construct. A classic example of poor biocompatibility is the formation over time of a thick capsule of dense connective tissue around implanted Millipore filter chambers. This effectively eliminates the passage of secreted products of the cells contained within the chambers to the host or the diffusion into the chambers of oxygen and nutrients needed by the contained cells.

The need for immunocompatibility is the same as for the transplantation of intact tissues and organs. Immunologic rejection is not an issue with the implantation of autologous cells or tissue components. Interestingly, immunologic rejection is also minimal in implants of foreign extracellular matrix material as long as associated cells are removed (Badylak, 2004). Particularly with xenogenic implants, removal of all cellular material is critical because of the presence in humans of natural antibodies to the terminal α-1,3 galactose epitope, which is a component of cell membranes on all mammals except for humans and old-world monkeys (Galili, 1993).

Despite the fact that if it occurs, immunologic rejection will effectively negate the positive influence of a tissue-engineering effort, it is important to note that, in mammals, the process of tissue regeneration per se proceeds quite vigorously despite a strong rejection response. Skeletal muscle from nonsibling rats and even rats of a different strain regenerates to the stage of cross-striated muscle fibers despite the concurrent presence of a vigorous cellular immune rejection response (see Fig. 11-2), which ultimately turns the regenerate into a thin sheath of connective tissue (Carlson, 1970b). Similarly, minced muscle tissue from a rat predictably regenerates to at least the multinucleated myotube stage when implanted beneath the skin of immunocompetent mice (Phillips et al., 1987).

Biocompatibility is a major technical issue for any manipulation involving the implantation of foreign materials into the body (Reis and Román, 2005), although most of the specific details are not directly relevant to the process of regeneration. Nevertheless, a significant feature of bioincompatibility is the presence of an inflammatory response at the interface between implant and host. As a general rule, such active inflammatory responses are detrimental to tissue regeneration. An acute inflammatory response can actually shut down tissue regeneration, whereas chronic inflammation of this type interferes with it to various degrees. In my experience with mammalian skeletal muscle, early muscle fiber regeneration is much less affected by immunologic rejection than by acute inflammation.

SUMMARY

A main goal of regenerative tissue engineering is to provide the means for regenerating a structure when natural restorative mechanisms prove to be insufficient. At a practical level, this means providing a source of cells, an appropriate substrate, and the local environmental signals (often growth factors) that stimulate or guide the regenerative process. Defining appropriate natural and artificial substrates has been a major focus of research. Some substrates are designed to allow the infiltration of cells from the host to drive regeneration. Others are combined with cells that are introduced into the body with the substrate. Most substrates are designed to support early regeneration, and then to be slowly resorbed as the regenerating tissue creates a natural substrate. Increasingly, the course of regeneration is being facilitated by the application of growth factors or molecules facilitating cell adhesion to the implanted substrates. Major challenges in tissue-engineering approaches to regeneration continue to be the issues of scale and of structural and functional integration of the engineered constructs into the body. Compatibility and immunologic issues must be dealt with in any new approach to tissue engineering.

CHAPTER 14

Stimulation of Regeneration

But if the above-mentioned animals, either aquatic or amphibians, recover their legs, even when kept on dry ground, how comes it to pass that other land animals, at least such as are commonly accounted perfect, and are better known to us, are not endowed with the same power? Is it to be hoped that they may acquire them by the same useful dispositions, and should the flattering expectation of obtaining this advantage for ourselves be considered entirely chimerical?

—Abbé Spallanzani (1769)

One of the major attractions for people entering the field of regeneration research is the prospect of stimulating the regeneration of structures that are not normally capable of natural regeneration. Particularly for those conducting research on limb regeneration, this has proved to be an elusive goal, but substantial progress has been made in other areas.

Both successes and failures in attempts to stimulate regeneration have been useful in further understanding many of the general principles underlying regenerative phenomena. By examining the rationale underlying these experiments, one can learn much about the evolution of thought in this complex field. Especially in the premolecular era of regeneration research, a common approach in attempts to stimulate regeneration in a number of systems was to manipulate a single variable that was thought to be deficient or inappropriate in the nonregenerating system. One of the hard-learned lessons from these attempts has been that regeneration is a complex phenomenon involving a whole host of variables, and that its absence in any system is likely to have a multifactorial basis. A basic question underlying attempts to stimulate regeneration in any system is whether the ability to regenerate was never present or has been irretrievably lost, or whether the fundamental capacity to regenerate exists but is suppressed in some manner or cannot be expressed because of some deficiency.

LIMB REGENERATION

Anuran amphibians highlight one of the real mysteries of limb regeneration. Their larvae (tadpoles) regenerate amputated limbs very well by a classic epimorphic mechanism that appears to differ little, if at all, from that occurring in the urodele limb. Yet, during the brief period of metamorphosis, the same limb that could regenerate perfectly in the tadpole stage quickly loses its natural capacity to regenerate along a proximodistal gradient (Schotté and Harland, 1943). Transplantation experiments have shown that anuran limbs that have lost their power to regenerate are still unable to regenerate, even if grafted to a host that is capable of regeneration (Guyénot, 1927). Conversely, limbs capable of regeneration can still regenerate if grafted to an anuran host that has lost its own capacity to regenerate (Liosner, 1931; Naville, 1927). These findings ruled out the possibility that the loss of regenerative power was due to the buildup of inhibitory substances in the body fluids as the animals matured. The early studies showed that, although it may be induced by changes in the hormonal status of the animal during metamorphosis, the loss of regenerative capacity becomes inherent in the cells and tissues of the limb and is not just due to suppression by the overall hormonal environment. They also showed that once the regenerative power is lost, it cannot be regained by simply altering the external hormonal environment.

In higher vertebrates, limb regenerative capacity appears to become extinguished during early stages of limb development. From early reports of regeneration in embryonic limbs, it was difficult to determine whether the restoration noted was true regeneration of pattern or merely embryonic regulation. It now appears that true regeneration can occur in the early embryonic limbs of both birds and mammals (rev. Han et al., 2005). The inability of the embryonic chick limb to regenerate is due to the inability of the apical ectoderm to re-form over the amputation surface. However, if the apical ectoderm is retained over the amputated mesodermal part of the limb, regeneration of the missing mesodermal tissue occurs (Muneoka and Sassoon, 1992). This finding led Muneoka to propose that the loss of regenerative capacity during ontogenesis involves sequential deficits (Muller et al., 1999). For example, in the chick limb, the first deficit is related to the absence of ectodermal wound healing, but in the presence of an apical ectoderm, the mesodermal cells can respond by regenerating. Later in development, the mesodermal cells themselves undergo a change that does not permit them to participate in regeneration even though they are supplied with a regeneration-competent apical ectoderm. Reginelli and colleagues (1995) report a strong correlation between the presence or absence of digital regeneration and the postamputational expression of Msx1 and Msx2 in the mesodermal tissues.

Although the nature of the regenerative process remains poorly understood, the success of human fingertip regeneration also declines with age. Clinical experience suggests that the degree of restoration is considerably less in individuals older than 12 years (Douglas, 1972; Illingworth, 1974), although successful cases of fingertip regeneration have also been reported in adults (Bossley, 1975; Holm and Zachariae, 1974).

Postamputational Differences between Regenerating and Nonregenerating Limbs

At the structural level, there is a remarkable similarity in the response of most nonregenerating limbs to amputation if the cross-sectional areas are similar. This is true for the amputated extremities of mammals, postmetamorphic frogs, and even limbs of salamanders, in which their natural regenerative ability has been suppressed by means such as denervation or x-irradiation.

The early epidermal wound healing response to amputation is quite similar, regardless of whether the extremity will regenerate. The presence of an eschar in amputated extremities of mammals, although a complicating factor, does not appear to alter significantly the fundamental dynamics of epidermal wound healing. The wound epidermis that covers the amputation surface is not underlain by a basal lamina in either urodele amphibians (Neufeld and Day, 1996) or mammals (Neufeld, 1989). One complicating factor in both frogs and newborn mammals is that the skin is quite loosely adherent to the underlying tissues, and that after amputation full-thickness skin has a tendency to seal off the amputation wound in a purse-string fashion.

One of the first clear-cut distinctions between regenerating and nonregenerating limbs occurs at the time when a regenerating limb enters the stage of dedifferentiation. The nonregenerating extremity exhibits much less loss of both skeletal and soft tissues near the level of amputation, and cells that had migrated between the cut end of the bone and the wound epidermis begin to differentiate into a layer of dense connective tissue (Fig. 14-1). As this is happening, tissue-regenerative changes are also seen in the distal part of the stump. The most prominent change is the formation of a cuff of periosteally derived cartilage around the distal end of the bone and some small areas of muscle fiber regeneration. In mammals, as well as in frogs, contraction of the original dermis is a prominent component of the healing response, and in rodent limbs, much of the original amputation surface becomes covered by stump skin. Beneath the relatively small area of remaining wound epidermis, a new dermis regenerates.

Interestingly, even in a nonregenerating limb, complex patterns of regeneration can occur. In amputated limbs of young mammals and humans, it is common to find elongation of the amputated bones past the original site of limb amputation. In humans, such outgrowth can interfere with the fitting of prosthetic devices. Speer (1981) attributes such outgrowth to intramembranous bone formation based on the orientation of the transected periosteum during wound contraction, whereas Libbin and Weinstein (1986) report the re-formation of growth plates at the ends of the bones after limb amputation in neonatal rats.

Methods Used in Attempts to Stimulate Limb Regeneration

Many methods and strategies have been used over the years in attempts to stimulate limb regeneration (Table 14-1). Often, they have reflected the state of knowledge at the time or the availability of new products or technology. More often, however, they have demonstrated the thought patterns prevailing in the overall field of developmental biology at the time.

Two main themes characterized the earliest attempts to stimulate limb regeneration. One was the pervasive influence of the concept of embryonic induction (Spemann, 1938). Under this influence, several investigators attempted to supply what were

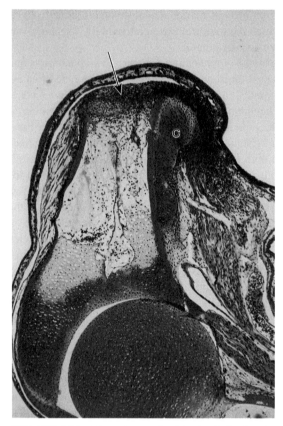

FIGURE 14-1   Photomicrograph of the forelimb of the frog, *Rana pipiens,* 20 days after amputation. Abundant subperiosteal cartilage (C) has regenerated around the shaft of the humerus, and a fibrocellular scar (arrow) is developing between the wound epidermis, which is underlain by dermis, and the remaining stump tissues.

TABLE 14-1

Summary of Major Methods Used in Attempts to Stimulate Regeneration in Normally Nonregenerating Limbs

Method	Selected References
Anuran Amphibians	
1. Replacing skin with tadpole skin	Gidge and Rose (1944)
2. Repeated skin removal	Rose (1944)
3. Mechanical trauma	Polezhaev (1933a, 1936)
	Tomlinson and colleagues (1985)
4. Chemical irritation	Polezhayev (1946)
a. NaCl	Rose (1942, 1944, 1945)
	Singer and colleagues (1957)
b. Na$_2$CO$_3$ and lactose	Polezhaev (1946)
c. Glucose, iodine, nitric acid, colchicine	Polezhaev (1945a, 1946)

TABLE 14-1

Summary of Major Methods Used in Attempts to Stimulate Regeneration in Normally Nonregenerating Limbs—cont'd

Method	Selected References
5. Augmentation of nerve supply	Konieczna-Marczynska and Skowron-Cendrzak (1958)
	Kudokotsev (1965)
	Kurabuchi (1992a,b)
	Singer (1954)
	Tomlinson and colleagues (1985)
6. Adrenal transplants	Schotté and Wilber (1958)
7. Hydrocortisone	Weis and Bleier (1973)
8. Ultraviolet irradiation	Rogal (1951)
9. Tissue implants or tissue extracts	Kudokotsev (1960)
	Malinin (1960)
	Malinin and Deck (1958)
	Polezhaev and Ramyenskaya (1950)
10. Electrical stimulation	Bodemer (1964)
	Borgens and colleagues (1977a)
	Smith (1967, 1974)
11. Electrical stimulation and NaCl	Bodemer (1964)
12. Heat shock	Rose and Rose (1947)
13. Removal of a bone	Goss (1953)
14. Ribonucleoprotein injection	Smith and Crawford (1969)
15. Nerve growth factor	Weis (1972)
16. Vitamin A	Cecil and Tassava (1986b)
17. Dimethyl sulfoxide and vitamin A	Cecil and Tassava (1986a)
18. Fibroblast growth factor	Yokoyama and colleagues (2001)
Lizards and mammals	
1. Repeated skin removal	Umansky and Kudokotsev (1948)
2. Nerve deviation	Kudokotsev (1962)
	Simpson (1961)
	Singer (1961)
3. Parathyroid hormone	Umansky and Kudokotsev (1951)
4. Adrenocorticotropic hormone and cortisone	Schotté and Smith (1961)
5. Trypsin and $CaCl_2$	Kudokotsev and Kuntsevich (1965)
	Scharf (1961, 1963)
6. Tissue extracts	Kudokotsev (1964)
7. Implants of brain and spinal ganglia	Mizell (1968)
	Mizell and Isaacs (1970)
	Taban (1971)
8. Injections of vitreous body	Polezhaev (1972b)
9. Electrical stimulation	Becker (1972)
	Becker and Spadaro (1972)
	Libbin and colleagues (1979)
	Sisken and colleagues (1979)
10. Nerve growth factor	Sisken and colleagues (1979)
11. Fibroblast growth factor	Neufeld (1996)
12. Nail organ transplantation	Mohammad et al. (1999)
	Zhao and Neufeld (1995)
13. Skin removal and NaCl	Neufeld (1980)

assumed to be missing or deficient regeneration-inducing factors by adding tissue implants or tissue extracts to the amputated limb (Kudokotsev, 1960; Malinin, 1960; Malinin and Deck, 1958; Polezhaev and Ramenskaya, 1950). In the same vein, Nassonov (1941) experimented with the implantation of many types of tissue implants to act as organizers in inducing the formation of supernumerary limbs in axolotls.

The second major early theme stemmed from the early research of Polezhaev (1933b), who recognized the importance of tissue damage and dedifferentiation in the early phases of normal limb regeneration in urodeles. He then traumatized amputated, normally nonregenerating hind limbs of metamorphosing tadpoles (Fig. 14-2) by repeated pricks with a needle and stimulated these limbs to regenerate (Polezhaev, 1933a). The basis behind this strategy was that the tissue trauma would increase the amount of subsequent dedifferentiation of mesodermal cells. Polezhaev (1945b) believed that dedifferentiation was a necessary early component of any epimorphic regenerative process.

Polezhaev's early experiments were followed by a variety of other experiments that involved either the infliction of mechanical trauma (e.g., Tomlinson et al., 1985) or a variety of types of chemical irritation applied to the wound surface (Rose, 1944) (see Table 14-1). In amputated mammalian limbs, the skeletal structures do not break down as they do in urodele limbs, leading Umansky and Kudokotsev (1951) to try to bring the destruction of the distal skeleton to a level consonant with that of the soft tissues by administering parathyroid hormone to the animals.

Early researchers, aware of the importance of a wound epidermis in supporting limb regeneration, noted that, in both frogs and mammals, there was a tendency for skin of the stump to contract over the amputation surface, thus reducing the area available for a true wound epidermis to form. This was dealt with by repeated skin removal (Rose, 1944; Umansky and Kudokotsev, 1948).

One of the most influential experiments in the field of limb regeneration was that of Singer (1954), who from a series of carefully performed experiments in newts (rev. Singer, 1952) had concluded that the basis for nonregeneration of postmetamorphic frog limbs was a quantitative deficiency in the number of nerve fibers or in total cross-sectional area of nerve fibers at the amputation surface. He increased the amount of innervation in amputated forelimbs of frogs by deviating the sciatic nerve from the hind limb into the forelimb (see Fig. 6-2). This resulted in the formation of some limb-like outgrowths; but in retrospect, it is now doubtful that these outgrowths were the result of true epimorphic regeneration. Nevertheless, this experiment inspired a number of follow-up experiments conducted not only on frogs, but also on lizards and mammals (see Table 14-1). These experiments produced similar results.

A somewhat different approach to testing the quantitative innervation hypothesis was used by Mizell (1968; Mizell and Isaacs, 1970), who implanted plugs of central nervous tissue into amputated limbs of newborn opossums and reported good regeneration of complex structures. Later, however, these results were placed in doubt by experiments of Fleming and Tassava (1981), who provided evidence that because of the disposition of the newborn opossum's limb, the level of amputation claimed by Mizell was probably considerably more distal than that which was actually reported.

A.

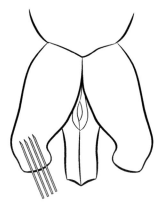

B.

FIGURE 14-2   Stimulation of limb regeneration in metamorphosing frogs. Both hind limbs were amputated below the knee. *(A)* One limb stump was traumatized by pricking it with a needle. The other was a
control. *(B)* The traumatized limb regenerated almost completely, whereas the control limb regenerated
only a short spike. Regenerated portions are outlined in red. (Based on Polezhaev's experiments
[1933a].)

Another approach in the stimulation of regeneration is based on the consistent observation of the formation of a mass of densely packed connective tissue between the amputated soft tissues and the overlying epidermis. According to one popular school of thought, limb regeneration in higher animals may be inhibited by the exuberant scarring response. By eliminating the scarring, conditions more amenable to the initiation of limb regeneration might be present. The discovery of the anti-inflammatory properties of cortisone in the early 1950s led some researchers to try to inhibit scar formation, and thus stimulate regeneration by the addition of adrenal cortical hormones through either adrenal transplantation (Schotté and Wilber, 1958) or the application of hydrocortisone (Weis and Bleier, 1973). Results of these treatments differed little from those reported for other methods.

The 1960s saw a resurgence of interest in electrical correlates of wound healing and amputation, and a number of investigators extrapolated from studies on normally regenerating limbs the design of techniques developed to stimulate limb regeneration by electrical means (rev. Borgens, 1989b) (see Table 14-1). The basic design of these experiments was to place a battery pack beneath the skin of the back and string an electrode down to the amputation surface of the limb and leave the other electrode beneath the skin of the back. With the exception of a couple of dubious cases in which complete regeneration of amputated frog limbs was reported, these attempts produced some forms of outgrowth, but their nature and the underlying mechanisms were difficult to pin down. One important finding was that limb outgrowth, whether in normal development or in regeneration, is associated with outflowing electrical current (Borgens et al., 1977b). Another was that a negative electrode (cathode) at the amputation surface is associated with the accumulation of considerable new tissue, including nerve, whereas the distal presence of a positive electrode (anode) results in tissue resorption (Borgens et al., 1977a).

In the early days of molecular biology, there was the sense that a deficiency of the appropriate RNA might account for a deficiency in regeneration. As a result, Polezhaev's laboratory (Polezhaev, 1972a, 1977a) tested the ability of the application of freshly prepared, highly polymerized RNA to stimulate regeneration in a variety of experimental systems, and Smith and Crawford (1969) also attempted to stimulate limb regeneration in anurans with ribonucleoprotein injections.

Although in recent decades published attempts to stimulate limb regeneration have been few, those that have been done largely involve the application of various growth factors to the limb stump. Yokoyama and colleagues (2001) applied fibroblast growth factor-10 (FGF-10) to nonregenerating *Xenopus* limbs and produced some regenerative outgrowth. Correlated with this was the expression of FGF-8 in the wound epidermis and an array of genes (*shh, Msx-1, FGF-10,* and *Hoxa-13*) associated with the mesodermal tissues during limb regeneration.

Stimulating the regeneration of a human limb remains the holy grail for those who work in limb regeneration, but to date, the results of attempts to stimulate limb regeneration have been largely disappointing. Most published attempts to stimulate regeneration have documented some degree of outgrowth past the amputation surface, but not all have documented the structural details of these regenerated structures. Longitudinal studies on the process of regeneration in these experiments are essentially lacking. To date, there is no convincing evidence that any of the above methods have succeeded in producing a regeneration blastema that is comparable with that seen in a regenerating salamander limb. Yet, there is no question that most of the manipulations have had some type of effect. Whether these are just exaggerated tissue regenerative responses or represent a more complex level of regenerative organization remains to be determined. For any study on the stimulation of limb regeneration, at least four elements should be well documented: (1) the exact structure of what was removed for each limb under experimentation; (2) the nature of the intervention; (3) the structure and ideally the function of the induced regenerate; and (4) the progression of developmental events leading to the production of the induced regenerate.

Possible Reasons for Nonregeneration of Mammalian Limbs

Implicit in Table 14-1 are reasons why, in the past, investigators have believed that limbs of higher vertebrates cannot regenerate. With the passage of time, certain possibilities seem less likely than others. Nevertheless, it has been difficult to definitively confirm or rule out most of them.

One of the first possibilities to be eliminated (through transplantation experiments, see p. 280) was the presence of a continuing inhibitory influence by the hormonal or humoral environment of the limb (Guyénot, 1927; Liosner, 1931; Naville, 1927). Nevertheless, the dramatic reduction of limb regenerative capacity during the metamorphosis of a tadpole into a frog provides strong evidence for the role of hormones as instigators or facilitators of a change in regenerative capacity in anurans.

One of the most often-cited reasons for the inability of limbs in higher vertebrates to regenerate is a rapid and excessive scarring response. There is no question that a substantial fibrocellular scar appears between the wound epidermis and underlying mesoderm in an amputated limb in species that do not normally regenerate, but investigators are faced with a classic "chicken or egg" question. Is the scar the reason the amputated limb does not regenerate, or is it simply evidence that the limb is not going to regenerate? One can cite evidence in favor of each of these possibilities. Certainly, the scar appears early, but in the critical first few days after amputation, is there insufficient scar tissue to block the presumably all-important interactions between wound epidermis and underlying tissues that are said to initiate and support dedifferentiation? In unpublished histologic studies in which I have examined large numbers of amputated limbs of newborn rats, a dense scar has not appeared within a time period that would preclude the initiation of dedifferentiation in amphibians. However, rats are not amphibians, and they may not necessarily follow a time schedule of events proportional to that in the regenerating salamander limb. In the stage at which wounds heal without scarring in the mouse embryo, limb buds amputated at a proximal level fail to regenerate (Wanek et al., 1989), again suggesting that scarring is certainly not the only cause of nonregeneration. Another discordance in time is that the amputation wound in newborn rats is not epithelialized as rapidly as it is in urodele amphibians with limbs of the same cross-sectional area. The lack of early epidermal signals may fail to trigger the requisite mesodermal cell responses.

On the other side of the coin, if an amputated limb of a newt is prevented from regenerating by denervation, irradiation, or some other means, it develops a terminal fibrocellular scar that is reminiscent in structure to that seen in a nonregenerating frog or mammalian limb. Viewed from that lens, one could say that the scar is the result, but not the cause, of nonregeneration. In the case of mammals, it is not impossible that the terminal pad of scar tissue could be both the cause and the result of nonregeneration. It is noteworthy, however, that experiments designed to retard the buildup of scar tissue after amputation have been no more successful than other methods used to stimulate limb regeneration (see Table 14-1).

Early in the study of limb regeneration, the lack of significant tissue destruction and absence of dedifferentiation was noted as a likely basis for nonregeneration in frogs (Polezhaev, 1933a, 1945b; Rose, 1944). In regenerating limbs of newts, evidence collected over many years (Grillo et al., 1968; Ju and Kim, 1998; Needham, 1952) has pointed to a strongly lytic environment, brought about principally by the secretion of matrix metalloproteinases, during the early days after amputation. Much less is known about the types and activities of similar enzymes in the amputated limbs of mammals. Experimental evidence (Carlson, 1982; Yokoyama et al., 2000) suggests that, in frogs at least, the missing element(s) lies in a deficient response of the underlying mesodermal tissues, rather than in the ability of the wound epidermis to stimulate tissue breakdown and dedifferentiation. Whether the wound epidermis of a mammal can support epimorphic regeneration has not been carefully tested, nor has the ability of a mammalian wound epidermis to produce regeneration-promoting FGFs. One of the molecular correlates of an epimorphic or other complex regenerative response is the expression of Msx and bone morphogenetic protein (BMP) signaling by the mesodermal tissues (rev. Han et al., 2005; Odelberg, 2005). Especially in view of the ability of Msx-1 to induce the dedifferentiation of mammalian muscle fibers (Odelberg et al., 2000) and its association with digital regeneration in mice (Han et al., 2005), understanding the limitations of Msx-1 expression in the amputated mammalian limb is of great importance. The finding that an extract derived from regenerating newt limbs can also induce the dedifferentiation of mammalian myotubes (McGann et al., 2001) suggests further approaches to the problem of stimulating dedifferentiation after amputation in mammalian limbs. A concomitant of stimulating dedifferentiation is the reentry of the dedifferentiated cells into the cell cycle. Especially in mammalian tissues, the role of the retinoblastoma (Rb) protein as an agent inhibiting entry into the cell cycle (Brockes, 1998) appears to be an important one.

Whether a deficient nerve supply (Singer, 1954) is the basis for the nonregeneration of limbs in higher animals remains an open question. The absence of a recognizable epimorphic regenerative response to the numerous nerve supplementation experiments certainly suggests that merely increasing the nerve supply by itself is not sufficient to bring about true limb regeneration. By the same token, it is still possible that the nerve supply in frog and mammalian limbs is sufficient to support epimorphic regeneration.

Current experimentation has not definitively established that a true neural insufficiency exists in these limbs, although quantitative extrapolations would suggest that this is the case (Rzehak and Singer, 1966). As in so many other biological processes, it is possible that, through compensatory mechanisms, a deficiency of one component of a process can be made up by another. A good example of this with respect to nerves is the well-known ability of transplanted limbs of urodeles to regenerate even though they are hypoinnervated.

One of the least understood aspects of epimorphic regeneration with respect to mammals is the role of morphogenetic signaling and boundary conditions. For many decades, the triumvirate of a wound epidermis, mesodermal damage, and an adequate nerve supply was considered to be the set of conditions that was absolutely required for the initiation of an epimorphic process (rev. Carlson, 1974c). Not until the flurry of experiments on morphogenetic control of the regenerating limb in the 1970s was it recognized that even in the presence of a wound epidermis, mesodermal damage, and nerves, epimorphic regeneration would not proceed in the absence of appropriate morphogenetic boundary conditions (Bryant, 1976; Lheureux, 1975b). This presents a real quandary for the mammalian experimentalist, because in the absence of a regenerative response it is difficult to test any hypothesis concerning morphogenetic control. This is due in large part to our utter lack of knowledge of the cellular and molecular basis for morphogenetic control, even in animals that regenerate well. The fundamental question in mammals is whether the cells of the limb have retained or lost critical morphogenetic properties, such as positional memory, and if they are retained, whether they can be unmasked.

Almost as obscure as the influence of the morphogenetic environment is the relation between the immune system and limb (or other forms of) regeneration. Although the relation between the immune system and mammalian regeneration has been examined particularly with respect to hypertrophy and tissue regeneration (Babaeva, 1985, 1989; Burwell, 1963), the appearance of complex regenerative responses in immunologically abnormal mice (Clark et al., 1998; Gawronska-Kozak, 2004) suggests that more experimentation is warranted in this long-neglected area. In their review, Harty and colleagues (2003) note the correlation between scarless healing in the mammalian fetus or limb regeneration in larval anurans and the presence of a poorly developed immune system during these ontogenetic stages. They then showed how the lack of good regeneration in frogs and the scarring response in adult mammals is associated with a higher degree of immune competence and a more complex inflammatory response after trauma. Similar ideas were put forth earlier to explain the absence of regeneration in the mammalian central nervous system.

Another factor that has been related to limb regeneration or its absence is the nature of the bioelectrical environment (Borgens et al., 1989). Whether expressed as differences in electrical potential or the flow of electrical current, certain electrical environments have been correlated with its regeneration, whereas inhibiting or reversing the favorable conditions has been correlated with the absence or elimination of the ability to regenerate (Becker, 1961; Borgens, 1989a). When treated as a single variable, the electrical environment alone has not been shown to be the sole determinant of regenerative capacity, but the weight of the cumulative evidence suggests that it constitutes part of the overall puzzle.

The sheer size of the cross-sectional area of a human limb in relation to that of an amputated salamander limb has often been thought of as a barrier to effective regeneration. However, a recent photograph of a large alligator that appears to be regenerating its tail (see Fig. 15-3) would suggest that size alone might not be an absolute barrier to regeneration. Nevertheless, large size would likely mean a long duration for a limb regenerative process.

Taken as a whole, the experimental data on limb regeneration strongly suggest that the lack of limb regeneration in mammals is due to much more than a deficiency of a single variable, but this is the way that most previous experimentation has been designed. A clever mix of both imaginative experimental approaches and contemporary technology will be necessary to generate significant insight into the historically intractable problem of stimulating limb regeneration in mammals.

STIMULATION OF REGENERATION IN THE CENTRAL NERVOUS SYSTEM

One of the most pressing problems in regenerative medicine is stimulating regeneration within the central nervous system (CNS). With an estimated quarter of a million Americans paralyzed by spinal cord trauma alone, the additional toll of disease-induced pathology further increases the stakes in this field. For many years, the CNS of mammals was considered for all practical purposes to be refractory to regeneration. One of the most enigmatic features of regeneration within the nervous system has been the relative ease by which transected axons within peripheral nerves can regenerate in comparison with the near lack of regeneration of injured axons within the CNS. Also, some fish and amphibians possess a substantial ability to regenerate both brain and spinal cord. During the 1960s and 1970s, a few isolated studies hinted at a greater degree of regenerative ability in the mammalian CNS than had been hitherto supposed, but it took much longer to change the prevailing attitude that it was nearly futile to attempt to stimulate regeneration within the CNS.

As was the case in research on limb regeneration, much early research in spinal cord regeneration consisted of attempts to reduce scarring at the site of the lesion. It was assumed that if scar formation could be prevented, central axons might then regenerate across the gap created by the lesion. The analysis of CNS regeneration in species capable of regenerating brain and spinal cord, together with lessons learned from analysis of regeneration in the peripheral nervous system (PNS), stimulated new approaches to a field that was previously dominated by attempts to stimulate regeneration by devising methods for reducing glial scarring at the site of the lesion.

The field of spinal cord regeneration, in particular, has historically been plagued by poorly designed experiments and grandiose claims for success. In an attempt to facilitate the evaluation of the results of research on spinal cord regeneration, a National Institutes of Health (NIH) committee (Guth et al., 1980) proposed the following criteria that need to be fulfilled to conclude that functional regeneration of spinal cord neurites has occurred:

1. There must be unquestioned proof that the experimental lesion has caused a disconnection of the nerve processes.
2. Processes of CNS neurons must cross the level of injury.
3. The regenerated fibers must make junctional contacts.
4. The regenerated fibers must generate responses from postjunctional cells.
5. Changes in function must be demonstrated to derive from regenerated connections.

Natural Regeneration within the Central Nervous System

Both phylogenetic and ontogenetic studies provide tantalizing clues regarding regeneration within the CNS. Most importantly, they have shown that good regeneration of both brain and spinal cord can occur naturally in adult cyclostomes and fish (Bernstein, 1967; Hibbard, 1963; Zupanc, 2001). Even in fish, however, not all types of central neurons regenerate well. Exceptions are the highly specialized Mauthner and Müller neurons, which regenerate poorly, if at all. However, even Mauthner cells can be stimulated to regenerate if dibutyryl cyclic adenosine monophosphate (cAMP) is applied to the cell bodies of the transected neurons (Bhatt et al., 2004). The regenerative capacity of the optic nerve in both fish and amphibians has been studied extensively and serves as the basis for a large body of experimental literature on cellular specificity in the restoration of functional connections in the visual system (Hunt and Jacobson, 1974).

Larval amphibians, both urodele and anuran, regenerate transected spinal cords or produce new spinal cord material after amputation of the tail (rev. Chernoff et al., 2002, 2003; Ferretti et al., 2003). However, transection lesions of the spinal cord in adult urodeles are repaired less successfully than those of larvae, and in anurans, spinal cord regeneration in postmetamorphic individuals is poor. A key element in the amphibian response to spinal cord trauma is the activation of the ependyma and formation of an ependymal tube containing proliferating ependymal cells. Under stable conditions, the ependymal cells contain long processes, similar to those of radial glial cells, that extend to the pial surface of the spinal cord and terminate in end feet. During tail regeneration in urodeles, cells of the ependymal tube form both central neurons and neurons of new spinal ganglia that form within the regenerating tail. The ontogenetic loss of regenerative capacity in the anuran spinal cord parallels that for the loss of limb regenerative capacity. The basis for this loss remains unclear.

True regeneration of neurites within the spinal cord has been demonstrated in both chick embryos (embryos up to 13 days) (Shimizu et al., 1990) and American opossums (up to 7–8 days after birth; a newborn opossum is roughly equivalent to a 14-day mouse embryo) (Saunders et al., 1998; Wang et al., 1998). Although the basis for the ontogenetic loss of regenerative capacity is not known, in both the chick and opossum, there is an interesting relation between the loss of regenerative capacity and the onset of myelination within the spinal cord (Ghooray and Martin, 1993; Macklin and Weill,

1985). In both these species, if myelination is experimentally delayed, the period during which spinal cord regeneration is possible is extended by a few days (rev. Ferretti et al., 2003). Whether this relation is more complex than the simple expression of Nogo and its receptor remains to be determined.

In most cases, regeneration of neural processes in the adult mammalian CNS is abortive. There are, however, a few notable exceptions (rev. Kiernan, 1979; Svendgaard et al., 1975). Within the olfactory system, new neurons are formed throughout life. Axons of the olfactory bulb, neurosecretory fibers in the neurohypophysis, and mono-aminergic neurons that are found in several regions of the adult CNS are able to regenerate.

Strategic Basis for Stimulating Regeneration within the Central Nervous System

A sea change in overall approaches to stimulating regeneration within the CNS has occurred in recent decades. The magnitude of this change can be appreciated by examining a well-done list of hypotheses and assumptions about the inability of the CNS to regenerate that reflected the state of the field in 1979 (Table 14-2). At that time, sufficient evidence did not exist to prove or disprove most of the hypotheses. As was the case in the field of limb regeneration, a dominant mindset that was reflected in the literature about stimulating regeneration in the CNS before that time was a focus on changing a single variable as the key. Since 1980, two broad changes have characterized the approach toward stimulation of regeneration in the CNS. One is a flood of new experimental findings that allow some clear-cut and well-founded approaches to be made. The second is a general recognition that stimulation of regeneration within the CNS involves many variables, and that the interventions must be applied in correct spatial and temporal order.

The first major breakthrough was the demonstration that central axons can regenerate if provided with a suitable environment (David and Aguayo, 1981; Richardson

TABLE 14-2

Hypotheses Regarding the Inability of the Central Nervous System to Regenerate as of 1979

1. Axons within the central nervous system (CNS) are intrinsically unable to regenerate.
2. The periaxonal environment of the CNS does not support axonal regeneration.
3. There is a scar tissue barrier to regeneration in the brain and spinal cord.
4. Inappropriate synapses are formed by transected neurons in the CNS.
5. There is an autoimmune inhibition of regeneration in the CNS.
6. Regenerating central axons must be surrounded by an extracellular fluid containing proteins derived from blood.
7. Endogenous electrical currents are not functioning optimally.

Adapted from Kiernan, J.A. 1979. Hypotheses concerned with axonal regeneration in the mammalian nervous system. *Biol Rev* 54:155–197, by permission.

et al., 1980). This was followed by the discovery of specific regeneration-inhibiting molecules within myelin (Caroni and Schwab, 1988). Shortly thereafter was the gradual realization that the glial scar was more than a physical barrier, and that present within the scar are also molecules that inhibit axonal extension. Finally, the resurgence of interest in stem cells and the belated recognition that proliferation is possible in the CNS has opened up a large field of cell-based therapies for diseases and trauma of the CNS. (Issues surrounding stem cells and neural regeneration are discussed in Chapter 12 and are not dealt with further here.)

A broad-based set of strategies for stimulating regeneration and restoring function in the injured CNS has evolved over time. The individual goals of these strategies are clear, but effecting these goals both individually and collectively is difficult because of the large number of variables to be dealt with in each category. In many cases, the state of knowledge is such that specific issues can be attacked at the molecular level; in others, much more remains to be learned before pinpoint strategies can be developed. The broad strategies (Fig. 14-3) are listed in the following paragraphs.

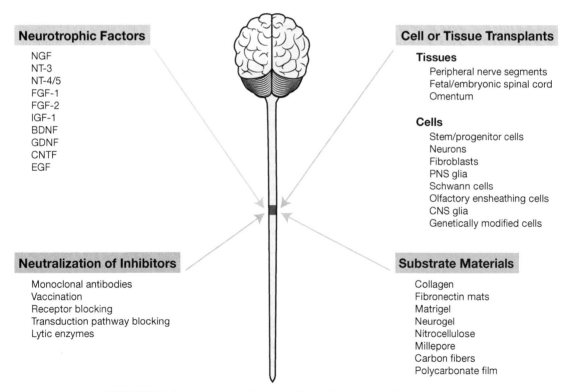

Neurotrophic Factors

NGF
NT-3
NT-4/5
FGF-1
FGF-2
IGF-1
BDNF
GDNF
CNTF
EGF

Cell or Tissue Transplants

Tissues
Peripheral nerve segments
Fetal/embryonic spinal cord
Omentum

Cells
Stem/progenitor cells
Neurons
Fibroblasts
PNS glia
Schwann cells
Olfactory ensheathing cells
CNS glia
Genetically modified cells

Neutralization of Inhibitors

Monoclonal antibodies
Vaccination
Receptor blocking
Transduction pathway blocking
Lytic enzymes

Substrate Materials

Collagen
Fibronectin mats
Matrigel
Neurogel
Nitrocellulose
Millepore
Carbon fibers
Polycarbonate film

FIGURE 14-3 A summary of strategies designed to promote spinal cord regeneration. The red area in the spinal cord represents a lesion. (After Tsai and Tator [2005].)

Reducing the Extent of Secondary Injury

Important functionally, but less glamorous, has been the realization that, in spinal cord lesions, secondary pathology, occurring after the primary lesion was inflicted, can cause a significant worsening of the individual's functional status. In the days after injury, cystic regions appear on either side of the primary lesion. These consolidate to form grossly visible cavitations, which destroy cells that were not injured by the primary traumatic event (rev. Profyris et al., 2004; Ramer et al., 2005). All of this results in a larger overall lesion to be regenerated and a larger and more complex scar that must be dealt with. There are three principal bases for the development of secondary injury: (1) vascular changes, such as hypoxia, ischemia, and edema; (2) toxic biochemical events, such as free radical formation and protease release; and (3) cellular responses, such as cell death, inflammatory reactions, and glial cell activation. Because much of the strategy involving secondary injury encompasses preventing degeneration rather than stimulating regeneration directly, this topic is not discussed further here.

Stimulating Axonal Growth and Restoring Myelination

With the demonstrations that central neurites do have the intrinsic capacity to regenerate, great effort has been placed on understanding the nature of the normal inhibitory environment and devising strategies to allow axonal elongation. The early work of Aguayo (David and Aguayo, 1981; Richardson et al., 1980) clearly showed that the environment of a peripheral nerve is compatible with central axonal regeneration. This led to two lines of research designed to determine (1) what factors in the PNS environment support axonal regeneration, and (2) what factors in the CNS environment inhibit it.

One of the earliest clues to the nature of the regeneration-promoting environment of the PNS was the correlation between the presence of laminin, and also fibronectin and collagen, as a substrate and the ability of neurites to elongate (rev. Grimpe and Silver, 2002). In addition, the importance of Schwann cells and their secretions of neurotrophic factors (NFs) was soon realized. This has led to a variety of therapies based on the introduction of NFs at the site of the lesion and cellular therapies involving the transplantation of Schwann cells into the regions where axonal regeneration was desired.

Many studies on inhibitory factors within the CNS have focused on the inhibitory proteins Nogo, MAG (myelin-associated glycoprotein), and OMGP (oligodendrocyte-myelin glycoprotein), all of which are bound to myelin and on the surface of oligodendrocytes (rev. David and LaCroix, 2003; He and Koprivica, 2004; Schwab, 2004). All three of these inhibitors bind to the same receptor, NgR (Nogo receptor), which is one of the many receptors located on the surface of growth cones of neurites (Fig. 14-4). In conjunction with p75, a coreceptor, the bound NgR activates a signal transduction pathway that involves Rho, which ultimately leads to the collapse of the growth cone by activating contraction through a mechanism involving myosin of the growth cone

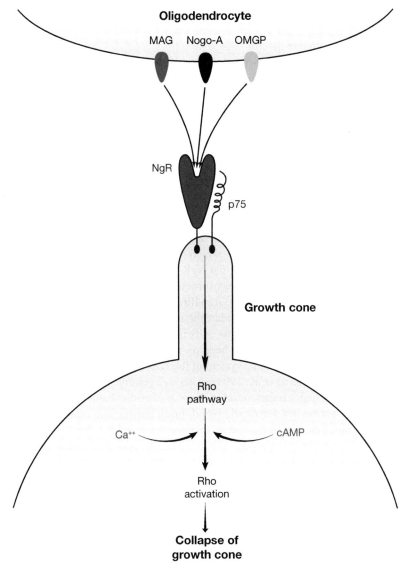

FIGURE 14-4   The Nogo pathway. Myelin-derived inhibitory proteins (MAG, Nogo-A, and OMGP) bind to a common receptor (NgR) on a growth cone. This stimulates the Rho pathway that ultimately results in the collapse of the growth cone. cAMP, cyclic adenosine monophosphate; MAG, myelin-associated glycoprotein; OMGP, oligodendrocyte-myelin glycoprotein.

itself and the actin filaments of the filopodia. This results in retraction of the filopodia and the inhibition of neurite elongation.

In addition to myelin-related inhibitors, another set of molecules with axonal growth-inhibiting properties is produced in the region of the glial scar. The most important of these are the chondroitin sulfate proteoglycans, which also serve to inhibit axonal growth in the embryo, and the tenascins. As a group, these molecules inhibit neurite outgrowth by binding to laminin, thus preventing the laminin from interacting with integrins on the surface of the growth cone. Laminin, however, is greatly reduced in the mature CNS. Interactions between neurites and the glial scar are complex because the early glial scar is permissive for neurite outgrowth, whereas the later scar is dominated by components that block neurite growth.

Several broad strategies have been developed to counteract the effects of these growth-inhibiting proteins, and they have been shown to improve neurite extension both *in vitro* and *in vivo* (rev. David and Lacroix, 2003). The earliest approach was to use monoclonal antibodies against the inhibitors. This was first done with the IN-1 monoclonal antibody, which recognizes Nogo-A (Schnell and Schwab, 1990). Later, with the recognition that NgR is a common receptor for the three known myelin-associated inhibitors, strategy turned toward attempts to block the receptor by antibodies or antagonist peptides (GrandPré et al., 2002). Common strategies for delivery of the antibodies have been either the use of osmotic pumps for direct installation or the transplantation of hybridoma cells into the CNS.

A second approach has been to vaccinate laboratory animals with purified myelin to stimulate the immune system of the host to produce antibodies against the inhibitory molecules (Huang et al., 1999). Although positive effects from this approach have been seen, two major issues stand in the way of clinical application. One issue is penetration of the antibodies across the blood–brain barrier; the other is the danger of producing autoimmune reactions against myelin.

A third broad approach involves the disruption of intracellular signal transduction mechanisms (rev. Kastin and Pan, 2005). This approach commonly involves the infusion of inhibitors of key elements of the pathways, a good example being the inactivation of Rho-A (see Fig. 14-4) by the C3 transferase enzyme derived from *Clostridium botulinum*. Along similar lines, Koprivica and colleagues (2005) focused on the activity of epidermal growth factor receptor (EGFR) on mediating the inhibitory action of both myelin by-products and chondroitin sulfate proteoglycans on regeneration within the CNS. By blocking the kinase function of EGFR by low-molecular-weight inhibitory compounds, they were able to stimulate significant outgrowth of optic nerve fibers.

A complementary strategy to inactivating inhibitors is to supply the area of regeneration with neurotrophic compounds. Neurotrophins are highly effective in the enhancement of neurite extension in gray matter, but by themselves they do not appear to be effective in overcoming the inhibitory properties of white matter. Effective neurotrophins within the CNS are NT-3 and brain-derived neurotrophic factor. The efficacy of these is reflected in an increased expression of growth-associated protein 43 (GAP-43) in the elongating axons. Glial-derived neurotrophic factor enhances neurite outgrowth in dorsal roots, but not in the CNS itself.

Another major molecular strategy for enhancing CNS regeneration is to attempt to neutralize the inhibitory components of the glial scar. Many of the early attempts to stimulate regeneration of the spinal cord involved the application of trypsin and other lytic enzymes to the area of trauma (Matinyan and Andreasyan, 1973; Nesmeyanova, 1977; Pettigrew and Windle, 1976). Although this was done to try to eliminate the scar, in retrospect, some of the positive effects might have been caused by the inactivation of inhibitory molecules within the forming scar. More recent attempts have involved the infusion of specific enzymes, such as chondroitinase, to remove the chondroitin sulfate glycosaminoglycan chains (Moon et al., 2001).

Bridging Gaps by Providing Permissive Substrates for Elongating Neurites

In almost all cases of spinal cord injury, one of the main barriers to regeneration is the gap created by both primary and secondary injury. Left alone, the space will be filled by a glial scar, which is inimical to the regeneration of axons across the gap. After the early demonstrations that axons can regenerate across a gap if provided with peripheral nerve conduits, more recent treatments have involved implants of cultured Schwann cells or olfactory ensheathing cells, both of which normally provide an environment compatible with neurite regeneration (rev. Ramer et al., 2005). These cells can be obtained from the patient and expanded in culture. Then they are introduced into the lesion, often in association with some type of artificial biodegradable substrate material. Both types of cells support the regeneration of neurites, but they differ in the reactions of the regenerating neurites when they reach the cell implant/spinal cord interface. When central axons have regenerated through spinal cord bridges lined by Schwann cells, they do not readily penetrate into the tissue of the spinal cord (Plant et al., 2001), although the use of olfactory ensheathing cells facilitates this transition (Raisman, 2004; Ramon-Cueto et al., 1998).

Another approach to bridging the gap is based on the known ability of neurites in mammalian embryos to regenerate in tracts that have not yet become myelinated. Even in normal development, the appearance of myelin in specific tracts is probably timed to allow pathways to be completed before cementing them in place with insulating material designed to promote stability. A number of animal experiments and clinical trials have involved the implantation of fetal spinal tissue into sites of spinal cord injury (Anderson et al., 1995; Lakatos and Franklin, 2002). Although most of the grafted cells become glia, some also turn into neurons. A major practical and ethical issue in this approach to human spinal cord repair has been the source of material.

A further approach to cellular therapy is to engineer some or all of the cells to produce NFs to further facilitate neurite elongation. Another variant is to construct biosynthetic substrate material so that it will release NF on a steady, controlled basis.

Achieving Appropriate Neuronal Connectivity and Enhancing Axonal Transmission

Experience has shown that even when axons have regenerated across a gap in the spinal cord, they do not become well myelinated. This reduces the functional effectiveness of the regeneration because transmission of impulses along the poorly myelinated axons is compromised. In a different domain of pathology, the demyelinating diseases also leave a functional deficit in their wake because of interference with the transmission of neural impulses. In endeavors to deal with both of these conditions, researchers have attempted transplants of various central and peripheral myelin-forming cells into the lesions. Commonly introduced cells have been glial precursor cells from the CNS or Schwann cells and olfactory ensheathing cells from the periphery. Of the peripheral cells, Schwann cells have proved to be the more effective myelinators, and even when retrovirally labeled olfactory ensheathing cells were introduced into a spinal cord lesion, Schwann cells from the host have been shown to produce most of the myelin (Boyd et al., 2004). Other effective sources of myelin are CNS precursor cells and a variety of stem cells.

Establishing appropriate neuronal connectivity is a long-range goal that first depends on obtaining sufficient axonal and dendritic elongation to make such connections even possible. Many studies on stimulated spinal cord regeneration have provided evidence of some degree of functional return, but there are a number of possible explanations for this in addition to the establishment of anatomically appropriate synaptic connections.

STIMULATION OF BONE REGENERATION

In most cases, fractured or damaged bone regenerates well, but there are circumstances, whether on an individual case basis or because of the type of bone, when natural healing or regeneration of bone does not occur. According to some estimates, well more than 100,000 cases of such nonunions per year occur in the United States alone. Especially in long bones of the leg, a chronically nonhealing fracture that forms a pseudarthrosis is highly debilitating, and treatment of this condition has long been a priority in the orthopedic surgery community. Other bones (e.g., those of the cranial vault) do not naturally fill in defects by regenerating new bone. The treatment of cranial defects by methods of inducing new bone has been discussed earlier in this book (see p. 21). This section concentrates on methods designed to stimulate bone formation in nonhealing fractures.

Bone Grafting

Bone grafting was one of the first methods used to stimulate the regeneration of bone in either large defects or nonhealing fractures (Bier, 1923). Although it was initially thought that the grafted bone survived, it soon became apparent that the grafted bone served principally as a scaffold on which new bone ultimately formed. Many types of bone, ranging from strips of dense cortical or spongy bone to bony fragments or even pastes, have been used for grafting (Habal and Reddi, 1992; Vinogradova and Lavrishcheva, 1974). Grafts of spongy bone not only provide an immediate measure of mechanical stability, but they are also a rich source of osteoprogenitor cells that are present within the red marrow contained in the graft (Block, 2005). Although in the short term, grafts of dense cortical bone provide mechanical stability in the case of fractured long bones, the main mechanism of success underlying bone grafting was the induction of new bone by materials of the graft.

Chemical Inducers

The experimental and clinical induction of new bone has been performed experimentally for more than a half century, starting with Huggins's (1931) report of bone induction by the urinary bladder. Initially, the inducing material consisted of ground bone or bone matrix (Polezhaev, 1951, 1977b; Urist, 1965). An early variant on the induction theme was Belous's (Belous et al., 1974) research on the induction of bone by exogenous RNA. This work was stimulated by the earlier reports of Niu (Niu et al., 1961), who claimed to get liver-tissue–specific inductions by RNA preparations derived from liver.

The contemporary era of bone induction began when Urist and colleagues (Urist et al., 1979, 1984) first named and then purified BMP, the active principle of the earlier biologic inductive preparations (see p. 88 for further details). Currently, the application of growth factors for inducing new bone is an integral part of many tissue-engineering approaches to bone formation (Seeherman and Wozney, 2005) (see p. 270). The addition of BMP is now an adjunct for many reconstructive procedures, ranging from spinal fusions to surgery to correct facial clefting.

Electrical Stimulation

The clinical application of electrical stimulation to promote bone regeneration is based on the natural production of electrical signals by mechanically stressed bone and the experimental findings that new bone forms around the cathode if electrodes are placed on a bone (see p. 237). A considerable body of research has shown that through implanted electrodes, new bone can be stimulated to form by the action of direct current, pulsed direct current, or low-frequency alternating current (rev. McGinnis, 1989). In addition, a number of systems have been developed for delivering an electrical stimulus to bone through externally applied electrodes, usually two insulated metal electrodes placed on either side of the fracture site, or by inducing electrical fields by current flowing through coils of wire wrapped around the skin covering the bony defect.

Clinical applications of electrically induced bone healing were first undertaken in the early 1800s, but it was not until 1979 that the U.S. Food and Drug Administration approved the use of electrical treatment for healing nonunions (McGinnis, 1989). Many types of devices have been fabricated, but they fall into three main categories: (1) invasive (electrodes and power supply embedded), (2) semi-invasive (embedded electrodes, but external power supply), and (3) noninvasive (externally placed coils or capacitative plates). Despite the solid experimental evidence that bioelectrical stimulation can induce the formation of bone, the overall clinical literature has proved to be difficult to evaluate because of a paucity of carefully controlled studies.

Stem Cells and Tissue-Engineering Approaches

With the recognition of the importance of the role of stem cells in the formation of new bone (Friedenstein and Lalikina, 1973) and the rapid development of the field of tissue engineering, many new approaches to the stimulation of regenerative bone formation have incorporated the three fundamental strategies of tissue engineering: (1) providing a source of cells, (2) implanting an appropriate substrate, and (3) adding inductive or stimulatory factors to the mix (Hollinger et al., 2005; Rose et al., 2004). Because of the rich stores of stem cells capable of making bone in the normal body, many tissue-engineering approaches to bone formation concentrate on implanting an appropriate substrate, either natural or artificial (Boskey, 2005; Hollister et al., 2005), and supplying an effective dose and distribution of BMP or one of the other growth factors that are known to stimulate bone formation (e.g., transforming growth factor-β, FGF, or insulin-like growth factor). A number of controlled clinical trials have shown improved and accelerated bone formation in nonhealing fractures (Geesink et al., 1999; Govender et al., 2002) and after spinal fusions (Boden et al., 1995, 2002) after application of BMP to the site.

Genetic Engineering

One of the most recent attempts to stimulate bone regeneration has been to implant genetically modified cells into an area requiring the formation of bone. Typically, the cells are modified to produce BMP or some other bone-stimulating factor, and combinatorial gene therapy is currently being proposed (Franceschi, 2005). In animal experiments, this has been used to facilitate the repair of defects in long bones (Baltzer et al., 2000; Lieberman et al., 1999) or to facilitate spinal fusions (Alden et al., 1999). Gene therapy approaches, particularly those that involve viral vectors, remain beset by technical issues, such as efficiency and immune reactions, and their application to clinical situations is limited.

CARDIAC MUSCLE

Mammalian cardiac muscle cells, together with neurons in the CNS and just a few other specialized cell types, have traditionally been considered to be postmitotic cells with no capacity to divide, and cardiac muscle as a tissue has similarly been categorized as nonregenerating. The recent burst of activity in stem-cell biology has placed cardiac muscle regeneration in a different light, with recent reports claiming amazing feats of regeneration within the damaged heart, although these reports are not without their detractors (see p. 249). Yet, some fish and amphibians possess a substantial capacity for cardiac muscle regeneration. It is useful to examine cardiac muscle regeneration in these animals in relation to the limited natural capacity for regeneration in the mammalian heart.

Heart Regeneration in Lower Vertebrates

In an era when the dominant mode of thought was that highly differentiated cells have permanently lost their capacity to re-enter the mitotic cycle (Holtzer, 1963), Rumyantsev (1961, 1973) and Oberpriller and Oberpriller (1971) clearly showed that differentiated cardiac myocytes in both frogs and newts are capable of undergoing mitosis after damage to the heart. This reparative reaction involves a partial cellular dedifferentiation (i.e., the loss of most contractile filaments) of the myocytes before the initiation of mitosis. In contrast to limb regeneration, the repair process after amputation of the tip of the ventricle involves mitosis of cells in the remaining ventricular wall near the wound, but a blastema does not form (Oberpriller and Oberpriller, 1974). The original wound is first filled with a blood clot, which is later replaced by a connective tissue scar. The mitotic response is delayed, with a peak 2 to 3 weeks after injury. The reparative response in the newt is so strong that if the tip of the ventricle is minced and replaced, it re-forms a new miniventricle that beats asynchronously with the remnant of the original ventricle (Bader and Oberpriller, 1978, 1979). An important characteristic of ventricular myocytes in the newt is that more than 98% of the nuclei are diploid (Oberpriller et al., 1988).

Although it has been known for years that some fish have a limited capacity for myocardial regeneration (Kolosova, 1961), only recently has it been shown that zebrafish possess a strong natural capacity for repair of the amputated myocardium (Poss et al., 2002; Raya et al., 2003). The regenerative process involves a substantial amount of mitosis of nearby cardiomyocytes. In contrast to the amputated ventricle of newts, similarly treated zebrafish hearts regenerate without a scar, through mitosis in the ventricular wall and the apparent ingrowth of myocytes into the area of the wound. The ploidy of zebrafish cardiomyocytes has not been reported. Interestingly, when ventricular regeneration was inhibited in mutants of *mps1,* a mitotic checkpoint kinase, the amputation surface became covered with scar tissue (Poss et al., 2002), again suggesting that scar tissue may be a consequence of, as well as a cause of, the lack of regeneration.

Natural Regeneration of Mammalian Heart Muscle

For much of the twentieth century, mammalian cardiac muscle was viewed as a tissue completely lacking in regenerative capacity. According to the prevailing dogma, after infarction or injury, the remaining cardiomyocytes could undergo hypertrophy, but could not reenter the mitotic cycle. It was not until the publication of careful electron microscopic and autoradiographic studies in the 1970s that unquestioned evidence of DNA synthesis and mitosis of cardiomyocytes appeared in the literature (rev. Borisov, 1998; Rumyantsev, 1977).

Rumyantsev and Kassem (1976) produced large left-ventricular infarcts in adult rats by ligating the left coronary artery. After single injections of H^3-thymidine a week after the lesion was created, they found a low (<1%) labeling index among perinecrotic ventricular myocytes. Further experiments involving multiple injections of tritiated thymidine (rev. Rumyantsev, 1991) showed considerably higher labeling indices in atrial myocytes than in the perinecrotic area and a slight increase in cells of the conducting system. Labeled myocytes in the right ventricle were rare. This low mitotic response to cardiac damage in mammals may be related to the state of ploidy of the cardiomyocytes. In rodents, within a few days after birth, most cardiomyocytes undergo various degrees of polyploidization, becoming either binucleated or single-nucleated polyploid (Brodsky et al., 1980). This could interfere with the entry of the cells into a conventional mitosis and may explain why cellular hypertrophy is a more common reaction to cardiac pathology in mammals. In addition to ploidy issues, mitotic blocks at specific checkpoints in the cell cycle—specifically at the G1/S and G2/M boundaries—appear to exist. Attempts to remove these blocks in cultured rat cardiomyocytes have met with some degree of success (Agah et al., 1997; Dätwyler et al., 2003; Field, 2004).

The classic view of the limited regenerative capacity of the mammalian heart has been challenged by controversial recent research based on the assumption that the mature heart contains a population of stem cells (rev. Leri et al., 2005). Initially reported by two different laboratories (Beltrami et al., 2003; Hierlihy et al., 2002), cardiac stem cells have been described as existing in clusters throughout the heart, although they are concentrated in the atria and in the apex of the heart (Anversa et al., 2003; Beltrami et al., 2003). According to Leri and colleagues (2005), there is 1 cardiac stem cell per 30,000 to 40,000 cardiomyocytes. They are characterized by possessing three stem-cell antigens (c-kit, MDR1, and Sca-1–like) in 65% of the cases, two of these antigens in 20% of the cases, and only one of the three antigens in the remaining 15% of cases. Other authors (rev. Laflamme and Murry, 2005) have also identified putative populations of intracardiac stem cells, but each appears to be characterized by different antigenic properties. Leri and colleagues (2005) envision a scenario in which cardiac stem cells, ensconced and nourished in protective niches, participate in a slow cycle of death and renewal of cells in the heart over the long lifetime of a human. The appearance of regenerative activity in the perinecrotic zone in human infarcts (Urbanek et al., 2003) has been attributed to the contribution of these local stem cells.

Many questions, both technical and biologic, remain to be answered before we achieve any real understanding of the true natural regenerative capacities of the human heart. The first question is how many new cardiomyocytes actually do form after various forms of injury; the second one is where any new cells come from.

Stimulation of Heart Regeneration in Mammals

Attempts to stimulate cardiac muscle regeneration in mammals can be categorized into two phases: a pre–stem-cell phase and a post–stem-cell phase. In the pre–stem-cell era, attempts to stimulate the regeneration of cardiac muscle by treatments as diverse as the injection of scar tissue inhibitors, RNA, tissue hydrolysates, vitamins, or various inducing compounds were largely unsuccessful (rev. Polezhaev et al., 1965).

In the short period since the advent of stem cell–based treatments, an enormous literature on the stimulation of regeneration in the mammalian heart has been built up (rev. Laflamme and Murry, 2005; Leri et al., 2005). In general, stem cell–based treatments fall into two broad categories: those that involve stem cells of bone marrow origin, and those that involve cardiac stem cells (see Chapter 12 for details). Although in this field consensus on any topic is rare, the current literature suggests that locally derived, rather than marrow-based, stem cells are more effective in restoring cardiac function. There also appears to be the sense that injection of the stem cells into perinecrotic areas or into the local coronary circulation, rather than introduction into the bloodstream outside the heart, is the most effective method of application. However, it has been proposed that immediately after an infarct, local molecular pathways are upregulated that promote circulating stem cells to home to the damaged regions of the heart (Ma et al., 2005; Penn et al., 2004). Tissue-engineering approaches have involved implanting constructs of cardiac myocytes adherent to substrates (e.g., collagen strings, biodegradable gels, or cell sheets) into an infarcted area in the ventricle, mainly in laboratory animals (Zammaretti and Jaconi, 2004).

A remarkable number of clinical trials of stem-cell therapy already have been undertaken, and improvements in symptoms have been noted in several of these trials, even though the mechanism underlying any of the effects remains obscure (Couzin and Vogel, 2004; Hassink et al., 2004). A major issue is the extent to which the introduced stem cells actually contribute directly to the formation of new cardiomyocytes versus their contribution to or stimulation of an enhanced local vascular response, which may act as a supportive environment for regeneration (Orlic, 2004). The collective evidence to date suggests that exogenous cells produce relatively few cardiomyocytes, but that some types contribute directly to the endothelial cell populations, especially near infarcted areas (Laflamme and Murry, 2005). The use of embryonic stem cells in facilitating cardiac muscle regeneration is still in its early stages. Many issues, including a propensity for some implanted embryonic stem cells to form teratomas in the heart, remain to be worked out.

Another new player in the field of cardiac regeneration is the MRL mouse, which appears to show an enhanced regenerative capacity in a number of tissues and organs (Heber-Katz et al., 2004). Leferovich and colleagues (2001) report that after cryoinjury, the region of damage, which initially is an early focus of immature scar tissue formation, is soon filled in by regenerating cardiac muscle cells. Unfortunately, quantification of this effect has been incomplete. Two follow-up studies on MRL mice used ischemic, rather than cryogenic, models of injury (Abdullah et al., 2005; Oh et al., 2004). Both of these studies conclude that ischemic myocardial injury is not followed by scarless recovery. Certainly more research will be necessary before it can be determined whether the MRL mouse will be a useful adjunct in studies of the stimulation of regeneration in the heart.

SUMMARY

The stimulation of regeneration, especially of limbs or the CNS, has always been the holy grail of regenerationists, and many largely unsuccessful attempts have been made in the past. One of the main conceptual defects inherent in most of these older attempts has been the focus on adjusting a single variable, the absence of which inhibits natural regeneration. It is becoming increasingly recognized that any regenerative process is a complex multifactorial event, and that the absence of regeneration is most likely associated with many factors. The amazing progress in understanding the basis for the lack of regeneration in the CNS and in attempting to stimulate CNS regeneration is due to a detailed understanding of some of the factors that allow regeneration in the PNS and that inhibit regeneration in the brain and spinal cord. To date, there remains insufficient understanding of the biology of normal limb regeneration to undertake serious attempts at stimulating the regeneration of mammalian limbs. The use of stem cells in the stimulation of mammalian tissues, such as cardiac muscle, is still in its infancy, and to date, it has been difficult to determine the efficacy of the procedures that have been used.

CHAPTER 15

What Have We Learned and Where Are We Going?

It is difficult to say what is impossible,
for the dream of yesterday
is the hope of today
and the reality of tomorrow.

—R. Goddard, rocket scientist

Thomas Hunt Morgan (1866–1945) is said to have left regeneration research in favor of the newly emerging field of genetics because he believed that the complexity of regeneration was too great to be realistically solved in his lifetime. A century later, his prescience has been amply borne out. If anything is now clear, it is that regeneration at any level of organization is much more complex than most researchers in the field had previously anticipated. Looking back to the level of optimism that prevailed when I was a beginning graduate student, I was seriously concerned that the problem of stimulating human limb regeneration would be solved before I finished my graduate studies!

Even though progress in the field of regeneration historically has advanced by fits and starts, we have collectively learned a great deal about regenerative phenomena. Probably as important as what we do know is a clearer recognition of what we do not know and what needs to be done to intellectually understand regeneration and to realize its great potential for medical applications. This chapter summarizes some impressions that I have gained after having spent many years trying to unlock some of the mysteries of regeneration.

WHAT HAVE WE LEARNED?

Regeneration Is a Complex Phenomenon

The more we learn about regeneration in any system, the more we realize how complex the phenomenon of regeneration really is. A classic example is regeneration of the liver (Fausto et al., 2006). When studies on liver regeneration were in their infancy, it was generally assumed that the action of a single stimulator or inhibitor was the key to understanding liver regeneration. Later, attention was focused on single pathways that would set off the regenerative response. Now, reading the literature on liver regeneration immerses one in a sea of cytokines, growth factors, receptors, signal transduction pathways, and nuclear responses, all on top of reports of the effects of specific components of metabolic pathways. It appears likely that regeneration of many individual organs and tissues will depend on activation and interactions among networks of signaling pathways that ultimately result in an appropriate nuclear response, which, in turn, affects other interacting networks.

In complex structures, such as extremities or significant parts of the body, epimorphic regeneration requires not only the well-described tissue interactions, but also molecular signaling that will likely prove to be every bit as complex as that which underlies the regeneration of isolated tissues. On top of that is the reactivation of the developmental patterning mechanisms that operated earlier in ontogenesis. For phenomena such as morphallaxis, we have hardly scratched the surface.

The recognition of the complexity of regeneration should not be viewed as discouraging so much as illustrating that we now have many of the tools needed to dissect a regenerative process to the point where it can be meaningfully manipulated.

What Do We Mean by Regeneration?

Many different types of regenerative phenomena exist, and they often appear to operate according to their own sets of rules. There has been a strong historical tendency in the field to try to assign new phenomena to categories with existing names, but sometimes they just do not fit. Although a taxonomy of regenerative processes can be highly useful, at the current state of knowledge, there is insufficient information for accurately placing many types of regeneration into their own boxes or lumping them into other boxes and then arranging them into a hierarchy. A danger with inappropriate categorization is that the name given to the phenomenon can blind the investigator to clues provided by the system. A good example of this concerns the term *epimorphic regeneration*. Morgan (1901) coined this term and contrasted it to morphallaxis:

> At present there are known two general ways in which regeneration may take place, although the two processes are not sharply separated, and may even appear combined in the same form. In order to distinguish broadly these two modes I propose to call those cases of regeneration in which a proliferation of material precedes the development of a new part, "epimorphosis." The other mode, in which a part is transformed directly into a new organism, or part of an organism without proliferation at the cut surfaces, "morphallaxis." (p. 23)

The terms "epimorphosis" and "morphallaxis" were meant to encompass the knowledge of regeneration at the time. In the context of current knowledge, these terms are much less meaningful. According to Morgan's original definition, *epimorphosis* would encompass virtually all of the regenerative phenomena mentioned in this book. Yet, many authors, including myself, have used it in a much more restricted manner, typically referring to a type of regeneration that involves the formation of a blastema and includes complex morphogenetic signaling. Even though my inclusion of Figure 1-1 in this book may appear contradictory to the cautionary note expressed earlier, the real key is to categorize carefully and not become wedded to terminology that is often inadequate for the job at hand.

One Size Does Not Fit All

Nature has devised an amazing variety of means by which to restore lost or damaged structures. In some cases, the end result is an almost perfect replica of what was lost; but in others, the product of regeneration is both structurally and functionally incomplete. Even among seemingly similar processes, experimentation has shown some profound differences. An excellent example of this is intercalary regeneration in the limbs of insects and amphibians. Both insects and salamanders have the ability to fill in a discontinuity created by a graft of a distal structure to a proximal stump by intercalary regeneration (see Fig. 7-6). In each case, the regenerated segment possesses appropriate polarity. Yet, insects go a step further. Intercalary regeneration in insects appears to operate on the basis of arrays of information contained in each proximodistal segment, so that if a proximal part of a distal segment is grafted onto the distal part of a proximal segment, intercalary regeneration also occurs, but the polarity of the intercalary regenerate is reversed (as indicated by the orientation of the bristles). Amphibian limbs do not appear to possess this degree of segmental control, but it remains unknown whether the fundamental basis for intercalary regeneration transcends broad taxonomic categories or whether intercalation is an example of a convergent evolutionary process.

Another example involves regeneration of the urodele limb and tail. For many years, these have been considered to be classic examples of epimorphic regeneration. In both systems, amputation is followed by epidermal wound healing. This is followed by a phase of dedifferentiation at both the tissue and muscle fiber level. The dedifferentiated cells then accumulate as a distal blastema, which in both instances is covered by a thickened apical epidermis. The blastema grows, and from it appears a new limb or tail that is grossly similar to that which was amputated. Despite the superficial similarities of these regenerative processes, detailed experimental and molecular analyses have demonstrated a number of profound differences between them.

In contrast to the dependence of the regenerating limb on peripheral nerves (Singer, 1952), the regenerating tail requires the presence of the spinal cord, rather than its peripheral nerve supply (Goldfarb, 1909). The ependymal tube emanating from the spinal cord induces the basal cartilaginous rod in the regenerating tail (Holtzer, 1956), whereas there appears to be no direct relation between the innervation of the regenerating limb and formation of the skeleton. When the spinal cord is rotated 180 degrees, the cartilaginous rod in the regenerating tail forms dorsally rather than ventrally (Holtzer, 1959). The switching of cell lineages has been demonstrated in both the limb and the tail, but in different cell types. In the regenerating limb, muscle-derived cells have been found in cartilage (Kumar et al., 2000), whereas in the tail, cells from the spinal cord have been shown to differentiate into muscle and cartilage (Echeverri and Tanaka, 2002a).

Significant differences in morphogenetic properties of skin and muscle became apparent when Dinsmore (1979, 1981b) rotated both tail skin and muscle with no effect on morphogenesis, in contrast to the formation of complex supernumerary regenerates when stump skin or muscle was rotated in amputated limbs (Carlson, 1974a, 1975a; Lheureux, 1972, 1975a). Similarly, when strips of stump skin on limbs are turned 90 degrees so that the amputation surface is confronted by all dorsal or all ventral skin, regeneration is inhibited (Lheureux, 1975b), whereas the same manipulation in the tail produces no effect on the course of regeneration (Dinsmore, 1981b). Lastly, unilateral ablation of both muscle and skin over half the circumference of the limb results in the regeneration of half limbs (Goss, 1957), but when the same operation was done on the tail, regeneration was perfect (Dinsmore, 1981a).

The response to retinoic acid is particularly striking. The amputated limb responds to retinoic acid treatment by producing serial (proximodistal) or parallel duplications, or both (Maden, 1982; Niazi and Saxena, 1978), but the regenerating tail produces homeotically duplicated limb structures (Maden, 1993; Mohanty-Hejmadi et al., 1992). There has been insufficient research to make meaningful molecular comparisons between limb and tail regeneration. It would not be surprising if there are similarities in the type of program that leads to the initiation of regeneration. However, in view of the profound differences in the molecular patterning of the body axis versus the extremities during embryonic development, it is highly likely that the results of the tissue manipulation studies cited earlier will be found to be based on significantly different molecular control mechanisms.

Different Ways to Accomplish the Same Goal: Double Assurance, or "There Are Several Ways to Skin a Cat"

Regeneration is an opportunistic phenomenon. Not infrequently, repair of a structure can be accomplished in different ways. In some cases, different species regenerate the same structure differently; in others, a single structure may be restored through different mechanisms even within the same individual. This was recognized early in the history of regeneration research when it was found that an amphibian limb regenerate forms a complete skeleton even after the skeleton of the stump had been removed (Fritsch, 1911). This type of biologic backup has been called *double assurance* in the regeneration literature. Although the term initially was used in the context of gross morphogenetic phenomena, the concept has utility at many levels. Contemporary literature on embryonic development includes many reports of single-gene knockouts that have had no discernible effect on development or only minimal disruption. However, when a close relative of that gene is simultaneously knocked out, devastating developmental defects are often seen. This is usually interpreted to mean that in the case of a single-gene knockout, the second, closely related gene and its products can take over some of the functions usually performed by the first gene.

Different ways of regenerating the same structure sometimes involve the use of different precursor cells as the source of the regenerate. In the current era of the stem cell, it would appear that most organs may have cellular backup systems in the form of locally or systemically derived stem cells. In some cases, entirely different tissues have the capacity to regenerate a structure. A good example is the amphibian lens. Under normal circumstances, the regenerating lens in the newt arises from the dorsal iris, and that in the frog comes from the cornea. Yet, in experimental situations, lenses can also regenerate from pieces of neural retina or retinal pigment epithelium (Eguchi, 1998). Table 15-1 summarizes some of the different cellular origins and backup systems that have been found to operate during regeneration in vertebrates.

Especially among the invertebrates, the same structure can form as a result of seemingly different processes, often involving asexual reproduction or somatic embryogenesis, as well as posttraumatic regeneration. Many examples are cited in the monographs of Ivanova-Kazac (1977), Tokin (1959, 1972), and Vorontsova and Liosner (1960). Among the protozoa, many ciliates regenerate well after resection. These same forms often reproduce asexually by fission. Interestingly, a form of cellular dedifferentiation occurs in some of these species before and during the natural fission process, whereas in experimentally resected fragments, dedifferentiation occurs after the fact. In addition to forming a new individual by fission, a number of species, particularly among

TABLE 15-1

Alternative Cellular Origins of Regenerates in Different Systems

Regenerating Structure	Cellular Origins
Lens	Dorsal iris (urodeles)
	Cornea (anurans)
	Neural retina (experimental)
	Pigment epithelium of retina (experimental)
Retina	Marginal zone of retina
	Pigment epithelium of retina
Liver	Hepatocytes
	Oval cells
	Stem cells
Bone	Periosteum
	Endosteum
	Stem cells in marrow
	Stem cells in soft tissues
Skeletal muscle	Dedifferentiation of muscle fibers (epimorphic)
	Satellite cells
	Postsatellite cells (urodeles)
	Stem cells
Cardiac muscle	Cardiomyocytes (mitosis)
	Intracardiac stem cells
	Circulating stem cells (?)
Spinal cord	Ependymal cells (fish and amphibians)
	Stem cells

the ameboid forms, use budding to reproduce new unicellular individuals. At a multi-cellular level, sponges (Porifera) and coelenterates use similar modes to reproduce asexually, although some sea anemones have also evolved the mode of fragmentation to produce multiple individuals (Fig. 15-1). Fragmentation involves breaking off small pieces from the base, which then proceed to regenerate new anemones.

Planaria and other platyhelminths have added a few new evolutionary twists to the process of reproduction. Whereas some species reproduce by straight fission, which requires each fragment to regenerate, others reproduce by paratomy, in which much of the new individual forms before fission detaches it from the parent (Fig. 15-2). Paratomy is also seen in some annelids (Bely and Wray, 2001), and certain of the genes expressed during fission (e.g., *engrailed* and *orthodenticle* homologs) are also expressed during regeneration. This process introduces the phenomenon of morphallaxis, which involves reorganization of the body to achieve a morphogenetic end result (see Fig. 1-12). These themes, which phylogenetically appeared in some of the most primitive animals, are repeated many times among the invertebrates. For some reason, verte-

A.

B.

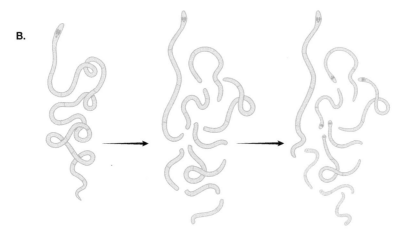

FIGURE 15-1   Asexual reproduction by fragmentation. *(A)* Detachment of pieces from the basal part of the body in the anemone *Aiptasia lacerta* and their transformation into daughter organisms. (After Andres. From Vorontsova and Liosner, 1960. p. 4.) *(B)* Fragmentation in the nemertean worm, *Lineus socialis.* Each fragment reorganizes into a new organism. (After Coe [1929].)

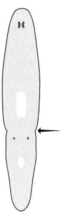

FIGURE 15-2 Reproduction by paratomy in the planarian *Planaria fussipara*. Before fission at the level indicated by the arrow, the posterior part of the planarian has already begun to transform into a new individual. (After Cattell. From Vorontsova and Liosner, 1960. p. 16)

brates have lost most of these exotic modes of reproduction and have to be satisfied with more prosaic backup mechanisms, usually involving different cellular sources for accomplishing the same regenerative response (see Table 15-1). One of the tasks for the future will be to determine whether, despite their different modes of initiation, the processes of regeneration and asexual reproduction involve the same or similar mechanisms.

The Environment as a Determinant of Regenerative Success

One of the most important realizations to come out of the past few decades of regeneration research on mammals is that the local environment is a critical determinant of regenerative success. Much of the experimentation to date has approached environmental influences at the tissue level; for example, peripheral nerve channels as an environment that permits axonal extension or decellularized matrices as substrates for natural tissue regeneration. A next logical step is a dissection of the tissue level environment to identify the components that influence regeneration. Such attempts have begun for a number of types of regeneration, and specific molecules are being identified as inhibitors or stimulators of regeneration. It will be important to learn from past experience in this field that at the molecular level, as well as at the tissue level, regeneration of almost any structure is a complex enough phenomenon that there are probably many modulating influences rather than a single "magic bullet." Control of regeneration is likely to involve a host of environmental factors acting as a group. Different configurations or combinations of these factors are likely under different circumstances of injury. How the remaining cells of the damaged organ respond to these environmental variations will determine the success of regeneration.

Ultimately, the issue of environment and regeneration will hinge to a great extent on interactions between surface receptors on cells and the myriad environmental signals present in any environment, especially that which follows tissue damage. The huge number of receptors present on the surface of a single growth cone of an outgrowing axon is a reflection of the complexity of that microenvironment. The multiplicity of potential signals must then be matched with the variety of transduction and processing pathways within the cells and the way that an individual cell finally responds to the environmental soup in which it resides. From the standpoint of stimulating or directing regeneration, we can be encouraged in knowing that, in many instances, a change in environment can make a profound difference in the success of a regenerative process.

Stem Cells and Regeneration

Nothing in recent decades has influenced the field of mammalian regeneration more than the recognition that many tissues of the body contain stem cells that can be activated to repair damage. One of the fascinating contemporary questions is how stem cells are involved in normal processes of tissue growth, maintenance, and adaptation, and how they fit into the overall hierarchy of regenerative responses to tissue damage. By now stem cells have been described, whether correctly or incorrectly, in almost every major tissue and organ in the body. The big question is, what are they doing?

Some investigators envision stem cells as the cellular basis for the growth of tissues, and then after overall growth ceases, as the source of cells to replace those that have died. Stem cells have also been considered to be the source of new tissue when the mass is increased because of heightened functional demand. One big issue is the extent to which growth and hypertrophy are based on the proliferation of parenchymal versus stem cells. Certainly in tissues such as skeletal muscle and bone, constraints on the differentiated cells make local stem cells (satellite cells in muscle and cells of the periosteum and endosteum in bone) the logical participants, although in the case of bone, there could be different contributions. For peripheral appositional growth, periosteal cells are likely the main source of new bone, whereas for internal remodeling, stem cells accompanying the vasculature are more likely to be dominant. In other tissues, parenchymal cells are the primary players. Hyaline cartilage is the clearest example, where interstitial growth is accomplished by the division of chondrocytes already surrounded by matrix.

In regeneration of damaged organs, one of the most important questions is whether there is a hierarchy of cellular responses, so that if the first level fails or is insufficient for any reason, then cells in the next level pick up the slack. This does appear to be the case in the liver, where the first level of response is proliferation of the hepatocytes. The next level of defense is the oval cells, and finally, if both of these fail, stem cells are called into action (Fausto, 2004). In the case of the mammalian heart, it has been suggested that the weak mitotic response of the cardiomyocytes is supplemented by contributions of local stem cells. Whether marrow-derived stem cells play a significant role in cardiac repair currently is being debated.

Regardless of the organ, many fundamental questions on the role of stem cells and regeneration remain to be answered. The first question is whether stem cells exist in all the tissues in which they have been claimed—in particular, whether there is a population of resident local stem cells, such as the satellite cells of skeletal muscle, or whether stem cells that have been identified are systemically derived transients. A second-level question is the extent to which tissue-specific stem cells actually participate in growth, maintenance, adaptation, or regeneration. If our experience with skeletal muscle, a relatively straightforward example, is any indication, this question will not always be easy to answer. An important question in situations of tissue equilibrium is what it is that maintains stem cells in a relatively inactive state. Is there a tissue-specific active negative feedback, or is the ground state maintained simply in the absence of positive stimuli? In the case of a need for more cells, whether for normal growth or repair, it will be important to understand the basis for activation of stem cells. Once an activating signal is received, is there a generic sequence of activation that awakens the stem cells of all types from their metabolic repose, or do stem cells in each tissue apply their own unique strategy? In cases such as the liver, how does the next level down in the hierarchy of repair (e.g., the oval cells) know when to become activated when the primary level of repair proves to be insufficient?

The above questions are just the beginning. Once stem cells are recruited and become involved in a repair process, how do they find their appropriate place in the damaged tissue, and what stimulates their differentiation into the required cell types? Because most tissues contain multiple cell types, it will be important to know whether a single type of stem cell produces only one type of cell (e.g., parenchyma), or whether it could also contribute to the endothelial cell population, for example. At the end of the process, knowing the basis for final functional differentiation and functional integration into the organ is critical, especially in complex organs such as the central nervous system.

The above questions are merely some of the biological inquiries. For the tissue engineer, prominent questions address finding sources of stem cells, expanding them and possibly genetically altering them before implantation, and properly associating cells and implantable substrate. Devising efficient methods for introducing the stem cells into the body is an important challenge. Experience in the implantation of both skeletal and cardiac myoblasts or stem cells has shown that within just a few hours, well more than 90% of the implanted cells have either died or otherwise been eliminated from the site of tissue damage. Then there are the chronic issues of biocompatibility and immunocompatibility.

Despite many challenges, stem-cell therapy remains one of the most important options for stimulating regeneration and reversing pathology in numerous clinical situations. A practical but important question is when to move from bench to bedside and on what basis. To date, the stem-cell field has sometimes shortchanged controlled clinical trials in favor of moving quickly ahead. In the long run, this could prove to be counterproductive. When human experimentation slides imperceptibly into clinical treatment without the tedious intervening proof-of-concept steps, long-term problems can arise.

Is There a Molecular Regeneration Program?

One of the most unusual and intriguing aspects of regeneration is that by definition the process is tied to the remaining part of a normally fully differentiated body. One of the fundamental questions is how the body knows that there is a loss and how tissue loss sets in motion a regenerative response. In many epimorphic systems, the question must be asked differently, because it is possible to induce the formation of supernumerary structures (e.g., limbs) in the absence of amputation (see Fig. 7-2). In this case, one must ask what set of circumstances sets the stage for the initiation of the epimorphic response (i.e., dedifferentiation, blastema formation, pattern formation, and morphogenesis of a complex structure). In the limb, these conditions are generally considered to be tissue damage, the presence of a wound epidermis, a nerve supply, and appropriate morphogenetic boundary conditions. The research from Agata's laboratory (Kato et al., 2001) on planarian regeneration strongly suggests that the differentiated tissues bordering the wound provide important morphogenetic cues that guide the initiation and the progression of the regenerative response.

A generic molecular regeneration program has yet to be discovered, but if one exists, it would have to accommodate several features common to many regenerating systems. The first is the necessity to remove tissue debris from the site of damage and, in some cases, to break down persisting original structures that could interfere with the regenerative response. This response can be viewed as preparing the interface between the future regenerate and the remaining body. It is commonly accomplished through the actions of both cells and enzymes. The most prominent cell is the macrophage, which not only physically engulfs tissue debris, but also secretes a wide variety of cytokines and growth factors that can move a purely phagocytic response into an early regenerative response.

Many complex systems of regeneration are characterized by the early up-regulation of a variety of lytic enzymes—the matrix metalloproteinases (MMPs; see Table 4-3) and their associated inhibitors, the tissue inhibitors of metalloproteinases—that break down components of the existing extracellular matrix in preparation for the initiation of the regenerative phase. If MMP activity is inhibited in species as diverse as hydra (Leontovich et al., 2000), sea cucumbers (Quinones et al., 2002), and newts (Vinarsky et al., 2005), regeneration fails to occur. The breakdown of existing scaffolding appears to be important for the initiation of dedifferentiation in classic epimorphic regeneration, whereas in mammalian tissue regeneration, skeletal muscle being a good example, it is often important to preserve elements of the extracellular matrix, for example, basal laminae of old muscle fibers, which serve as a scaffolding on which the formation of new muscle fibers takes place.

The appearance of Msx-1 as a precursor or accompaniment of dedifferentiation is beginning to emerge as a common molecular theme in regeneration (Akimenko et al., 1995; Han et al., 2005; Odelberg, 2005). Its association with predifferentiative stages in embryonic development (e.g., the limb bud) reinforces its importance in somehow maintaining cells in an undifferentiated state. Another common theme in systems based on interactions between epithelia and mesodermal tissues is the production of fibroblast growth factors and MMPs by the epithelium. The necessity of FGF secretion by the apical ectodermal ridge for outgrowth of the embryonic limb bud and by the thickened apical epidermal cap in the regenerating amphibian limb has been documented extensively. More research is required to determine whether FGF secretion by the wound epidermis plays a major role in initiating the dedifferentiative process in the amputed limb. Through differential screening, Wolfe and colleagues (2004) have identified 13 genes that are up-regulated in the early regeneration of both the hind limb and lens in *Xenopus*.

The events of the early destructive phase lead into a phase that causes the precursor cells for regeneration to reenter the mitotic cycle. One recurring theme is the role of a thrombin-activated serum factor (Straube et al., 2004) that triggers reentry into the mitotic cycle, possibly by hyperphosphorylating the Rb (retinoblastoma) protein. Rb actively inhibits the G1-S transition in the cell cycle, and phosphorylation renders Rb into an inactive form, thus relieving this block to mitosis (Brockes, 1998). Such a mechanism operates in systems as diverse as skeletal muscle, hair cell, and lens regeneration (Imokawa et al., 2004; Sage et al., 2005; Tanaka et al., 1999; Thitoff et al., 2003).

In epimorphic regeneration, the formation of the blastema coincides with the reexpression of many genes that earlier played a prominent role in the embryonic development of that organ or organism (Alvarado, 2004; Bryant et al., 2002). This includes both transcription factors, such as the *Hox* clusters, and representatives of major signaling factors, such as sonic hedgehog and Wnt. In regenerating tails of *Xenopus* tadpoles, Slack and coworkers (2004) found different signaling pathways from those seen in the urodele limb, with Notch signaling proving to be essential for outgrowth. The recent identification (Morais da Silva et al., 2002) of the CD59 ortholog, Prod 1, in the regenerating limb gives one of the first clues concerning the molecular basis of proximodistal positional identity in limb regeneration. The ramifications of this discovery are considerable, as seen in the marking experiments of Echeverri and Tanaka (2005; see p. 149). It is too early to know whether this molecule will be a key to unlocking an overall program for morphogenetic control in epimorphic regeneration.

The Role of Scarring

One thing that is now abundantly clear is that where regeneration does not occur, scar tissue forms. As mentioned earlier in this book, whether scar tissue formation is a cause of no regeneration, a default reaction to the absence of regeneration, or both remains an open question. In favor of a default explanation is the formation of a cap of dense scar tissue over the end of an amputated limb stump in urodeles after experimental manipulations (e.g., denervation, skin flaps, excess vitamin A, hypophysectomy, digital implants) that block regeneration. In favor of a regeneration-blocking action is the early appearance of a similar scar in amputated limbs of frogs and mammals. What is not known is whether in these latter cases, some early step leading to regeneration has failed to occur, allowing scar tissue formation to take place. In both the regenerating urodele limb (Vinarsky et al., 2005) and ear holes in the MRL mouse (Gourevitch et al., 2003), high levels of MMP activity have been credited with reducing scarring and allowing blastema formation, whereas in the same experimental models, scar tissue forms when MMP levels are lower.

The field of skin wound healing is confronted with the same conundrum. Skin regeneration occurs in the fetus, only to be replaced by a scarring response after birth. Ferguson and O'Kane (2004) refer to dermal scarring as a morphogenetic problem, that is, failure of the regeneration of normal skin structure. They have focused on differing levels of TGF-β between embryos (low TFG-β_1 and -β_2; high TGF-β_3) and adults (high TGF-β_1 and -β_2; low TGF-β_3) and the difference that these environments make on fibroblast behavior early in the healing process. Early treatment of adult skin wounds with TGF-β_3 appears to reprogram the secretory activity of macrophages and alter the migratory behavior of fibroblasts so that they lay down extracellular matrix in a normal, basket-weave formation. Because skin wounds in different locations or different types of skin wound in the same location can elicit either regeneration or a scarring response, Ferguson and O'Kane (2004) postulate that local differences, mainly in growth factor and/or cytokine activity, rather than systemic conditions, determine the nature of the healing response.

WHERE ARE WE HEADING?

For most of its history, the field of regeneration has progressed in roughly 20-year cycles of great optimism, followed by long stretches of seemingly little progress. With the availability of discriminating molecular techniques, the resurgence of interest in stem cells, and the great interest of the tissue-engineering community, the phenomenon of regeneration has been thrust into greater prominence than at any time in its history since its initial heyday in the 1700s. Almost every major biomedical research institution has recently established some sort of center or institute for regenerative medicine, and impressive resources currently are being poured into the field. There is no lack of major problems to attack. From my perspective as a long-time researcher in the field, the following sections present some of the most important issues that need to be addressed.

Epimorphic (Blastema-Based) Regeneration

A fundamental issue in epimorphic regeneration is whether from a mechanistic basis all blastema-based systems of regeneration operate according to the same general principles. We know that there are substantial differences, for example, the way in which blastema cells are recruited (from neoblasts or stem cells in planaria and local dedifferentiation in amphibians) or the morphogenetic controls in regenerating limbs versus tails, but beneath these is there some fundamental unity?

How Is Epimorphic Regeneration Tied to the Remaining Body, and Are These Influences from the Body Fundamental to the Process or Merely Modulators?

Epimorphic regeneration occurs against the backdrop of an amazing number of influences from the remaining parts of the body. These include sealing the amputation surface with an epithelium, the bioelectrical environment, the vascular and nerve supply, the humoral (both hormonal and immune) environment, and morphogenetic information on cells at the amputation surface, which may play important roles in both dedifferentiation and pattern formation. Collectively, these influences affect many stages in the regenerative process, although their direct effect is usually an early one. A final integrating influence occurs when local tissue regeneration unites the blastema-derived regenerate with the remaining body. How all these influences from the body act both to stimulate and integrate the epimorphic process on a collective basis has received little attention.

What Does the Wound Epidermis Really Do?

The necessity of a wound epidermis is a hallmark of epimorphic systems, but exactly how it exerts its effect continues to be poorly understood. From studies in amphibians, we know that there is a close relation between the wound epidermis and the appearance of the full dedifferentiative phenomenon. Whether this effect, which occurs during a period when little appears to be happening at a histologic level, can be explained purely by the production of lytic enzymes or growth factors remains to be determined.

What Is the Basis for the Stabilization and Destabilization of the Differentiative State?

Many decades of painstaking research were required before the concept of dedifferentiation could be verified by well-controlled experimental data. The most important problem now is to identify the conditions that allow stable differentiated cells to become reprogrammed by reentering the mitotic cycle and reacquiring a degree of developmental flexibility. The ability of extracts from regenerating limbs or certain types of degenerating tissues to stimulate dedifferentiation can provide important clues to this question.

What Is the Molecular Basis for the Requirement of Morphogenetic Opposites at the Wound Surface for Epimorphic Regeneration to Proceed?

Experimental data from planaria and amphibian limbs have highlighted the requirement for the presence of morphogenetic opposites as a precondition for blastema formation to occur. Yet, there are some circumstances—for example, the amputated urodele tail and some types of double half limbs—that may not completely follow this rule. We know virtually nothing about the molecular nature of morphogenetic information (positional information), especially along the anteroposterior and dorsoventral axes, although the recent identification of Prod 1 as a player along the proximodistal axis has already provided major insight into potential mechanisms of morphogenetic control. The tail could be an excellent model for testing the hypothesis that the meeting of morphogenetic opposites is a generic requirement for blastema formation or whether that characteristic is limb specific.

What Is the Basis of the Neural Requirement for Epimorphic Regeneration?

Abundant experimental data support the requirement for a quantitatively adequate nerve supply during the early blastemal phases of regeneration, and the evidence in favor of transferrin's playing a major role as a mediator of the neural effect is quite strong (Mescher et al., 1997). The nature of the assay systems makes it difficult to rule out other factors that could affect regeneration at different times from when the major transferrin effect occurs or that might function concurrently with transferrin. Particularly in the formation of accessory limbs after nerve deviation, one wonders whether there is an effect of nerves earlier than at the level of blastemal cell proliferation.

Is There a Common Molecular Program Underlying Epimorphic Regeneration?

The question of a common molecular program underlying epimorphic regeneration has been brought up with greater frequency in recent years, especially with respect to the question of how the scale is tipped from a wound healing response to a regenerative response. Fundamental to this question is whether the scattered phylogenetic appearances of what is currently referred to as *epimorphic regeneration* (see Fig. 1-5) represent the perpetuation of a fundamental property of living beings, or whether these appearances in taxa as different as flatworms, annelids, echinoderms, fish, and amphibians represent independent evolutionary adaptations. However, even if the latter were true, it would not *de facto* rule out the existence of a common program, because similar pathways could be independently coopted. As Alvarado (2004) notes (p. 760), "It is quite possible that the study of regeneration will not uncover novel molecular pathways. Instead, the most likely initial outcome will be the identification of pleiotropic functions for known genetic pathways in the milieu of regeneration." Large-scale analyses of genes involved in a variety of regenerating systems (e.g., Katogi et al., 2004; King et al., 2003; Reddien et al., 2005) are beginning to produce the type of data required to test this possibility.

What Is the Relation between Immune Function and Epimorphic Regeneration?

Whether there is any relation between immune function and limb regeneration or its absence is a topic that has received remarkably little attention. Earlier speculation

(Sicard, 1985) centered mostly on possible stimulatory or antifibrotic influences of the immune system in urodeles. More recent treatment has focused on the relative immaturity of the immune system in urodeles and larval anurans in relation to that of vertebrates incapable of regeneration. Mescher and Neff (2005) raise the interesting, but untested, possibility that, in nonregenerating limbs, cytotoxic T cells or natural killer cells might recognize the antigens on the surface of dedifferentiated cells as foreign and destroy them, thus eliminating any possibility for the formation of a blastema. That prod 1, an important part of the proximodistal recognition system in the urodele limb (Morais de Silva et al., 2002), is an ortholog of CD59, a member of the complement family, suggests an intriguing relation between immune recognition and morphogenetic control.

What Is the Basis for Positional Memory?

One of the least understood aspects of epimorphic regeneration is the nature of the morphogenetic information that characterizes the position rather than the tissue type of a group of cells within an epimorphic field. This information is highly stable on differentiated cells, and in muscle, at least, it survives the complete destruction and subsequent regeneration of that tissue (Carlson, 1975b). With the identification of prod 1 as a major morphogenetic player along the proximodistal axis, one would similarly expect that cell-surface glycoproteins would play an important role in morphogenetic control along the transverse axes. Whether the limb possesses a system of cross-sectional positional information as well ordered and as complex as that which specifies the cross-sectional organization of the embryonic central nervous system remains to be investigated. One of the most important questions concerning the lack of regeneration in the limbs of higher vertebrates is whether or not positional memory has been preserved on the cells within the limbs.

The Issue of Cross-Sectional Area

One of the most intriguing questions concerning epimorphic regeneration is the relation between cross-sectional area and the ability to construct a blastema. The limbs of most animals that regenerate well are very small, as are embryonic limb buds. The critical question is whether the types of communications and signaling required for blastema formation and development can take place in limbs with a large cross-sectional area. In amphibian larvae and even many small adults, the early to mid-blastema is roughly the same cross-sectional area as that of the limb. In larger urodeles, for example, axolotls in the 250+ mm range, the cone-shaped blastema occupies only part of the total cross section of the limb. Even when morphogenesis is complete, the regenerate is much smaller than the original limb (Tank et al., 1976). A prolonged period of later growth brings the regenerate to normal size. Of interest is the ability of the giant Japanese salamander, *Andrias (Megalobatrachus) japonicus,* to regenerate limbs, albeit quite slowly (Tochimoto, 1997).

The recent discovery of an adult alligator that is apparently regenerating a tail, which has the cross-sectional area of the human limb (Fig. 15-3), engenders some hope for the possibility of success in stimulating regeneration in limbs of large mammals. A significant biologic question, however, is whether differences in morphogenetic control of tail regeneration would permit a blastema to be built on a larger stump than one that could support blastema formation in the limb.

Morphallaxis

Morphallaxis, the reshaping of the remainder of the body during regeneration, has received little attention in recent years. Because the phenomenon occurs principally in invertebrates, it has been regarded as something of a biologic curiosity. Yet, these processes, which range from the reorganization of some remaining body segments in regenerating sabellid worms to the maintenance of normal body proportions during major degrowth in starving planarians (Alvarado, 2004), are reflections of powerful morphogenetic control mechanisms. In an era in which discussions of phylogenetically conserved molecular pathways are commonplace, a detailed analysis of invertebrate systems that can accomplish such amazing morphogenetic feats is almost certain to provide clues of relevance to those trying to understand the basis for morphogenesis in vertebrates.

FIGURE 15-3   Regeneration in the tail of an alligator living in a preserve in Louisiana. The 3- to 4-foot-long alligator in the foreground of this group had, according to park rangers, suffered an amputation injury and was in the process of regenerating its tail (see inset). The cross-sectional diameter of the amputated tail was estimated to be approximately that of a human arm or leg. (Reprinted from Han, M., et al. 2005. Limb regeneration in higher vertebrates: Developing a roadmap. *Anat Rec* 287B:14–24, by permission of the publisher and Carol A. Burdsal, the photographer.)

Regeneration of Mammalian Tissues and Organs
and Tissue Engineering

The regeneration of individual tissues and organs in mammals is subject to a diverse set of conditions and control mechanisms. Nevertheless, certain broad questions apply to many individual situations.

Stem Cells: How Many Tissues Contain Them and What Do They Do?

Many aspects of stem-cell biology and their application to regeneration are covered earlier in this chapter, but of fundamental biologic interest is whether a substrate of stem cells is present in almost all tissues and organs. If so, do they represent members of a single stem-cell population, or are they tissue specific in terms of their origins and properties? This is of more than theoretic interest, because if there is any unity to their origin, there may be similar unity in the molecular basis for their activation and differentiation. Another important biologic question is the relation between tissue-specific and marrow-derived or circulating stem cells. Are they all part of the same system, or do circulating stem cells represent a case of cellular double or triple assurance that is called into action under only the most desperate of circumstances? Studies on the development of skeletal muscle now suggest that the satellite cells, which are the tissue-specific stem cells of muscle, are derived from a different cellular lineage from that which creates primary and secondary myoblasts. For many tissues, understanding the developmental origins of these stem cells will depend on more definitive characterization of the stem-cell phenotype of that tissue. How stem cells remain stem cells is also almost totally unknown. Do they occupy protected niches within their respective tissues, and if so, what is the nature of the microenvironment that keeps them stem cells rather than differentiating? Do stem cells embedded in tissues use the same types of control mechanisms that have been discovered in the hematopoietic stem-cell lineage?

Different Strategies for Different Insults

Tissue damage can be the result of actual loss of tissue mass, usually caused by trauma, or it can result from the actions of factors such as ischemia or toxins, which leave in their wake devitalized cells with varying numbers of survivors. Posttraumatic loss is typically harder to deal with, because tissue regeneration is normally not a phenomenon that is confined to the edge of a wound. Here, the classic approach of tissue engineering comes to the fore, with the implantation of a matrix, usually seeded with cells and juiced up with a cocktail of growth factors.

Some long-standing problems still face the tissue engineer. The issue of scale remains one of the most important. Wonderful constructs can often be made *in vitro*, but when produced in a volume or mass large enough to be effective in humans, such an implanted construct cannot be maintained, usually because of a deficient vascular supply early after implantation. Much remains to be done on the development of layering techniques, the concomitant production of vascular channels in the constructs before implantation, or the sequential addition of cellular components after implantation. Stimulating ingrowth of host cells into an implanted acellular substrate involves a number of important biologic issues, but appears to be a quite promising approach.

For ischemic or toxic lesions, one of the most significant problems is removal of the dead cells and debris and the stimulation of parenchymal cell division before the scarring response sets in. From our studies on muscle transplantation, it is ironic that, in rodents, an area of ischemic necrosis can remain in a state of suspended animation for many weeks without the formation of scar tissue before the necrotic area becomes revascularized and regenerates, but in primates, the scarring response occurs before muscle fiber regeneration can occur. Devising ways of stimulating rapid angiogenesis into necrotic areas is an important approach for ischemic lesions.

The Immune System, Scarring, and Regeneration: What Is the Relation?

Studies on the healing of skin wounds in both anuran amphibians and mammals have shown a close correlation between the time when fetal or larval skin regeneration converts to a scarring response and maturation of the immune system. The big question is whether they are related causally (Mescher and Neff, 2005). Experiments involving transplantation of pieces of skin from old (scarring) to earlier (regenerating) ontogenetic stages have shown clearly that the nature of the response depends on the developmental state of the transplanted tissue, rather than the overall environment of the host (rev. Ferguson and O'Kane, 2004). However, the factors that determine the conditioning of the old transplant toward a scarring response are not completely clear, although differing levels of TGF-β isoforms certainly appear to be major influences. The earlier reports on the effect of spleen cells on hepatocyte proliferation in the resected liver (see Fig. 11-3) are worthy of further attention. Another approach to the relation between immune function and regeneration is careful investigation of the healing and regeneration of tissues in immunodeficient animals, whether genetic strains such as the MRL mouse or after the administration of immunosuppressive drugs.

The Microenvironment and Regeneration

It is now clear that the success of most tissue regenerative processes in mammals depends heavily on the environment in which regeneration occurs. At one level, the environment must stimulate proliferation of the local parenchymal or stem cells, or it must be able to attract immigrant cells. Whether that microenvironment consists of phagocytic cells and their secretions or growth factors bound to the extracellular matrix, further definition of the specific features that are mitogenic will be important for optimizing the regenerative response.

At a later stage in regeneration, the microenvironment affects cellular differentiation. Especially where stem cells are involved, identifying the factors that steer differentiation down specific channels is a major task for the future, especially when there may be more than one option. Finally, the role of structural or mechanical factors in guiding morphogenesis of the regenerating tissue needs to be further clarified and controlled. For skeletal muscle and bone, this is dealt with at some length in Chapter 11. Ferguson and O'Kane (2004) pose an interesting twist to the issue of the local mechanical environment and morphogenesis. They attribute the regenerative type of healing of skin wounds after treatment with TGF-β_3 to an early invasion of the fibrin clot by highly motile fibroblasts, which deposit matrix in a basket-weave orientation. This results in much less scarring, whereas in normal wound healing, the fibroblasts lay down collagen in abnormally organized parallel bundles oriented around the interface between the fibrin clot and the wound margin. To date, subtle interactions of this sort during tissue regeneration have received little attention and merit further investigation. Such studies will be of particular importance in regenerating cardiac muscle, where the entire architecture of the heart is based on mechanical needs.

Much of the tissue-engineering approach to regeneration is based on devising implantable matrices that will enhance tissue regeneration. These matrices represent the temporary substrate on which regeneration will begin. One of the major issues for the future will be to design artificial matrices that will anticipate or accommodate the changing overall environment as regeneration progresses in the days and weeks after implantation.

CONCLUSION

The study of regeneration is an intellectual addiction that is finally being recognized as a field worthy of serious scientific study. Looking back at the long history of the field and the complexity of most regenerative processes, it is remarkable that so much useful information was learned through early experiments that relied on tricking a regenerating system into revealing some of its secrets. We have now entered an era in which it is possible to technologically overpower many biologic systems and to dissect their essence at a molecular level. Many aspects of regeneration are sufficiently complex that they are resistant to revealing their secrets through either low-technology experimental tricks or high-technology analytic methods alone. The ability to combine these approaches lends hope that some of the questions in regeneration that have been puzzling investigators for decades will be answered, and that these answers may provide the basis for the application of some of the fundamental principles of regeneration to the healing of broken bodies.

REFERENCES

Abdullah, I., et al. 2005. MRL mice fail to heal the heart in response to ischemia-perfusion injury. *Wound Repair Regen* 13:205–208.

Aboody, K.S., et al. 2000. Neural stem cells display extensive tropism for pathology in adult brain: Evidence from intracranial gliomas. *Proc Natl Acad Sci USA* 97:12846–12851.

Achermann, J., and T. Sugiyama. 1985. Genetic analysis of developmental mechanisms in Hydra. X. Morphogenetic potentials of a regeneration-deficient strain (reg-16). *Dev Biol* 107:13–27.

Adams, J.C., and F.M. Watt. 1993. Regulation of development and differentiation by the extracellular matrix. *Development* 117:1183–1198.

Addis, T., and W. Lew. 1940. The restoration of lost organ tissue: The rate and degree of restoration. *J Exp Med* 71:325–334.

Adova, A.N., and A.M. Feldt. 1939. Biochemical properties of different regions of the body of the axolotl, connected with morphogenesis under the effect of organizers (Russian). *Compt Rend Acad Sci URSS* 25:43.

Aebischer, P., et al. 1996. Intrathecal delivery of CNTF using encapsulated genetically modified xenogeneic cells in amyotrophic lateral sclerosis patients. *Nat Med* 2:696–699.

Agah, R., et al. 1997. Adenoviral delivery of E2F-1 directs cell cycle reentry and p53-independent apoptosis in postmitotic adult myocardium in vivo. *J Clin Invest* 100:2722–2728.

Agata, K., et al. 2003. Intercalary regeneration in planarians. *Dev Dyn* 226:308–316.

Akimenko, M.-A., et al. 1995. Differential induction of four *msx* homeobox genes during fin development and regeneration in zebrafish. *Development* 121:347–357.

Akimoto, T., et al. 2005. Mechanical stretch inhibits myoblast-to-adipocyte differentiation through Wnt signaling. *Biochem Biophys Res Comm* 329:381–385.

Alden, T.D., et al. 1999. Percutaneous spinal fusion using bone morphogenetic protein-2 gene therapy. *J Neurosurg* 90 (1 suppl):109–114.

Alitzer, A.M., et al. 2002. Skin flaps inhibit both the current of injury at the amputation surface and regeneration of that limb in newts. *J Exp Zool* 293:467–477.

Alonso, L., and E. Fuchs. 2003. Stem cells of the skin epithelium. *Proc Natl Acad Sci USA* 100 (suppl 1):11830–11835.

Altman, J., and G.D. Das. 1965. Autoradiographic and histological evidence of postnatal hippocampal neurogenesis in rats. *J Comp Neurol* 124:319–335.

Altman, J., and G.D. Das. 1967. Postnatal neurogenesis in the guinea-pig. *Nature* 214:1098–1101.

Alvarado, A.S. 2004. Regeneration and the need for simpler model organisms. *Philos Trans R Soc Lond B* 359:759–763.

Alvarez-Buylla, A., and D.A. Lim. 2004. For the long run: Maintaining germinal niches in the adult brain. *Neuron* 41:683–686.

Anderson, D.K., et al. 1995. Fetal neural grafts and repair of injured spinal cord. *Brain Pathol* 5:451–457.

Anderson, J.E. 2000. A role for nitric oxide in muscle repair: NO-mediated satellite cell activation. *Mol Biol Cell* 11:1859–1874.

Anderson, J.E., and A.G. Wozniak. 2004. Satellite cell activation on fibers: Modeling events *in vivo*—An invited review. *Can J Physiol Pharmacol* 82:300–310.

Andrade, J., et al. 2005. Predominant fusion of bone marrow-derived cardiomyocytes. *Cardiovasc Res* 68:387–393.

Anonymous. 1948. *The situation in biological science*. Moscow: Foreign Languages Publishing House.

Anversa, P., et al. 2003. Primitive cells and tissue regeneration. *Circ Res* 92:579–582.

Armstrong, J.R., and M.W.J. Ferguson. 1995. Ontogeny of the skin and the transition from scar-free to scarring phenotype during wound healing in the pouch young of a marsupial, *Monodelphis domestica*. *Dev Biol* 169:242–260.

Arsanto, J.-P., et al. 1992. Patterns of dystrophin expression in developing, adult and regenerating tail skeletal muscle of Amphibian urodeles. *Int J Dev Biol* 36:555–565.

Ashcroft, G.S., et al. 1995. The effects of aging on cutaneous wound healing in mammals. *J Anat* 187:1–26.

Ashhurst, D.E. 1986. The influence of mechanical conditions on the healing of experimental fractures in the rabbit: A microscopical study. *Philos Trans R Soc Lond B* 313:271–302.

Aspiz, M.E. 1954. On the regeneration of skin glands and hair in certain mammals (Russian). In: *Problems of restoration of organs and tissues in vertebrates animals* (Russian). Trudy Inst. Morphol. Zhivotnikh im. A.N. Severtsov. 11:92–113.

Athenstaedt, H. 1974. Pyroelectric and piezoelectric properties of vertebrates. *Ann NY Acad Sci* 238:68–94.

Atkinson, K.H., et al. 1976. Morphogenetic sequences during tadpole tail regeneration. *Can J Zool* 54:1314–1325.

Auerbach, R., et al. 1985. Expression of organ-specific antigens on capillary endothelial cells. *Microvasc Res* 29:401–411.

Avel, M. 1961. L'influence du système nerveux sur la régénération chez les urodèles et les oligochètes. *Bull Soc Zool France* 86:464–483.

Babaeva, A.G. 1972. *Immunological mechanisms of regulation of restorative processes* (Russian). Moscow: Meditsina.

Babaeva, A.G. 1985. *Regeneration and the system of immunogenesis* (Russian). Moscow: Meditsina.

Babaeva, A.G. 1989. Immune system and regeneration. In: V. Kiortsis et al., eds. *Recent trends in regeneration research*. New York: Plenum Press. 121–128.

Babaeva, A.G., and E.A. Zotnikov. 1987. *The immunology of processes of adaptive growth, proliferation and their disturbance* (Russian). Moscow: Nauka.

Bader, D., and J.O. Oberpriller. 1978. Repair and reorganization of minced cardiac muscle in the adult newt *(Notophthalmus viridescens)*. *J Morphol* 155:349–358.

Bader, D., and J.O. Oberpriller. 1979. Autoradiographic and electron microscopic studies of minced cardiac muscle regeneration in the adult newt *Notophthalmus viridescens*. *J Exp Zool* 208:177–194.

Badylak, S.F. 2002. The extracellular matrix as a scaffold for tissue reconstruction. *Semin Cell Dev Biol* 13:377–383.

Badylak, S.F. 2004. Xenogenic extracellular matrix as a scaffold for tissue reconstruction. *Transplant Immunol* 12:367–377.

Badylak, S.F., et al. 2000. Resorbable bioscaffold for esophageal repair in a dog model. *J Pediatr Surg* 35:1097–1103.

Bak, B., and T.T. Andreassen. 1989. The effect of aging on fracture healing in the rat. *Calcif Tissue Int* 45:292–297.

Balabaud, C., et al. 2004. The role of hepatic stellate cells in liver regeneration. *J Hepatol* 40:1023–1026.

Balsam, L.B., et al. 2004. Haemopoietic stem cells adopt mature haematopoietic fates in ischaemic myocardium. *Nature* 428:668–673.

Baltzer, A.W., et al. 2000. Genetic enhancement of fracture repair: Healing of an experimental segmental defect by adenoviral transfer of the BMP-2 gene. *Gene Ther* 7:734–739.

Bandyopadhyay, B., et al. 2006. A "traffic control" role for TGFβ3: Orchestrating dermal and epidermal cell motility during wound healing. *J Cell Biol* 172:1093–1105.

Bar-Maor, J.A., and G. Gitlin. 1961. Attempted induction of forelimb regeneration by augmentation of nerve supply in young rats. *Transplant Bull* 27:460–461.

Bart, A. 1971. Morphogenèse surnuméraire au niveau de la pattes u phasme *Carausius morosus* Br. *Arch f Entw-mech* 166:331–364.

Barton-Davis, E.R., et al. 1998. Viral-mediated expression of insulin-like growth factor I blocks the aging-related loss of skeletal muscle function. *Proc Natl Acad Sci USA* 95:15603–15607.

Bassett, C.A. 1972. A biophysical approach to craniofacial morphogenesis. *Acta Morphol Neerl Scand* 10:71–86.

Bassett, C.A.L., et al. 1982. Treatment of therapeutically resistant non-unions with bone grafts and pulsing electromagnetic fields. *J Bone Joint Surg* 64-A:1214–1220.

Bateson, R.G., et al. 1967. Circulating cell as a source of myoblasts in regenerating injured mammalian skeletal muscle. *Nature* 213:1035–1036.

Bateson, W. 1894. *Materials for the study of variation.* London: MacMillan.

Bauduin, B., et al. 2000. Stimulation of axon growth from the spinal cord by a regenerating limb blastema in newts. *Brain Res Dev Brain Res* 119:47–54.

Beck, C.W., et al. 2003. Molecular pathways needed for regeneration of spinal cord and muscle in a vertebrate. *Dev Cell* 5:429–439.

Becker, R.O. 1961. The bioelectric factors in amphibian limb regeneration. *J Bone Joint Surg* 43-A:643–656.

Becker, R.O. 1972. Stimulation of partial limb regeneration in rats. *Nature* 235:109–111.

Becker, R.O., and J.A. Spadaro. 1972. Electrical stimulation of partial limb regeneration in mammals. *Bull NY Acad Med* 48:627–641.

Beigel C. 1912. Regeneration der Barteln bei Siluroiden. *Arch f Entw-mech* 34:363–370.

Bell, E., et al. 1991. Recipes for reconstituting skin. *J Biomech Eng* 113:113–119.

Belous, A.M., et al. 1974. *Exogenous nucleic acids and restorative processes* (Russian). Moscow: Meditsina.

Beltrami, A.P., et al. 2003. Adult cardiac stem cells are multipotent and support myocardial regeneration. *Cell* 114:763–776.

Bely, A.E., and G.A. Wray. 2001. Evolution of regeneration and fission in annelids: Insights from *engrailed-* and *orthodenticle-*class gene expression. *Development* 128:2781–2791.

Ben Ze'ev, A., et al. 1988. Cell-cell and cell-matrix interactions differentially regulate the expression of hepatic and cytoskeletal genes in primary cultures of rat hepatocytes. *Proc Natl Acad Sci USA* 85:2161–2165.

Benoit, D.S.W., and K.S. Anseth. 2005. The effect on osteoblast function of colocalized RGD and PHSRN epitopes on PEG surfaces. *Biomaterials* 26:5209–5220.

Bernal, G.M., and D.A. Peterson. 2004. Neural stem cells as therapeutic agents for age-related brain repair. *Aging Cell* 3:345–351.

Bernstein, B.E., et al. 2006. A bivalent chromatin structure marks key developmental genes in embryonic stem cells. *Cell* 125:315–326.

Bernstein, J.J. 1967. The regenerative capacity of the telencephalon of the goldfish and rat. *Exp Neurol* 17:44–56.

Berrill, N.J. 1931. Regeneration in *Sabella pavonina* (Sav.) and other sabellid worms. *J Exp Zool* 58:495–523.

Berrill, N.J. 1978. Induced segmental reorganization in sabellid worms. *J Embryol Exp Morphol* 47:85–96.

Berry, M. 1982. Post-injury myelin-breakdown products inhibit axonal growth: An hypothesis to explain the failure of axonal regeneration in the mammalian central nervous system. *Bibl Anat* 23:1–11.

Bhatt, D.H., et al. 2004. Cyclic AMP-induced repair of zebrafish spinal circuits. *Science* 305:254–258.

Bier, A. 1923. Über Knochenregeneration, über Pseudarthrosen und über Knochentransplantate. *Arch f klin Chir* 127:1–136.

Bischler, V. 1926. L'influence du squelette dans la régénération, et les potentialités des divers territoires du membre chez *Triton cristatus*. *Rev Suisse Zool* 33:430–560.

Black, D., et al. 2004. Molecular and cellular features of hepatic regeneration. *J Surg Res* 117:306–315.

Block, J.E. 2005. The role and effectiveness of bone marrow in osseous regeneration. *Med Hypotheses* 65:740–747.

Bode, H.R. 2003. Head regeneration in *Hydra*. *Dev Dyn* 226:225–236.

Bode, P.M., and H.R. Bode. 1980. Formation of pattern in regenerating pieces of *Hydra attenuata*. I. Head-body proportion regulation. *Dev Biol* 78:484–496.

Bodemer, C.W. 1964. Evocation of regrowth phenomena in anuran limbs by electrical stimulation of the nerve supply. *Anat Rec* 148:441–457.

Boden, S.D., et al. 1995. The use of an osteoinductive growth factor for the spinal lumbar fusion. 2. Study of dose, carrier, and species. *Spine* 20:2633-2644.

Boden, S.D., et al. 2002. Use of recombinant human bone morphogenetic protein-2 to achieve posterolateral lumbar spine fusion in humans: A prospective, randomized clinical pilot trial: 2002 Volvo Award in clinical studies. *Spine* 27:2662–2673.

Bohler, K.R., et al. 2002. Differentiation of pluripotent embryonic stem cells into cardiomyocytes. *Circ Res* 91:189–201.

Bohn, H. 1970. Interkalare Regeneration und segmentale Gradienten bei den Extremitäten von *Leucophaea*-Larven (Blattaria). II. Coxa und Tarsus. *Dev Biol* 23:355–379.

Bohn, H. 1972. The origin of the epidermis in the supernumerary regenerates of triple legs in cockroaches (Blatteria). *J Embryol Exp Morphol* 28:185–208.

Bohn, H. 1976. Tissue interactions in the regenerating cockroach leg. In: P.A. Lawrence, ed. *Insect development*. Oxford, UK: Blackwell. 170–185.

Boilly, B., et al. 1985. Control of the blastemal cell cycle by the peripheral nervous system during newt limb regeneration: Continuous labeling analysis. *Biol Cell* 55:107–112.

Boilly-Marer, Y. 1969. Induction expérimentale de parapodes surnuméraires chez l'Annélide Polychète *Nereis pelagica*. *CR Acad Sci Paris* 268:1300–1302.

Boilly-Marer, Y. 1971. Rôle du système nerveux parapodial dans l'induction de parapodes surnuméraires par greffes hétérologues chez *Nereis pelagica*. *CR Acad Sci Paris* 272:261–264.

Borgens, R.B. 1982. Mice regrow the tips of their foretoes. *Science* 217:747–750.

Borgens, R.B. 1989a. Natural and applied currents in limb regeneration and development. In: R.B. Borgens et al., eds. *Electric fields and vertebrate repair.* New York: Alan R. Liss. 27–75.

Borgens, R.B. 1989b. Artificially controlling axonal regeneration and development by applied electrical fields. In: R.B. Borgens et al., eds. *Electric fields and vertebrate repair.* New York: Alan R. Liss. 117–170.

Borgens, R.B., et al. 1977a Bioelectricity and regeneration. I. Initiation of frog limb regeneration by minute currents. *J Exp Zool* 200:403–416.

Borgens, R.B., et al. 1977b. Bioelectricity and regeneration: Large currents leave the stumps of regenerating newt limbs. *Proc Natl Acad Sci USA* 74:4528–4532.

Borgens, R.B., et al. 1979. Reduction of sodium dependent stump currents disrupts urodele limb regeneration. *J Exp Zool* 209:377–386.

Borgens, R.B., et al., eds. 1989. *Electric fields in vertebrate repair.* New York: Alan R. Liss.

Borisov, A.B. 1998. Cellular mechanisms of myocardial regeneration. In: P. Ferretti and J. Géraudie, eds. *Cellular and molecular basis of regeneration.* Chichester, UK: Wiley. 335–354.

Bosch, T.C.G. 2003. Ancient signals: Peptides and the interpretation of positional information in ancestral metazoans. *Comp Biochem Physiol B* 136:185–196.

Boskey, A.L. 2005. The organic and inorganic matrices. In: J.O. Hollinger et al., eds. *Bone tissue engineering.* Boca Raton, FL: CRC Press. 91–123.

Bossley, C.J. 1975. Conservative treatment of digit amputations. *N Z Med J* 82:379–380.

Bottai, D., et al. 2003. Neural stem cells in the adult nervous system. *J Hematother Stem Cell Res* 12:655–670.

Boyd, J.G., et al. 2004. LacZ-expressing olfactory ensheathing cells do not associate with myelinated axons after implantation into the compressed spinal cord. *Proc Natl Acad Sci USA* 101:2162–2166.

Bray, G.M., et al. 1987. Regeneration of axons from the central nervous system of adult rats. *Prog Brain Res* 71:373–379.

Brazel, C.Y., and M.S. Rao. 2004. Aging and neuronal replacement. *Ageing Res Rev* 3:465–483.

Breedis, C. 1954. Regeneration of hair follicles and sebaceous glands from the epithelium of scars in the rabbit. *Cancer Res* 14:575–579.

Brighton, C.T., et al. 1977. Treatment of nonunion with constant direct current. *Clin Orthopaed* 124:106–122.

Brockes, J.P. 1984. Mitogenic growth factors and nerve dependence of limb regeneration. *Science* 225:1280–1287.

Brockes, J.P. 1998. Regeneration and cancer. *Biochem Biophys Acta* 1377:M1–M11.

Brockes, J.P., and A. Kumar. 2002. Plasticity and reprogramming of differentiated cells in amphibian regeneration. *Nat Rev Mol Cell Biol* 3:566–574.

Brockes, J.P., and A. Kumar. 2005. Appendage regeneration in adult vertebrates and implications for regenerative medicine. *Science* 310:1919-1923.

Brockes, J.P., et al. 2001. Regeneration as an evolutionary variable. *J Anat* 199:3–11.

Brodsky, W.Ya. 1991. Cell ploidy in the mammalian heart. In: J.O. Oberpriller et al., eds. *The development and regenerative potential of cardiac muscle.* London: Harwood Academic Publishers. 253–292.

Brodsky, V.Ya., et al. 1988. Polyploidization of transplanted cardiac myocytes. *Cell Differ Dev* 25:177–184.

Brodsky, W.Ya., et al. 1980. Mitotic polyploidization of mouse heart myocytes during the first postnatal week. *Cell Tissue Res* 210:133–144.

Brooks, S.V., and J.A. Faulkner. 1994. Skeletal muscle weakness in old age: Underlying mechanisms. *Med Sci Sports Exer* 26:432–439.

Brown, M.D., and O. Hudlicka. 2003. Modulation of physiological angiogenesis in skeletal muscle by mechanical forces: Involvement of VEGF and metalloproteinases. *Angiogenesis* 6:1–14.

Bryant, S.V. 1976. Regenerative failure of double half limbs in *Notophthalmus viridescens*. *Nature* 263:676–679.

Bryant, S.V., and B.A. Baca. 1978. Regenerative ability of double–half and half upper arms in the newt, *Notophthalmus viridescens*. *J Exp Zool* 204:307–324.

Bryant, S.V., and L.E. Iten. 1976. Supernumerary limbs in amphibians: Experimental production in *Notophthalmus viridescens* and a new interpretation of their formation. *Dev Biol* 50:212–234.

Bryant, S.V., et al. 1981. Distal regeneration and symmetry. *Science* 212:993–1002.

Bryant, S.V., et al. 1987. Limb development and regeneration. *Am Zool* 27:675–696.

Bryant, S.V., et al. 2002. Vertebrate limb regeneration and the origin of limb stem cells. *Int J Dev Biol* 46:887–896.

Bubenik, G.A., and A.B. Bubenik, eds. 1990. *Horns, Pronghorns, and Antlers*. Springer-Verlag, New York.

Bucher, N.L.R., and A.D. Glinos. 1950. The effect of age on regeneration of rat liver. *Cancer Res* 10:324–332.

Bucher, N.L.R., and R.A. Malt. 1971. *Regeneration of liver and kidney*. Boston: Little, Brown & Co.

Bucher, N.L.R., et al. 1964. Influence of age upon incorporation of thymidine-2-C[14] into DNA of regenerating rat liver. *Cancer Res* 24:509–512.

Bullough, W.S. 1962. The control of mitotic activity in adult mammalian tissues. *Biol Rev* 37:307–402.

Bullough, W.S. 1967. *The evolution of differentiation*. Academic Press: London.

Bullough, W.S., and E.B. Lawrence. 1960. The control of epidermal mitotic activity in the wound. *Proc Roy Soc* (London), *Series B* 151:517-536.

Bunting, K.D., and R.G. Hawley. 2003. Integrative molecular and developmental biology of adult stem cells. *Biol Cell* 95:563–578.

Burden, S.J., et al. 1979. Acetylcholine receptors in regenerating muscle accumulate at original synaptic sites in the absence of the nerve. *J Cell Biol* 82:412-425.

Burger, E.H. 2001. Experiments on cell mechanosensitivity: Bone cells as mechanical engineers. In: S.C. Cowin, ed. *Bone mechanics handbook*. 2nd ed. Boca Raton, FL: CRC Press. 28-1–28-16.

Burnett, A.L. 1962. The maintenance of form in Hydra. In: D. Rudnick, ed. *Regeneration*. New York: Ronald Press. 27–52.

Burwell, R.G. 1963. The role of lymphoid tissue in morphostasis. *Lancet* 2:69–74.

Buss, R.R., and R.W. Oppenheim. 2004. Role of programmed cell death in normal neuronal development and function. *Anat Sci Int* 79:191–197.

Butler, E.G. 1935. Studies on limb regeneration in x-rayed *Amblystoma* larvae. *Anat Rec* 62:295-307.

Butler, E.G. 1951. The mechanics of blastema formation and regeneration in urodele limbs of reversed polarity. *Trans N Y Acad Sci Ser II* 13:164–167.

Butler, E.G., and O.E. Schotté. 1941. Histological alterations in denevated non-regenerating limbs of urodele larvae. *J Exp Zool* 88:307–341.

Cai, J., et al. 2004. In search of "stemness." *Exp Hematol* 32:585–598.

Caldwell, C.J., et al. 1990. Role of the basement membrane in the regeneration of skeletal muscle. *Neuropath Appl Neurobiol* 16:225-238.

Camargo, F.D., et al. 2003. Single hematopoietic stem cells generate skeletal muscle through myeloid intermediates. *Nat Med* 9:1520–1527.

Canalis, E., ed. 2000. *Skeletal growth factors.* Philadelphia: Lippincott Williams & Wilkins.

Caplan, A.I. 2006. Mesenchymal stem cells. In: R. Lanza, ed. *Essentials of stem cell biology.* Amsterdam: Elsevier. 205–210.

Caporaso, G.L., et al. 2003. Telomerase activity in the subventricular zone of adult mice. *Mol Cell Neurosci* 23:693–702.

Carleton, A., et al. 2003. Becoming a new neuron in the adult olfactory bulb. *Nat Neurosci* 6:507–518.

Carlson, B.M. 1967. Studies on the mechanism of implant-induced supernumerary limb formation in urodeles. I. The histology of supernumerary limb formation in the adult newt, *Triturus viridescens. J Exp Zool* 164:227–242.

Carlson, B.M. 1968. Regeneration of the completely excised gastrocnemius muscle in the frog and rat from minced muscle fragments. *J Morphol* 125:447–472.

Carlson, B.M. 1970a. The regeneration of a limb muscle in the axolotl from minced fragments. *Anat Rec* 166:423–436.

Carlson, B.M. 1970b. Regeneration of the rat gastrocnemius muscle from sibling and non-sibling fragments. *Am J Anat* 128:21–32.

Carlson, B.M. 1972a. Muscle morphogenesis in Urodele limb regenerates following removal of the stump musculature. *Dev Biol* 28:487–497.

Carlson, B.M. 1972b. *The regeneration of minced muscles.* Basel, Switzerland: S. Karger.

Carlson, B.M. 1974a. Morphogenetic interactions between rotated skin cuffs and underlying stump tissues in regenerating axolotl forelimbs. *Dev Biol* 39:263–285.

Carlson, B.M. 1974b. Regeneration from short stumps of the gastrocnemius muscle. *Experientia* 30:275–276.

Carlson, B.M. 1974c. Factors controlling the initiation and cessation of early events in the regenerative process. In: G.V. Sherbet, ed. *Neoplasia and cell division.* Basel, Switzerland: S. Karger. 60–105.

Carlson, B.M. 1975a. The effects of rotation and positional change of stump tissues upon morphogenesis of the regenerating axolotl limb. *Dev Biol* 47:269–291.

Carlson, B.M. 1975b. Multiple regeneration from axolotl limb stumps bearing cross-transplanted minced muscle regenerates. *Dev Biol* 454:203–208.

Carlson, B.M. 1979. The relationship between the tissue and epimorphic regeneration of muscle. In: A. Mauro, ed. *Muscle regeneration.* New York: Raven Press. 57–71.

Carlson, B.M. 1982. The regeneration of axolotl limbs covered by frog skin. *Dev Biol* 90:435–440.

Carlson, B.M. 1983. Positional memory in vertebrate limb development and regeneration. In: J.F. Fallon et al., eds. *Limb development and regeneration.* New York: Alan R. Liss. 433–443.

Carlson, B.M. 1995. Factors influencing the repair and adaptation of muscles in aged individuals: Satellite cells and innervation. *J Gerontol Ser A* 50A:96–100.

Carlson, B.M. 2004. *Human embryology and developmental biology.* 3rd ed. Philadelphia: Mosby. 178–180.

Carlson, B.M., and J.A. Faulkner. 1989. Muscle transplantation between young and old rats: Age of host determines recovery. *Am J Physiol Cell Physiol* 256:C1262–C1266.

Carlson, B.M., and J.A. Faulkner. 1996. The regeneration of noninnervated muscle grafts and Marcaine-treated muscles in young and old rats. *J Gerontol (Biol Sci)* 51:B43–B49.

Carlson, B.M., and E. Gutmann. 1972. Development of contractile properties of minced muscle regenerates in rats. *Exp Neurol* 36:239–249.

Carlson, B.M., and E. Gutmann. 1974. Transplantation and "cross-transplantation" of free muscle grafts in the rat. *Experientia* 30:1292–1294.

Carlson, B.M., and S.L. Rogers. 1976. Satellite cells in the limb musculature of the axolotl. *Folia Morph (Praha)* 24:359–361.

Carlson, B.M., et al. 1974. Nerve interactions and regenerative processes occurring in newt limbs fused end-to-end. *Dev Biol* 37:248–262.

Carlson, B.M., et al. 1979a. The life history of a free muscle graft. In: A. Mauro, ed. *Muscle regeneration.* New York: Raven Press. 493–507.

Carlson, B.M., et al. 1979b. Retention of hormonal sensitivity in free grafts of the levator ani muscle. *Exp Neurol* 63:94–107.

Carlson, B.M., et al. 1981. Comparison between grafts with intact nerves and standard free grafts of the rat extensor digitorum longus muscle. *Physiol Bohemoslovaca* 30:505–513.

Carlson, B.M., et al. 2001. Skeletal muscle regeneration in very old rats. *J Gerontol* 56A: B224–B233.

Carlson, E.C., and B.M. Carlson. 1991. A method for preparing skeletal muscle fiber basal laminae. *Anat Rec* 230:325–331.

Carlson, M.R.J., et al. 2001. Expression of Hoxb13 and Hoxc10 in developing and regenerating axolotl limbs and tails. *Dev Biol* 229:396–406.

Carmeliet, P. 2000. Mechanisms of angiogenesis and arteriogenesis. *Nat Med* 6:389–395.

Caroni, P., and M.E. Schwab, 1988. Antibody against myelin-associated inhibitor of neurite growth neutralizes nonpermissive substrate properties of CNS white matter. *Neuron* 1:85–96.

Cecil, M.L., and R.A. Tassava. 1986a. Forelimb regeneration in the postmetamorphic bullfrog: Stimulation by dimethyl sulfoxide and retinoic acid. *J Exp Zool* 239:57–63.

Cecil, M.L., and R.A. Tassava. 1986b. Vitamin A enhances forelimb regeneration in juvenile leopard frogs *Rana pipiens. J Exp Zool* 237:57–61.

Cederna, P.S., et al. 2001. Motor unit properties of nerve-intact extensor digitorum longus muscle grafts in young and old rats. *J Gerontol (Biol Sci)* 556A:B254–B258.

Celeste, A.J., et al. 1990. Identification of transforming growth factor β family members present in bone-inductive protein purified from bovine bone. *Proc Natl Acad Sci USA* 87:9843–9847.

Chalkley, D.T. 1954. A quantitative histological analysis of forelimb regeneration in *Triturus viridescens. J Morphol* 94:21–70.

Chandebois, R. 1957. Recherches expérimentales sur la régénération de la planaire marine *Procerodes lobata* O Schmidt *Bull Biol Fr Belg* 91:1–94.

Chandebois, R. 1976. *Histogenesis and morphogenesis in planarian regeneration.* Basel, Switzerland: S. Karger.

Chandebois, R. 1979. The dynamics of wound closure and its role in the programming of planarian regeneration. *Dev Growth Differ* 21:195–204.

Chargé, S.B.P., and M.A. Rudnicki. 2004. Cellular and molecular regulation of muscle regeneration. *Physiol Rev* 84:209–238.

Chen, X., et al. 2005. Dedifferentiation of adult human myoblasts induced by ciliary neurotrophic factor. *Mol Biol Cell* 16:3140–3151.

Cherkasova, L.V. 1982. Postsatellites in muscular tissue in adult tailed amphibian (Russian). *Dokl Akad Nauk SSSR* 267:1235–1236.

Chernoff, E.A.G., and S. Robertson. 1990. Epidermal growth factor and the onset of epithelial wound healing. *Tissue Cell* 22:123–135.

Chernoff, E.A.G., et al. 2002. Spinal cord regeneration: Intrinsic properties and emerging mechanisms. *Semin Cell Dev Biol* 13:361–368.

Chernoff, E.A.G., et al. 2003. Urodele spinal cord regeneration and related processes. *Dev Dyn* 226:295–307.

Chevallier, A., et al. 1977. Limb-somite relationship: Origin of the limb musculature. *J Embryol Exp Morphol* 42:275–292.

Chiakulas, J.J. 1952. The role of tissue specificity in the healing of epithelial wounds. *J Exp Zool* 121:383–417.

Child, C.M. 1941. *Patterns and problems of development*. Chicago: Univ. of Chicago Press.

Christen, B., and J.M. Slack. 1997. FGF-8 is associated with anteroposterior patterning and limb regeneration in *Xenopus*. *Dev Biol* 192:455–466.

Christensen, R.N., and R.A. Tassava. 2000. Apical epithelial cap morphology and fibronectin gene expression in regenerating axolotl limbs. *Dev Dyn* 217:216–224.

Christensen, R.N., et al. 2002. Expression of fibroblast growth factors 4, 8, and 10 in limbs, flanks and blastemas of *Ambystoma*. *Dev Dyn* 223:193–203.

Christophers, E. 1972. Kinetic aspects of epidermal healing. In: H.I. Maibach and D.T. Rovee, eds. *Epidermal wound healing*. Chicago: Year Book Medical Publishers. 53–69.

Čihák, R. 1972. Ontogenesis of the skeleton and intrinsic muscles of the human hand and foot. *Adv Anat Embryol Cell Biol* 46:1–194.

Clark, L.D., et al. 1998. A new murine model for mammalian wound repair and regeneration. *Clin Immunol Immunopathol* 88:35–45.

Coe, W.R. 1929. Regeneration in nemerteans. *J Exp Zool* 54:411–459.

Colucci, V. 1891. Sulla rigenerazione parziale dell-occhio nei tritoni. Istogenesi e sviluppo. Studio sperimentale. *Mem R Accad Sci Ist Bologna Ser* 5:593–629.

Conboy, I.M., et al. 2003. Notch-mediated restoration of regenerative potential to aged muscle. *Science* 302:1575–1577.

Conboy, I.M., et al. 2005. Rejuvenation of aged progenitor cells by exposure to a young systemic environment. *Nature* 433:760–764.

Corbel, S.Y., et al. 2003. Contribution of hematopoietic stem cells to skeletal muscle. *Nat Med* 9:1528–1532.

Coulombre, J.L., and A.J. Coulombre. 1963. Lens development: Fiber elongation and lens orientation. *Science* 142:1489–1490.

Couzin, J., and G. Vogel. 2004. Renovating the heart. *Science* 304:192–194.

Cowin, S.C., and M.L. Moss. 2000. Mechanosensory mechanisms in bone. In: R.P. Lanza et al., eds. *Principles of tissue engineering*. 2nd ed. San Diego: Academic Press. 723–737.

Cox, P.G. 1969. Some aspects of tail regeneration in the lizard, *Anolis carolinensis*. II. The role of the peripheral nerves. *J Exp Zool* 171:151–159.

Crawford, K., and D.L. Stocum. 1988. Retinoic acid coordinately proximalizes regenerate pattern and blastemal differential affinity in axolotl limbs. *Development* 102:687–698.

Cummings, S.G., and H.R. Bode. 1984. Head regeneration and polarity reversal occur in the absence of DNA synthesis in *Hydra attenuata*. *Roux' Arch Dev Biol* 194:79–86.

Curtis, A.S.G., et al. 2004. Cells react to nanoscale order and symmetry in their surroundings. *IEEE Trans Nanobiosci* 3:61–65.

d'Albis, A., et al. 1989. Myosin isoform transitions in regeneration of fast and slow muscles during postnatal development of the rat. *Dev Biol* 135:320–325.

Daniel, C.W., et al. 1968. The *in vivo* life span of normal and preneoplastic mouse mammary glands: A serial transplantation study. *Proc Natl Acad Sci USA* 61:53–60.

Dätwyler, D.A., et al. 2003. Reactivation of the mitosis-promoting factor in postmitotic cardio-myocytes. *Cells Tissues Organs* 175:61–71.

David, S., and A.J. Aguayo. 1981. Axonal elongation into peripheral nervous system "bridges" after central nervous system injury in adult rats. *Science* 214:931–933.

David, S., and S. Lacroix. 2003. Molecular approaches to spinal cord repair. *Annu Rev Neurosci* 26:411–440.

Deasy, B.M., et al. 2004. Tissue engineering with muscle-derived stem cells. *Curr Opin Biotechnol* 15:419–423.

del Rio-Tsonis, K., and G. Eguchi. 2004. Lens regeneration. In: F.J. Lovicu and M.L. Robinson, eds. *Development of the ocular lens.* Cambridge, UK: Cambridge Univ. Press. 290–311.

del Rio-Tsonis, K., et al. 1998a. Regulation of lens regeneration by fibroblast growth factor receptor 1. *Dev Dyn* 213:140–146.

del Rio-Tsonis, K., et al. 1998b. Expression of the third component of complement, C3, in regenerating limb blastema cells of urodeles. *J Immunol* 616:6819–6824.

del Rio-Tsonis, K., et al. 1999. Regulation of Prox 1 during lens regeneration. *Invest Ophthalmol Vis Sci* 40:2039–2045.

Dent, J.N. 1962. Limb regeneration in larvae and metamorphosing individuals of the South African clawed toad. *J Morphol* 110:61–77.

Devor, S.T., and J.A. Faulkner. 1999. Regeneration of new fibers in muscles of old rats reduces contraction-induced injury. *J Appl Physiol* 87:750–756.

Dezawa, M., et al. 2005. Bone marrow stromal cells generate muscle cells and repair muscle degeneration. *Science* 309:314–317.

Dhawan, J., and T.A. Rando. 2005. Stem cells in postnatal myogenesis: Molecular mechanisms of satellite cell quiescence, activation and replenishment. *Trends Cell Biol* 15:665–673.

Dijkstra, C. 1933. Die De- und Regeneration der sensiblen Endkörperchen des Entenschnabels (Grandry- und Herbst-Körperchen) nach Durchschneidung des Nerven, nach Fortnahme der ganzen Haut und nach Transplantation des Hautstückchens. *Z micro-anat Forsch* 34:75–158.

Ding. S., and P.G. Schultz. 2004. A role for chemistry in stem cell biology. *Nat Biotechnol* 22:833–840.

Dinsmore, C.E. 1979. Morphogenetic control during tail regeneration in *Plethodon cinereus:* The role of skeletal muscle. *Dev Biol* 72:244–253.

Dinsmore, C.E. 1981a. Regulative ability of the regenerating urodele tail: The effect of unilat-ereateral soft tissue ablation. *Dev Biol* 82:186–191.

Dinsmore, C.E. 1981b. Morphogenetic effect of rotated skin cuffs on tail regeneration in *Plethodon cinereus. J Exp Zool* 215:151–161.

Dinsmore, C.E., ed. 1991. *A history of regeneration research.* Cambridge, UK: Cambridge Univ. Press.

DiPietro, L.A. 2000. Thrombospondin and angiogenesis. In: S.A. Mousa, ed. *Angiogenesis inhibitors and stimulators.* Georgetown, TX: Landes Bioscience. 52–60.

DiPietro, L.A., and N.N. Nissen. 1998. Angiogenic mediators in wound healing. In: M.E. Maragoudakis, ed. *Angiogenesis: Models, modulators and clinical applications.* New York: Plenum Press. 121–128.

Discher, D.E., et al. 2005. Tissue cells feel and respond to the stiffness of their substrate. *Science* 310:1139–1143.

Djojosubroto, M.W., et al. 2003. Telomeres and telomerase in aging, regeneration and cancer. *Mol Cells* 15:164–175.

Doetsch, F. 2003. The glial identity of neural stem cells. *Nat Neurosci* 6:1127–1134.

Doherty, T.J. 2003. Invited review: Aging and sarcopenia. *J Appl Physiol* 95:1717–1727.

Doletsky, S.Ya., et al. 1976. Injuries of terminal phalanges of fingers in children (Russian). *Khirurgia* 10–16.

Donaldson, D.J., and J.T. Mahan. 1988. Keratinocyte migration and the extracellular matrix. *J Invest Dermatol* 90:623–628.

Doolabh, A.B., et al. 1996. The role of conduits in nerve repair: A review. *Rev Neurosci* 7:47–84.

Douglas, B.S. 1972. Conservative management of guillotine amputation of the finger in children. *Aust Paediatr J* 8:86–89.

Dow, D.E. 2002. Electrical stimulation of denervated EDL muscles of rats: Maintenance of mass, contractile properties and recovery following grafting. PhD Thesis. University of Michigan, Ann Arbor.

Drahota, Z., and E. Gutmann. 1961. The influence of age on the course of reinnervation of muscle. *Gerontologia* 5:88–109.

Drescher, U., et al. 1995. In vitro guidance of retinal ganglion cell axons by RAGS, a 25kDa tectal protein related to ligands for Eph receptor tyrosine kinases. *Cell* 82:359–370.

Dreyfus, P.A., et al. 2004. Adult bone marrow-derived stem cells in muscle connective tissue and satellite cell niches. *Am J Pathol* 164:773–779.

Driesch, H. 1908. *The science and philosophy of the organism.* London: Black.

Dubois, F. 1949. Contribution à l'étude de la régénération chez planaires dulcicoles. *Bull Biol* 83:213–283.

Duckmanton, A., et al. 2005. A single-cell analysis of myogenic dedifferentiation induced by small molecules. *Chem Biol* 12:1117–1126.

Durchon, M., and R. Marcel. 1962. Influence du cerveau sur la régénération postérieure chez *Nereis diversicolor* O.F. Müller (Annélide Polychète). *CR Séanc Soc Biol* 156:661–663.

Echeverri, K., and E. Tanaka. 2002a. Ectoderm to mesoderm lineage switching during axolotl tail regeneration. *Science* 298:1993–1996.

Echeverri, K., and E.M. Tanaka. 2002b. Mechanisms of muscle dedifferentiation during regeneration. *Semin Cell Dev Biol* 13:353–360.

Echeverri, K., and E.M. Tanaka. 2005. Proximodistal patterning during limb regeneration. *Dev Biol* 279:391–401.

Echeverri, K.E., et al. 2001. In vivo imaging indicates muscle fiber dedifferentiation is a major contributor to the regenerating tail blastema. *Dev Biol* 236:151–164.

Efimov, M.I. 1931. Materials for the investigation of the mechanics of the regenerative process. I. The role of the skin in the process of organ regeneration in the axolotl (Russian). *Zh Exp Biol* 7:352.

Efimov, M.I. 1933. Materials for the investigation of the mechanics of the regenerative process. II. The role of the skin in the process of organ regeneration in the axolotl (Russian). *Biol Zhur* 2:214–219.

Egar, M., and M. Singer. 1972. The role of ependyma in spinal cord regeneration in the urodele, *Triturus. Exp Neurol* 37:422–430.

Eglitis, M.A., and E. Mezey. 1997. Hematopoietic cells differentiate into both microglia and macroglia in the brains of adult mice. *Proc Natl Acad Sci USA* 94:4080–4085.

Eguchi, G. 1998. Transdifferentiation as the basis of eye lens regeneration. In P. Ferretti and J. Géraudie, eds. *Cellular and molecular basis of regeneration*. Chichester, UK: Wiley, 207–229.

Eisen, A.Z., and J. Gross. 1965. The role of the epithelium and mesenchyme in the production of a collagenolytic enzyme and a hyaluronidase in the anuran tadpole. *Dev Biol* 121:408–418.

Endo, T., et al. 2000. Analysis of gene expressions during *Xenopus* forelimb regeneration. *Dev Biol* 220:296–306.

Engel, A.G., and G. Biesecker. 1982. Universal involvement of complement in muscle fiber necrosis. In: D.L. Scotland, ed. *Disorders of the motor unit*. New York: Wiley. 535–546.

Epstein, W.L., and D.J. Sullivan. 1964. Epidermal mitotic activity in wounded human skin. In: W. Montagna and R.E. Billingham. *Advances in biology of skin*. Vol. 5. New York: Pergamon Press. 68–75.

Faulkner, J.A., et al. 1995. Muscle atrophy and weakness with aging: Contraction-induced injury as an underlying mechanism. *J Gerontol (Biol Sci)* 50A:124–129.

Fausto, N. 2004. Liver regeneration and repair: Hepatocytes, progenitor cells and stem cells. *Hepatology* 39:1477–1487.

Fausto, N., et al. 2006. Liver regeneration. *Hepatology* 43:S45–S53.

Fawcett, J.W., and R.J. Keynes. 1990. Peripheral nerve regeneration. *Annu Rev Neurosci* 13:43–60.

Fekete, D.M., and J.P. Brockes. 1987. A monoclonal antibody detects a difference in the cellular composition of developing and regenerating limbs of newts. *Development* 99:589–602.

Feldman, J.I., and F.E. Stockdale. 1992 Temporal appearance of satellite cells during myogenesis. *Dev Biol* 153:217–226.

Ferguson, M.W.J., and S. O'Kane. 2004. Scar-free healing: From embryonic mechanisms to adult therapeutic intervention. *Philos Trans R Soc Lond B* 359:839–850.

Ferrari, G., and F. Mavilio. 2002. Myogenic stem cells from the bone marrow: A therapeutic alternative for muscular dystrophy? *Neuromuscul Disord* 12:S7–S10.

Ferrari, G., et al. 1998. Muscle regeneration by bone marrow-derived myogenic precursors. *Science* 279:1528–1530.

Ferretti, P., and J.P. Brockes. 1988. Culture of newt cells from different tissues and their expression of a regeneration-associated antigen. *J Exp Zool* 247:77–91.

Ferretti, P., et al. 2003. Changes in spinal cord regenerative ability through phylogenesis and development: Lessons to be learnt. *Dev Dyn* 226:245–256.

Ferry, N., and M. Hadchouel. 2002. Liver regeneration: With a little help from marrow. *J Hepatol* 36:695–697.

Field, L.J. 2004. Modulation of the cardiomycete cell cycle in genetically altered animals. *Ann N Y Acad Sci* 1015:160–170.

Filbin, M.T. 2003. Myelin-associated inhibitors of axonal regeneration in the adult mammalian CNS. *Nat Rev Neurosci* 4:1–11.

Fine, E.G., et al. 2000. Nerve regeneration. In: R.P. Lanza et al., eds. *Principles of tissue engineering*. 2nd ed. San Diego: Academic Press. 785–798.

Fleming, M.W., and R.A. Tassava. 1981. Preamputation and postamputation histology of the neonatal opossum hindlimb: Implications for regeneration experiments. *J Exp Zool* 215:143–149.

Flickinger, R.A. 1959. A gradient of protein synthesis in planaria and reversal of axial polarity of regenerates. *Growth* 23:251–271.

Folkman, J., and C. Haudenschild. 1980. Angiogenesis in vitro. *Nature* 288:551–556.

Fraisse, P. 1885. *Die Regeneration von Geweben und Organen bei den Wirbelthieren, besonders Amphibien und Reptilien.* Cassel und Berlin: Verlag von Theodor Fischer.

Franceschi, R.T. 2005. Biological approaches to bone regeneration by gene therapy. *J Dent Res* 84:1093–1103.

Frankel, J. 1974. Positional information in unicellular organisms. *J Theor Biol* 47:439–481.

Freeman, G. 1963. Regeneration from the cornea in *Xenopus laevis. J Exp Zool* 154:39–66.

French, V. 1976. Leg regeneration in the cockroach, *Blatella germanica.* II. Regeneration from a non-congruent tibial graft/host junction. *J Embryol Exp Morphol* 35:267–301.

French, V., et al. 1976. Pattern regulation in epimorphic fields. *Science* 193:969–981.

Friedenstein, A.Ya., and K.S. Lalikina. 1973. *Induction of bony tissue and osteogenic progenitor cells* (Russian). Moscow: Meditsina.

Fritsch, C. 1911. Experimentelle Studien über Regenerationsvorgänge des Gliedmassenskelets der Amphibien. *Zool Jahrb Abt allg Zool u Physiol* 30:377–472.

Frost, H.M. 1973. *Orthopaedic biomechanics.* Springfield, IL: Charles C. Thomas.

Frost, H.M. 1989. The biology of fracture healing: An overview for clinicians. Part I. *Clin Orthopaed* 248:283–293.

Fugleholm, K., et al. 1994. Early peripheral nerve regeneration after crushing, sectioning, and freeze studied by implanted electrodes in the cat. *J Neurosci* 14:2659–2673.

Fukada, E., and I. Yasuda. 1957. On the piezoelectric effect of bone. *J Phys Soc Japan* 10:1158–1169.

Gage, F.H. 1998. Stem cells of the nervous system. *Curr Opin Neurobiol* 8:671–676.

Gage, F.H. 2000. Mammalian neural stem cells. *Science* 287:1433–1438.

Gailit, J., and R.A.F. Clark. 1994. Wound repair in the context of extracellular matrix. *Curr Opin Cell Biol* 6:717–725.

Galili, U. 1993. Interaction of the natural anti-Gal antibody with a-galactosyl epitopes: A major obstacle for xenotransplantation in humans. *Immunol Today* 14:480–482.

Galle, S., et al. 2005. The homeobox gene *Msx* in development and transdifferentiation of jellyfish striated muscle. *Int J Dev Biol* 49:961–967.

Gardiner, D.M., and S.V. Bryant. 1996. Molecular mechanisms in the control of limb regeneration: The role of homeobox genes. *Int J Dev Biol* 40:797–805.

Gardiner, D.M., et al. 1995. Regulation of *HoxA* expression in developing and regenerating axolotl limbs. *Development* 121:1731–1741.

Gardiner, D.M., et al. 1999. Towards a functional analysis of limb regeneration. *Semin Cell Dev Biol* 10:385–393.

Gardiner, D.M., et al. 2002. The molecular basis of amphibian limb regeneration: Integrating the old with the new. *Semin Cell Dev Biol* 13:345–352.

Gardner, S.E., et al. 1999. Effect of electrical stimulation on chronic wound healing: A meta-analysis. *Wound Repair Regen* 7:495–503.

Garlick, J.A., and L.B. Taichman. 1994. Fate of human keratinocytes during reepithelialization in an organotypic culture model. *Lab Invest* 70:916–924.

Gartner, L.P., and J.L. Hiatt. 1997. *Color textbook of histology.* Philadelphia: W.B. Saunders.

Gawronska-Kozak, B. 2004. Regeneration in the ears of immunodeficient mice: Identification and lineage analysis of mesenchymal stem cells. *Tissue Eng* 10:1251–1265.

Geesink, R.G.T., et al. 1999. Osteogenic activity of OP1 bone morphogenetic protein (BMP7) in a human fibular defect. *J Bone Joint Surg* 81 B:710–718.

Geiger, H., and G. van Zant. 2002. The aging of lympho-hematopoietic stem cells. *Nat Immunol* 3:329–333.

George, A.J.T., and R.I. Lechler. 2002. Xenotransplantation: Will pigs fly? In: J.M. Polak et al., eds. *Future strategies for tissue and organ replacement.* London: Imperial College Press. 215–236.

Géraudie, J., and P. Ferretti. 1998. Gene expression during amphibian limb regeneration. *Int Rev Cytol* 180:1–50.

Ghooray, G.T., and G.F. Martin. 1993. The development of myelin in the spinal cord of the North American opossum and its possible role in loss of rubrospinal plasticity: A study using myelin basic protein and galactocerebroside immunohistochemistry. *Brain Res Dev Brain Res* 72:67–74.

Giacometti, L. 1967. The healing of skin wounds in primates. I. The kinetics of cell proliferation. *J Invest Dermatol* 48:133–137.

Gibbins, J.R. 1978. Epithelial migration in organ culture: A morphological and time lapse cinematographic analysis of migrating stratified squamous epithelium. *Pathology* 10:207–218.

Gibson, M.C., and E. Schultz. 1983. Age-related differences in absolute numbers of skeletal muscle satellite cells. *Muscle Nerve* 6:574–580.

Gidge, N.M., and S.M. Rose. 1944. The role of larval skin in promoting limb regeneration in adult Anura. *J Exp Zool* 97:71–93.

Gierer, A., and H. Meinhardt. 1972. A theory of biological pattern formation. *Kybernetic* 12:30–39.

Gierer, A., et al. 1972. Regeneration of Hydra from cell aggregates. *Nat New Biol* 239:98–101.

Gillen, G., et al. 1997. Gene expression in nerve regeneration. *Neuroscientist* 3:112–122.

Globus, M. 1988. A neuromitogenic role for substance P in urodele limb regeneration. In: S. Inoue et al., eds. *Regeneration and development.* Maebashi: Okada Printing & Publ. 675–685.

Glowacki, J., et al. 1983. Cell shape and phenotypic expression in chondrocytes. *Proc Soc Exp Biol Med* 172:93–98.

Godlewski, E. 1928. Untersuchungen über Auslösung und Hemmung der Regeneration beim Axolotl. *Arch f Entw-mech* 114:108–143.

Goldfarb, A.J. 1909. The influence of the nervous system in regeneration. *J Exp Zool* 7:643–722.

Golding, D.W. 1967. Neurosecretion and regeneration in *Nereis.* I. Regeneration and the role of the supraesophageal ganglion. *Gen Comp Endocrinol* 8:348–355.

Goldman, S.A., and F. Nottebohm. 1983. Neuronal production, migration and differentiation in a vocal control nucleus of the adult female canary brain. *Proc Natl Acad Sci USA* 80:2390–2394.

Goldring, K., et al. 2002. Muscle stem cells. *J Pathol* 197:457–467.

Goodson, W.H., and T.K. Hunt. 1979. Wound healing and aging. *J Invest Dermatol* 73:88–91.

Goss, R.J. 1953. Regeneration in anuran forelimb following removal of the radio-ulna. *Anat Rec* 115:311 (abstract).

Goss, R.J. 1954. The role of the central cartilaginous rod in the regeneration of the catfish taste barbel. *J Exp Zool* 181–199.

Goss, R.J. 1956a. Regenerative inhibition following limb amputation and immediate insertion into the body cavity. *Anat Rec* 126:15–27.

Goss, R.J. 1956b. The regenerative responses of amputated limbs to delayed insertion into the body cavity. *Anat Rec* 126:283–298.

Goss, R.J. 1956c. An experimental analysis of taste barbel regeneration in the catfish. *J Exp Zool* 131:27–49.

Goss, R.J. 1957. The relation of skin to defect regulation in regenerating half limbs. *J Morphol* 100:547–564.

Goss, R.J. 1964a. The role of the skin in antler regeneration. *Adv Biol Skin* 5:194–207.

Goss, R.J. 1964b. *Adaptive growth.* London: Academic Press.

Goss, R.J. 1966. Hypertrophy vs. hyperplasia. *Science* 153:1615–1620.

Goss, R.J. 1969. *Principles of regeneration.* New York: Academic Press.

Goss, R.J. 1978. *The physiology of growth.* New York: Academic Press.

Goss, R.J. 1981. Tissue interactions in mammalian regeneration. In: R.O. Becker, ed. *Mechanisms of growth control.* Springfield, IL: C.C. Thomas. 12–26.

Goss, R.J. 1983. *Deer antlers: Regeneration, function, evolution.* New York: Academic Press.

Goss, R.J., and L.N. Grimes. 1972. Tissue interactions in the regeneration of rabbit ear holes. *Am Zool* 12:151–157.

Goss, R.J., and L.N. Grimes. 1975. Epidermal downgrowths in regenerating rabbit ear holes. *J Morphol* 146:533–542.

Goulet, F., et al. 2000. Tendons and ligaments. In: R.P. Lanza, et al., eds. *Principles of tissue engineering.* 2nd ed. San Diego: Academic Press. 711–722.

Gourevitch, D., et al. 2003. Matrix metalloproteinase activity correlates with blastema formation in the regenerating MRL mouse ear hole model. *Dev Dyn* 226:377–387.

Govender, S., et al. 2002. Recombinant human bone morphogenetic protein 2 for treatment of open fractures: A prospective, controlled, randomized study of four hundred fifty patients. *J Bone Joint Surg* 84 A:2123–2134.

Grand, D.S., et al. 1992. Interaction of endothelial cells with a laminin A chain peptide (SIKVAV) in vitro and induction of angiogenic behavior in vivo. *J Cell Physiol* 153:614–625.

GrandPré, T., et al. 2002. Nogo-66 receptor antagonist peptide promotes axonal regeneration. *Nature* 417:547–551.

Green, S.A. 2003. The Ilizarov method. In: B.D. Browner et al., eds. *Skeletal trauma: Basic science management, and reconstruction.* 3rd ed. Philadelphia: W.B. Saunders. 605–638.

Griffin, K.J.P., et al. 1987. A monoclonal antibody stains myogenic cells in regenerating newt muscle. *Development* 101:267–278.

Grillo, H.C., and J. Gross. 1967. Collagenolytic activity during mammalian wound repair. *Dev Biol* 15:300–317.

Grillo, H.C., et al. 1968. Collagenolytic activity in regenerating forelimbs of the adult newt *(Triturus viridescens). Dev Biol* 17:571–583.

Grim, M., and B.M. Carlson. 1974. A comparison of morphogenesis of muscles of the forearm and hand during ontogenesis and regeneration in the axolotl *(Ambystoma mexicanum). Z Anat Entw-Gesch* 1445:137–167.

Grim, M., et al. 1986. Enzymatic differentiation of arterial and venous segments of the capillary bed during the development of free muscle grafts in the rat. *Am J Anat* 177:149–159.

Grimpe, B., and J. Silver. 2002. The extracellular matrix in axon regeneration. *Prog Brain Res* 137:333–349.

Grisham, J.W. 1962. Morphologic study of deoxyribonucleic acid synthesis and cell proliferation in regenerataing rat liver: Autoradiography with thymidine-H^3. *Cancer Res* 22:842–849.

Grothe, C., and G. Nikkhah. 2001. The role of basic fibroblast growth factor in peripheral nerve regeneration. *Anat Embryol* 204:171–177.

Grove, G.L., and A.M. Kligman. 1983. Age-associated changes in human epidermal cell renewal. *J Gerontol* 38:137–142.

Gulati, A.K. 1985. Basement membrane component changes in skeletal muscle transplants undergoing regeneration or rejection. *J Cell Biochem* 27:337–346.

Gulati, A.K., and M.S. Swamy. 1991. Regeneration of skeletal muscle in streptozotocin-induced diabetic rats. *Anat Rec* 229:298–304.

Gulati, A.K., et al. 1983. Changes in the basement membrane zone components during skeletal muscle fiber degeneration and regeneration. *J Cell Biol* 97:957–962.

Gunderson, K., and T. Eken. 1992. The importance of frequency and amount of electrical stimulation for contractile properties of denervated rat muscles. *Acta Physiol Scand* 145:49–57.

Gussoni, E., et al. 2002. Long-term persistence of donor nuclei in a Duchenne muscular dystrophy patient receiving bone marrow transplantation. *J Clin Invest* 110:807–814.

Guth, L. 1958. Taste buds on the cat's circumvallate papilla after reinnervation by glossopharyngeal, vagus, and hypoglossal nerves. *Anat Rec* 130:25–37.

Guth, L., et al. 1980. Criteria for evaluating spinal cord regeneration experiments. *Exp Neurol* 69:1–3.

Gutmann, E., and B.M. Carlson. 1975. Contractile and histochemical properties of regenerating cross-transplanted fast and slow muscles in the rat. *Pflügers Arch* 353:227–239.

Gutmann, E., et al. 1942. The rate of regeneration of nerves. *J Exp Biol* 19:14–44.

Gutowska, A., et al. 2001. Injectable gels for tissue engineering. *Anat Rec* 263:342–349.

Guyénot, E. 1927. La perte du pouvoir régénérateur des Anoures, étudiée par les hétérogreffes, et la notion de territories. *Rev Suisse de Zool* 343:1–54.

Guyénot, E., et al. 1948. L'exploration du territoire de la patte antérieure du Triton. *Rev Suisse de Zool* 55:1–120.

Habal, M.B., and A.H. Reddi, eds. 1992. *Bone grafts and bone substitutes.* Philadelphia: W.B. Saunders.

Ham, A.W., and W.R. Harris. 1956. Repair and transplantation of bone. In: G.H. Bourne, ed. *The biochemistry and physiology of bone.* New York: Academic Press. 475–506.

Hämmerling, J. 1963. Nucleo-cytoplasmic interactions in *Acetabularia* and other cells. *Ann Rev Plant Physiol* 14:65–92.

Han, M., et al. 2003. Digit regeneration is regulated by Msx1 and BMP4 in fetal mice. *Development* 130:5123–5132.

Han, M., et al. 2005. Limb regeneration in higher vertebrates: Developing a roadmap. *Anat Rec* 287B:14–24.

Han, M.-J., et al. 2001. Expression patterns of FGF-8 during development and limb regeneration of the axolotl. *Dev Dyn* 220:40–48.

Hansen-Smith, F.M. 1986. Formation of acetylcholine receptor clusters in mammalian sternohyoid muscle regenerating in the absence of nerves. *Dev Biol* 118:129–140.

Hansen-Smith, F.M., et al. 1980. Revascularization of the freely grafted extensor digitorum longus muscle in the rat. *Am J Anat* 158:65–82.

Harrisingh, M.C., et al. 2004. The Ras/Raf/ERK signaling pathway drives Schwann cell dedifferentiation. *EMBO J* 23:3061–3071.

Hartley, R.S., et al. 1992. Skeletal muscle satellite cells appear during late chicken embryogenesis. *Dev Biol* 153:206–216.

Hartwell, S.W. 1955. *The mechanisms of healing in human wounds.* Springfield, IL: C.C. Thomas.

Harty, M., et al. 2003. Regeneration or scarring: An immunologic perspective. *Dev Dyn* 226:268–279.

Hašek, M., and V. Hašková. 1953. *Biologie: Učební Text pro Zdravotnické Školy*. Praha: Státní Pedagogické Nakladatelství, Praha.

Hassink, R.J., et al. 2004. Human stem cells shape the future of cardiac regeneration research. *Int J Cardiol* 95(suppl 1):S20–S22.

Hawke, T.J., and D.J. Garry. 2001. Myogenic satellite cells: Physiology to molecular biology. *J Appl Physiol* 91:534–551.

Hay, E.D. 1959. Microscopic observations of muscle dedifferentiation in regenerating *Amblystoma* limbs. *Dev Biol* 1:555–585.

Hay, E.D. 1970. Regeneration of muscle in the amputated amphibian limb. In: A. Mauro et al., eds. *Regeneration of striated muscle, and myogenesis*. Amsterdam: Excerpta Medica. 3–24.

Hay, E.D., and D. Fischman 1961. Origin of the blastema in regenerating limbs of the newt *Triturus viridescens*. *Dev Biol* 3:26–59.

Hayashi, T., et al. 2004. FGF2 triggers iris-derived lens regeneration in newt eye. *Mech Dev* 121:519–526.

He, Z., and V. Koprivica. 2004. The Nogo signaling pathway for regeneration block. *Annu Rev Neurosci* 27:341–368.

Heber-Katz, E. 1999. The regenerating mouse ear. *Semin Cell Dev Biol* 10:415–419.

Heber-Katz, E., et al. 2004. Spallanzani's mouse: A model of restoration and regeneration. In: E. Heber-Katz, ed. *Regeneration: Stem cells and beyond*. Berlin: Springer. 165–189.

Hellmich, W. 1930. Untersuchungen über Herkunft und Determination des regenerativen Materials bei Amphibien. *Arch f Entw-mech* 21:135–203.

Henry, G., et al. 2003. Migration of human keratinocytes in plasma and serum and wound re-epithelialisation. *Lancet* 361:574–576.

Herbst, C. 1896. Über die Regeneration von antennenähnlichen Organen an Stelle von Augen. *Arch f Entw-mech* 3:544–558.

Herbst, C. 1902. Über die Regeneration von antennenähnlichen Organen an Stelle von Augen. *Arch f Entw-mech* 13:436–447.

Herlant-Meewis, H. 1964. Regeneration in annelids. *Adv Morphogen* 4:155–215.

Heumann, R., et al. 1987. Changes of nerve growth factor synthesis in non-neuronal cells in response to sciatic nerve transection. *J Cell Biol* 104:1623–1631.

Hibbard, E. 1963. Regeneration of the severed spinal cord of chordate larvae of *Petromyzon marinus*. *Exp Neurol* 7:175–185.

Hicklin, J., and L. Wolpert. 1973. Positional information and pattern regulation in Hydra: The effect of radiation. *J Embryol Exp Morphol* 30:741–752.

Hierlihy A.M., et al. 2002. The post-natal heart contains a myocardial stem cell population. *FEBS Lett* 530:239–243.

Higgins, G.M., and R.M. Anderson. 1931. Experimental pathology of the liver. I. Restoration of the liver of the white rat following partial surgical removal. *Arch Pathol* 12:186–202.

Hill, S.D. 1970. Origin of the regeneration blastema in polychaete annelids. *Am Zool* 10:101–112.

Hill, S.D. 1972. Caudal regeneration in the absence of a brain in two species of sedentary polychaetes. *J Embryol Exp Morphol* 28:667–680.

Hobmayer, B., et al. 2000. WNT signaling molecules act in axis formation in the diploblastic metazoan Hydra. *Nature* 407:186–189.

Hodde, J. 2002. Naturally occurring scaffolds for soft tissue repair and regeneration. *Tissue Eng* 8:295–308.

Höke, A. 2005. Proteoglycans in axonal regeneration. *Exp Neurol* 195:273–277.

Holder, N., and R. Klein. 1999. Eph receptors and ephrins: Effectors of morphogenesis. *Development* 126:2033–2044.

Hollinger, J.O., et al. 2005. *Bone tissue engineering.* Boca Raton, FL: CRC Press.

Hollister, S.J., et al. 2005. Design and fabrication of bone tissue engineering scaffolds. In: J.O. Hollinger et al., eds. *Bone tissue engineering.* Boca Raton, FL: CRC Press. 167–192.

Holm, A., and L. Zachariae. 1974. Fingertip lesions in evaluation of conservative treatment vs. free skin grafting. *Acta Orthop Scand* 45:382–392.

Holtzer, H. 1959. The development of mesodermal axial structures in regeneration and embryogenesis. In: C.S. Thornton, ed. *Regeneration in vertebrates.* Chicago: Univ. Chicago Press. 15–33.

Holtzer, H. 1963. Mitosis and cell transformations. In: D. Mazia and A. Tyler, eds. *General physiology of cell specialization.* New York: McGraw-Hill. 80–90.

Holtzer, H., et al. 1955. An experimental analysis of the development of the spinal column. IV. Morphogenesis of tail vertebrae during regeneration. *J Morph* 96:145–172.

Holtzer, S.W. 1956. The inductive activity of the spinal cord in urodele tail regeneration. *J Morphol* 99:1–39.

Horackova, M., et al. 2004. Cell transplantation for treatment of acute myocardial infarction: Unique capacity for repair by skeletal muscle satellite cells. *Am J Physiol Heart Circ Physiol* 287:H1599–H1608.

Hornsby, P.J. 2001. Cell proliferation in mammalian aging. In: E.J. Masaro and S.N. Austad, eds. *Handbook of the biology of aging.* 5th ed. San Diego: Academic Press. 207–245.

Horton, D.L. 1967. The effect of age on hair growth in the CBA mouse: Observations on transplanted skin. *J Gerontol* 22:43–45.

Hsu, L. 1974. The role of nerves in the regeneration of minced skeletal muscle in the adult Anurans. *Anat Rec* 179:119–136.

Huang, D.W., et al. 1999. A therapeutic vaccine approach to stimulate axon regeneration in the adult mammalian spinal cord. *Neuron* 24:639–647.

Huang, J.K., et al. 2005. Glial membranes at the node of Ranvier prevent neurite outgrowth. *Science* 310:1813–1817.

Hubbell, J.A. 2000. Matrix effects. In: R.P. Lanza, et al., eds. *Principles of tissue engineering.* 2nd ed. San Diego: Academic Press, 237–250.

Hudson, T.W., et al. 1999. Engineering strategies for peripheral nerve repair. *Clin Plast Surg* 26:617–628.

Huggins, C.B. 1931. The formation of bone under the influence of epithelium of the urinary tract. *Arch Surg* 22:377–408.

Hunt, D., et al. 2002. The Nogo receptor, its ligands and axonal regeneration in the spinal cord: A review. *J Neurocytol* 31:93–120.

Hunt, R.K., and M. Jacobson. 1974. Neuronal specificity revisited. *Curr Top Dev Biol* 8:203–259.

Hynes, R.O. 1992. Integrins: Versatility, modulation and signaling in cell adhesion. *Cell* 69:11–25.

Iakova, P., et al. 2003. Aging reduces proliferative capacities of liver by switching pathways of C/EBPα growth arrest. *Cell* 113:495–506.

Ide, C. 1996. Peripheral nerve regeneration. *Neurosci Res* 25:101–121.

Ihara, S., et al. 1990. Ontogenetic transition of wound healing pattern in rat skin occurring at the fetal stage. *Development* 110:671–680.

Ikegami, Y., et al. 2002. Neural cell differentiation from retinal pigment epithelial cells of the newt: An organ culture model for the Urodele retinal regeneration. *J Neurobiol* 50:209–220.

Illingworth, C.M. 1974. Trapped fingers and amputated finger tips in children. *J Pediatr Surg* 9:853–858.

Imokawa Y, Yoshizato K. 1997. Expression of *sonic hedgehog* gene in regenerating newt limb blastemas recapitulates that in developing limb buds. *Proc Natl Acad Sci USA* 94:9159–9164.

Imokawa, Y., et al. 2004. A critical role for thrombin in vertebrate lens regeneration. *Philos Trans R Soc Lond B Biol Sci* 359:765–776.

Ingber, D.E., and J. Folkman. 1989a. Tension and compression as basic determinants of cell form and function: Utilization of a cellular tensegrity mechanism. In: W.D. Stein and F. Bronner, eds. *Cell shape: Determinants, regulation, and regulatory role.* San Diego: Academic Press. 3–31.

Ingber, D.E., and J. Folkman. 1989b. Mechanochemical switching between growth and differentiation during fibroblast growth factor-stimulated angiogenesis in vitro: Role of extracellular matrix. *J Cell Biol* 109:317–330.

Iten, L.E., and S.V. Bryant. 1973. Forelimb regeneration from different levels of amputation in the newt, *Notophthalmus viridescens:* Length, rate and stages. *Wilhelm Roux' Arch* 173:263–282.

Iten, L.E., and S.V. Bryant. 1975. The interaction between the blastema and stump in the establishment of the anterior-posterior and proximal-distal organization of the limb regenerate. *Dev Biol* 44:119–147.

Itescu, S. 2004. Strategies using cell therapy to induce cardiomyocyte regeneration in adults with heart disease. In: S. Sell, ed. *Stem cells handbook.* Totawa, NJ: Humana Press. 251–258.

Ito, M., et al. 1999. Lens formation by pigmented epithelial cell reaggregate from the dorsal iris implanted into the limb blastema in the adult newt. *Dev Growth Differ* 41:429–440.

Ivanova-Kazac, O.M. 1977. *Asexual reproduction of animals* (Russian). Leningrad: Izdatel. Leningrad Univ.

Jackson, K.A., et al. 2001. Regeneration of ischemic cardiac muscle and vascular endothelium by adult stem cells. *J Clin Invest* 107:1395–1402.

Jankowski, R.J., et al. 2002. Muscle-derived stem cells. *Gene Ther* 9:642–647.

Jockusch, H., et al. 1983. Beating heart muscle in a skeletal muscle bed. *Exp Neurol* 81:749–755.

Johnson, S.E., and R.E. Allen. 1990. The effects of bFGF, IGF-1, and TGF-β on RMo skeletal muscle cell proliferation and differentiation. *Exp Cell Res* 187:250–254.

Ju, B.-G., and W.-S. Kim. 1998. Upregulation of cathepsin D expression in the dedifferentiating salamander limb regenerates and enhancement of its expression by retinoic acid. *Wound Repair Regen* 6:349–357.

Kadi, F., et al. 2004. Satellite cells and myonuclei in young and elderly women and men. *Muscle Nerve* 29:120–127.

Kamińska, A., and A. Fidziańska. 1990. Basement membrane component changes during experimental skeletal muscle degeneration and regeneration. *Neuropatol Pol* 28:141–150.

Kamrin, R.P., and M. Singer. 1955a. The influence of the spinal cord in regeneration of the tail of the lizard, *Anolis carolinensis. J Exp Zool* 128:611–628.

Kamrin, R.P., and M. Singer. 1955b. The influence of the nerve on regeneration and maintenance of the barbel of the catfish, *Ameiurus nebulosus*. *J Morphol* 96:173–188.

Kanatani, H. 1958. Formation of bipolar heads induced by demecolcine in the planarian, *Dugesia gonocephala*. *J Fac Sci Univ Tokyo* Section 4.8(pt 2):254–270.

Kastin, A.J., and W. Pan. 2005. Targeting neurite growth inhibitors to induce CNS regeneration. *Curr Pharm Des* 11:1247–1253.

Kato, K., et al. 1999. The role of dorsoventral interaction in the onset of planarian regeneration. *Development* 126:1031–1040.

Kato, K., et al. 2001. Dorsal and ventral positional cues required for the onset of planarian regeneration may reside in differentiated cells. *Development* 233:109–121.

Kato, T., et al. 2003. Unique expression patterns of matrix metalloproteinases in regenerating newt limbs. *Dev Dyn* 226:366–376.

Katogi, R., et al. 2004. Large-scale analysis of the genes involved in fin regeneration and blastema formation in the medaka, *Oryzias latipes*. *Mech Dev* 121:861–872.

Katoh, M., et al. 2004. An orderly retreat: Dedifferentiation is a regulated process. *Proc Natl Acad Sci USA* 101:7005–7010.

Kempermann, G., et al. 2004. Functional significance of adult neurogenesis. *Curr Opin Neurobiol* 14:186–191.

Khrestchatisky, M., et al. 1988. Expression of muscle actin genes in early differentiation stages of tail regeneration of the urodele amphibian *Pleurodeles waltlii*. *Cell Differ Dev* 25:203–212.

Khrushchov, G.K., ed. 1954. *Questions about the restoration of organs and tissues in vertebrate animals* (Russian). Moscow: Izdatel. Akad. Nauk SSSR.

Kiernan, J.A. 1979. Hypotheses concerned with axonal regeneration in the mammalian nervous system. *Biol Rev* 54:155–197.

Kiffmeyer, W.R., et al. 1991. Axonal transport and release of transferrin in nerves of regenerating amphibian limbs. *Dev Biol* 147:392–402.

Kimura, Y., et al. 2003. Expression of complement 3 and complement 5 in newt limb and lens regeneration. *J Immunol* 170:2331–2339.

King, M.W., et al. 2003. Identification of genes expressed during *Xenopus laevis* limb regeneration by using subtractive hybridization. *Dev Dyn* 226:398–409.

Kiortsis, V., and M. Moraitou. 1965. Factors of regeneration in *Spirographis spallanzani*. In: V. Kiortsis and H.A.L. Trampusch, eds. *Regeneration in animals and related problems*. Amsterdam: North-Holland. 250–261.

Kirk, S., et al. 2000. Myostatin regulation during skeletal muscle regeneration. *J Cell Physiol* 184:356–363.

Kléber, M., and L. Sommer. 2004. Wnt signaling and the regulation of stem cell function. *Curr Opin Cell Biol* 16:681–687.

Kleeberger, W., et al. 2002. High frequency of epithelial chimerism in liver transplants demonstrated by microdissection and STR-analysis. *Hepatology* 35:110–116.

Kleinman, H.K., et al. 1982. Isolation and characterization of type IV procollagen, laminin, and heparin sulfate proteoglycan from EHS sarcoma. *Biochemistry* 221:6188–6193.

Klein-Nulend, J., et al. 1995. Pulsating fluid flow increases nitric oxide (NO) synthesis by osteocytes but not periosteal fibroblasts: Correlation with prostaglandin upregulation. *Biochem Biophys Res Commun* 217:640–648.

Kolosova, A.A. 1961. Reactive changes of fish heart tissue as a result of its injury (Russian). *Doklad Akad Nauk SSSR* 138:1443–1445.

Konieczna-Marczynska, B., and A. Skowron-Cendrzak. 1958. The effect of the augmented nerve supply on the regeneration in postmetamorphic *Xenopus laevis*. *Folia Biol* 6:37–46.

Kopen, G.C., et al. 1999. Marrow stromal cells migrate throughout forebrain and cerebellum, and they differentiate into astrocytes after injection into neonatal mouse brains. *Proc Natl Acad Sci USA* 96:10711–10716.

Koprivica, V., et al. 2005. EGFR activation mediates inhibition of axon regeneration by myelin and chondroitin sulfate proteoglycans. *Science* 310:106–110.

Körbling, M., and Z. Estrov. 2003. Adult stem cells for tissue repair: A new therapeutic concept? *N Engl J Med* 349:570–582.

Körbling, M., et al. 2002. Hepatocytes and epithelial cells of donor origin in recipients of peripheral-blood stem cells. *N Engl J Med* 346:738–746.

Korschelt, E. 1927. *Regeneration und transplantation.* Berlin: Gebrüder Borntraeger. B.M. Carlson, trans. 1931. *Regeneration and transplantation.* Vols. I, II-1, and II-2. Canton, MA: Science History Publication, Watson Publishing International.)

Kosnik, P.E., et al. 2001. Functional development of engineered skeletal muscle from adult and neonatal rats. *Tissue Eng* 7:573–584.

Krause, D.S., et al. 2001. Multi-organ, multi-lineage engraftment by a single bone marrow-derived stem cell. *Cell* 105:369–377.

Krawczyk, W.S. 1971. A pattern of epidermal cell migration during wound healing. *J Cell Biol* 49:247–263.

Kudokotsev, V.P. 1960. Stimulation of regeneration of limbs of tailless amphibians by the action of tissue extract (Russian). *Dokl Akad Nauk SSSR* 132:715–718.

Kudokotsev, V.P. 1962. Stimulation of the regenerative process in lizard limbs by the method of supplementary innervation (Russian). *Dokl Akad Nauk SSSR* 142:233–236.

Kudokotsev, V.P. 1964. Stimulation of the regenerative process of mammalian limbs by the action of tissue extract (Russian). *Nauch Dokl Vissh Skol (Biol Nauk)* 3:40–43.

Kudokotsev, V.P. 1965. The influence of supplementary innervation on the regeneration of limbs in tailless amphibians (Russian). *Vestn Kharkov Univ Ser Biol* 1:119–122.

Kudokotsev, V.P., and V.A. Kuntsevich. 1965. Stimulation of restorative processes by the method of treating the wound surface with trypsin and calcium chloride after amputation of external organs in mammals (Russian). *Byull Exp Biol Med* 9:106–109.

Kuhn, H.G., et al. 1996. Neurogenesis in the dentate gyrus of the adult rat: Age-related decrease of neuronal progenitor proliferation. *J Neurosci* 16:2027–2033.

Kumar, A., et al. 2000. Plasticity of retrovirus-labelled myotubes in the newt limb regeneration blastema. *Dev Biol* 218:125–136.

Kumar, A., et al. 2004. The regenerative plasticity of isolated urodele myofibers and its dependence on Msx1. *PLOS Biol* 2:E218.

Kurabuchi, S. 1992a. Effects of an augmented nerve supply on forelimb regeneration in the adult mud frog, *Rana rugosa. J Exp Zool* 264:75–81.

Kurabuchi, S. 1992b. Relationship between innervation and forelimb regenerative capacity in the postmetamorphic pond frog *Rana brevipoda porosa. Int J Dev Biol* 36:429–433.

LaBarge, M.A., and H.M. Blau. 2002. Biological progression from adult bone marrow to mononucleate muscle stem cell to multinucleate muscle fiber in response to injury. *Cell* 111:589–601.

Laflamme, M.A., and C.E. Murry. 2005. Regenerating the heart. *Nat Biotechnol* 23:845–856.

Laflamme, M.A., et al. 2002. Evidence for cardiomyocyte repopulation by extracardiac progenitors in transplanted human hearts. *Circ Res* 90:634–640.

Lakatos, A., and R.J. Franklin. 2002. Transplant mediated repair of the central nervous system: An imminent solution? *Curr Opin Neurol* 15:701–705.

Lanza, R.P., et al., eds. 2000. *Principles of tissue engineering.* 2nd ed. San Diego: Academic Press.

Larsson, L., and T. Ansved. 1995. Effects of aging on the motor unit. *Prog Neurobiol* 445:397–458.

Lash, J.W. 1955. Studies on wound closure in urodeles. *J Exp Zool* 128:13–26.

Lash, J.W. 1956. Experiments in epithelial migration during the closure of wounds in urodeles. *J Exp Zool* 131:239–256.

Laugwitz K.-L., et al. 2005. Postnatal isli+ cardioblasts enter fully differentiated cardiomyocyte lineages. *Nature* 433:647–653.

Lavine L.S., et al. 1977. Treatment of congenital pseudarthrosis of the tibia with direct current. *Clin Orthopaed* 124:69–74.

LeCouter, J., and N. Ferrara. 2002. EG-VEGF and the concept of tissue-specific angiogenic growth factors. *Semin Cell Dev Biol* 13:3–9.

Lee, T.I., et al. 2006. Control of developmental regulators by polycomb in human embryonic stem cells. *Cell* 125:301–313.

Leferovich, J., et al. 2001. Heart regeneration in adult MRL mice. *Proc Natl Acad Sci USA* 98:9830–9835.

Lender, T. 1974. La régénération animale. Presses Univ. de France, Vendôme, France.

Lenhoff, H.M., and S.G. Lenhoff. 1991. Abraham Trembley and the origins of research on regeneration in animals. In: C.D. Dinsmore, ed. *A history of regeneration research.* Cambridge, UK: Cambridge Univ. Press. 47–66.

Leobon, B., et al. 2003. Myoblasts transplanted into rat infarcted myocardium are functionally isolated from their host. *Proc Natl Acad Sci USA* 100:7808–7811.

Leontovich, A.A., et al. 2000. A novel hydra metalloproteinase (HMMP) functions in extracellular matrix degradation, morphogenesis and the maintenance of differentiated cells in the foot process. *Development* 127:907–920.

Lepeshinskaya, O.B. 1945. *The origin of cells from living substance and the role of living substance in the organism* (Russian). Moscow: Izdatel. Akad. Nauk SSSR.

Lepeshinskaya, O.B. 1951. *Über die Entstehung von Zellen.* Berlin: Verlag Kultur und Fortschritt.

Lepeshinskaya, O.B. 1952. *Extracellular forms of life* (Russian). Moscow: Izdatel. Akad. Pedagogicheskikx Nauk RSFSR.

Leri, A., et al. 2005. Cardiac stem cells and mechanisms of myocardial regeneration. *Physiol Rev* 85:1373–1416.

Levander, G. 1964. *Induction phenomena in tissue regeneration.* Baltimore: Williams & Wilkins.

Levenberg, S., et al. 2005. Engineering vascularized skeletal muscle tissue. *Nat Biotechnol* 23:879–884.

Lévesque, M., et al. 2005. Expression of heat-shock protein 70 during limb development and regeneration in the axolotl. *Dev Dyn* 233:1525–1534.

Levin, M. 2003. Bioelectromagnetics in morphogenesis. *Bioelectromagnetics* 24:295–315.

Lheureux, E. 1972. Contribution a l'étude du role de la peau et des tissues axiaux du member dans la déclenchement de morphogenesis régénératrices anormales chez le triton *Pleurodeles waltlii* Michah. *Ann Embryol Morphogen* 5:165–178.

Lheureux, E. 1975a. Nouvelles données sur les roles de la peau et des tissues internes dans la régénération du membre du triton, *Pleurodeles waltlii* Michah. (urodèle): Influence des qualities et orientation de greffons non-irradiés. *Roux Arch* 176:303–327.

Lheureux, E. 1975b. Régénération des members irradiés de *Pleurodeles waltlii* Micah. (Urodèle): Influence des qualities et orientations des greffons non irradies. *Arch f Entw-mech* 176:303–327.

Lheureux, E. 1977. Importance des associations de tissus du membre dans le développement des membres surnuméraires induits par deviation de nerf chez le triton *Pleurodeles waltlii* Micah. *J Embryol Exp Morphol* 38:151–173.

Li, C., and J.M. Suttie. 2001. Deer antlerogenic periosteum: A piece of postnatally retained embryonic tissue? *Anat Embryol* 204:375–388.

Li, C., et al. 2005. Histological examination of antler regeneration in the red deer *(Cervus elaphus)*. *Anat Rec* 282A:163–174.

Li, G., et al. 2003a. c-Jun is essential for organization of the epidermal leading edge. *Dev Cell* 4:865–877.

Li, W.-Y., et al. 2003b. Plasminogen activator/plasmin system: A major player in wound healing? *Wound Repair Regen* 11:239–247.

Libbin, R.M., and M. Weinstein. 1986. Regeneration of growth plates in the long bones of the neonatal rat limb. *Am J Anat* 177:369–383.

Libbin, R.M., et al. 1979. Partial regeneration of the above-elbow amputated rat forelimb. II. Electrical and mechanical facilitation. *J Morphol* 159:439–451.

Lieberman, J.R., et al. 1999. The effect of regional gene therapy with bone morphogenetic protein-2-producing bone-marrow cells on the repair of segmental femoral defects in rats. *J Bone Joint Surg* 81 A:905–917.

Limke, T.L., and M.S. Rao. 2003. Neural stem cell therapy in the aging brain: Pitfalls and possibilities. *J Hematol Stem Cell Res* 12:615–623.

Lindvall, O., et al. 1990. Neural transplantation in Parkinson's disease: The Swedish experience. *Prog Brain Res* 82:729–734.

Liosner, L.D. 1931. Concerning the mechanism of the loss of regenerative capacity during the time of development of *Rana temporaria* tadpoles (Russian). *Zhur Exp Biol* 7:163–171.

Liosner, L.D., ed. 1960. *The regeneration of organs in mammals* (Russian). Moscow: Medgiz.

Liosner, L.D., ed. 1972. *Conditions of organ regeneration in mammals* (Russian). Moscow: Izdatel. Meditsina.

Liozner, L.D. 1974. *Organ regeneration* (English translation of *Conditions of organ regeneration in mammals* [Russian], Moscow: Meditsina, 1972). New York: Consultants Bureau.

Litver, G.M., et al. 1961. Organic regeneration of skeletal muscles in rats (Russian). *Byull Exp Biol Med* 52:101–105.

Locatelli, P. 1929. Der Einfluss des Nervensystems auf die Regeneration. *Arch f Entw-mech* 114:686–770.

Lohmann, J.U., and T.C. Bosch. 2000. The novel peptide HEADY specifies apical fate in a simple radially symmetric metazoan. *Genes Dev* 14:2771–2777.

Lois, C., and A. Alvarez-Buylla. 1993. Proliferating subventricular zone cells in the adult mammalian forebrain can differentiate into neurons and glia. *Proc Natl Acad Sci USA* 90:2074–2077.

Lois, C., and A. Alvarez-Buylla. 1994. Long-distance neuronal migration in the adult mammalian brain. *Science* 264:1145–1148.

Longaker, M.T., and N.S. Adzick. 1991. The biology of fetal wound healing: A review. *Plast Reconstr Surg* 87:788–798.

Lorenz, H.P., et al. 1992. Scarless wound repair: A human fetal skin model. *Development* 114:253–259.

Lund, E.J. 1925. Experimental control of organic polarity by the electric current. V. The nature and control of organic polarity by the organic current. *J Exp Zool* 41:155–190.

Lund, E.J. 1947. *Bioelectric fields and growth*. Austin, TX: Univ. of Texas Press.

Luskin, M.B. 1993. Restricted proliferation and migration of postnatally generated neurons derived from the forebrain subventricular zone. *Neuron* 11:173–189.

Ma, J., et al. 2005. Time course of myocardial stromal cell-derived factor 1 expression and beneficial effects of intravenously administered bone marrow stem cells in rats with experimental myocardial infarction. *Basic Res Cardiol* 100:217–223.

MacKay, E.M., et al. 1932. The degree of compensatory renal hypertrophy following unilateral nephrectomy. I. The influence of age. *J Exp Med* 565:255–265.

Macklin, W.B., and C.L. Weill. 1985. Appearance of myelin proteins during development in the chick central nervous system. *Dev Neurosci* 7:170–178.

MacWilliams, H.K. 1983a. *Hydra* transplantation phenomena and the mechanism of *Hydra* head regeneration. I. Properties of head inhibition. *Dev Biol* 96:217–238.

MacWilliams, H.K. 1983b. *Hydra* transplantation phenomena and the mechanism of *Hydra* head regeneration. II. Properties of head activation. *Dev Biol* 96:239–257.

Maden, M. 1977. The regeneration of positional information in the amphibian limb. *J Theor Biol* 69:735–753.

Maden, M. 1982. Vitamin A and pattern formation in the regenerating limb. *Nature* 295:672–675.

Maden, M. 1993. The homeotic transformation of tails into limbs in *Rana temporaria* by retinoids. *Dev Biol* 159:379–391.

Maden, M., and M. Hind. 2003. Retinoic acid, a regeneration-inducing molecule. *Dev Dyn* 226:237–244.

Maden, M., and M. Hind. 2004. Retinoic acid in alveolar development, maintenance and regeneration. *Philos Trans R Soc Lond B* 359:799–808.

Madri J.A., et al. 1988. Phenotypic modulation of endothelial cells by transforming growth factor-β depends upon the composition and organization of the extracellular matrix. *J Cell Biol* 106:1375–1384.

Magavi, S.S., et al. 2000. Induction of neurogenesis in the neocortex of adult mice. *Nature* 405:951–955.

Mahan, J.T., and D.J. Donaldson. 1986. Events in the movement of newt epidermal cells across implanted substrates. *J Exp Zool* 237:35–44.

Makino, S., et al. 2005. Heat-shock protein 60 is required for blastema formation and maintenance during regeneration. *Proc Natl Acad Sci USA* 102:14599–14604.

Malinin, T.I. 1960. The effects of implantation of embryonic and tadpole tissues into adult frog limbs. II. Histological observations. *J Exp Zool* 143:1–20.

Malinin, T.I., and J.D. Deck. 1958. The effects of implantation of embryonic and tapole tissues into adult frog limbs. I. Regeneration after amputation. *J Exp Zool* 139:307–328.

Malt, R.A. 1983. Humoral factors in regulation of compensatory renal hypertrophy. *Kidney Int* 23:611–615.

Maréchal, G., et al. 1984. Isozymes of myosin in growing and regenerating rat muscles. *Eur J Biochem* 138:421–428.

Marino, A.A., ed. 1988. *Modern bioelectricity.* New York: Marcel Dekker Inc.

Markelova, I.V. 1953. Regeneration of external organs in mammals. Thesis. (Russian).

Markelova, I.V. 1960. On the question of the regeneration of external organs in mammals (Russian) (unpublished dissertation). In: I.N. Maiskii and L.D. Liosner, eds. *Questions of reparative and physiological regeneration* (Russian). Moscow: Medgiz. 122–141.

Marsh, D.R., et al. 1998. The force-frequency relationship is altered in regenerating and senescent rat skeletal muscle. *Muscle Nerve* 21:1265–1274.

Marsh, G., and H.W. Beams. 1952. Electrical control of morphogenesis in regenerating *Dugesia trigrina*. I. Relation of axial polarity to field strength. In: R.A. Flickinger, ed. *Developmental biology*. Dubuque, IA: Wm. C. Brown. 60–77.

Martin, C., and J.G. del Pino. 1998. Controversies in the treatment of fingertip amputations. *Clin Orthopaed Rel Res* 353:63–73.

Martin, G.R. 1998. The roles of FGFs in the early development of vertebrate limbs. *Genes Dev* 12:1571–1586.

Martin, P. 1997. Wound healing: Aiming for perfect skin regeneration. *Science* 276:75–81.

Massaro, G.D., and D. Massaro. 1997. Retinoic acid treatment abrogates elastase-induced pulmonary emphysema in rats. *Nat Med* 3:675–677.

Matinyan, L.A., and A.S. Andreasyan. 1973. *Enzymotherapy in organic damage of the spinal cord* (Russian). Yerevan, Armenia: Izdatel. Akad. Nauk Armyanskoi SSR.

Matsui, J.I., et al. 2005. Regeneration and replacement in the vertebrate inner ear. *Drug Discov Today* 10:1307–1312.

Matveeva, A.I. 1958. Dynamics of the process of regeneration of skull bone in the dog, elicited by the method of destruction (Russian). *Doklad Akad Nauk SSSR* 119:830–833.

Matveeva, A.I. 1959a. Regeneration of skull bones in dogs, elicited by the method of destruction (Russian). *Folia Biol* 7:239–257.

Matveeva, A.I. 1959b. Regeneration of skull bones in the dog through the homotransplantation of fresh, preserved and autoclaved bone fragments (Russian). *Dokl Akad Nauk SSSR* 129:460–463.

Mauro, A. 1961. Satellite cell of skeletal muscle fibers. *J Biophys Biochem Cytol* 9:493–495.

Mauro, A., ed. 1979. *Muscle regeneration*. New York: Raven Press.

Mauro, A., et al., eds. 1970. *Regeneration of striated muscle, and myogenesis*. Amsterdam: Excerpta Medica.

Maxwell, P.H., and P.J. Ratcliffe. 2002. Oxygen sensors and angiogenesis. *Semin Cell Dev Biol* 13:29–38.

Maynard, D.M. 1965. The occurrence and functional characteristics of heteromorphy antennules in an experimental population of spiny lobsters, *Panulirus argus*. *J Exp Biol* 43:79–106.

McCue, S., et al. 2004. Shear-induced reorganization and endothelial cell cytoskeleton and adhesion complexes. *Trends Cardiovasc Med* 14:143–151.

McGann, C.J., et al. 2001. Mammalian myotube dedifferentiation induced by newt regeneration extract. *Proc Natl Acad Sci USA* 98:13699–13704.

McGeachie, J.K., and M.D. Grounds. 1995. Retarded myogenic cell replication in regenerating skeletal muscles of old mice: An autoradiographic study in young and old BALBc and SJL/J mice. *Cell Tissue Res* 280:277–282.

McGinnis, M.E. 1989. The nature and effects of electricity in bone. In: R.B. Borgens, et al., eds. *Electric fields in vertebrate repair*. New York: Alan R. Liss. 225–284.

McKinney-Freeman, S.L., et al. 2002. Muscle-derived hematopoietic stem cells are hemapoietic in origin. *Proc Natl Acad Sci USA* 99:1341–1346.

McMinn, R.M.H. 1969. *Tissue repair*. New York: Academic Press.

McPherron, A.C., et al. 1997. Regulation of skeletal muscle mass in mice by a new TGF-beta superfamily member. *Nature* 387:83–90.

Medvedev, Z.A. 1969. *The rise and fall of T. D. Lysenko*. New York: Columbia Univ. Press.

Mehendale, F., and P. Martin. 2001. The cellular and molecular events of wound healing. In: V. Falanga, ed. *Cutaneous wound healing*. London: Martin Dunitz. 15–37.

Menkin, V. 1956. *Biochemical mechanisms in inflammation.* Springfield, IL: Charles C. Thomas.

Mercader, N., et al. 2005. Proximodistal identity during vertebrate limb regeneration is regulated by Meis homeodomain proteins. *Development* 132:4131–4142.

Merzkirch, C., et al. 2001. Engineering of vascular ingrowth matrices: Are protein domains an alternative to peptides? *Anat Rec* 263:379–387.

Mescher, A.L. 1976. Effects on adult newt limb regeneration of partial and complete skin flaps over the amputation surface. *J Exp Zool* 195:117–128.

Mescher, A.L. 1996. The cellular basis of limb regeneration in urodeles. *Int J Dev Biol* 40:785–795.

Mescher, A.L., and S.I. Muniam. 1988. Transferrin and the growth-promoting effect of nerves. *Int Rev Cytol* 110:1–26.

Mescher, A.L., and A.W. Neff. 2005. Regenerative capacity and the developing immune system. *Adv Biochem Eng Biotechnol* 93:39–66.

Mescher, A.L., and R.A. Tassava. 1975. Denervation affects DNA replication and mitosis during the initiation of limb regeneration in adult newts. *Dev Biol* 44:187–197.

Mescher, A.L., et al. 1997. Transferrin is necessary and sufficient for the neural effect on growth in amphibian regeneration blastemas. *Dev Growth Differ* 39:677–684.

Michalopoulos, G.K., and M.C. DeFrances. 1997. Liver regeneration. *Science* 276:60–66.

Migliaccio, A.R., et al. 1996. Molecular control of erythroid differentiation. *Int J Hematol* 64:1–29.

Milburn, A. 1976. The effect of the local anesthetic bupivacaine on the muscle spindle of rat. *J Neurocytol* 5:425–446.

Mire-Sluis, A., and R. Thorpe, eds. 1998. *Cytokines.* San Diego: Academic Press.

Mizell, M. 1968. Limb regeneration: Induction in the newborn opossum. *Science* 161:283–286.

Mizell, M., and J.J. Isaacs. 1970. Induced regeneration of hindlimbs in the newborn opossum. *Am Zool* 10:141–155.

Mizuno, N., et al. 1999a. Lens regeneration in *Xenopus* is not a mere repeat of lens development, with respect to crystallin gene expression. *Differentiation* 64:143–149.

Mizuno, N., et al. 1999b. Pax-6 and prox 1 expression during lens regeneration from *Cynops* iris and *Xenopus* cornea: Evidence for a genetic program common to embryonic lens development. *Differentiation* 65:141–149.

Mohammad, K.S., and D.A. Neufeld. 2000. Denervation retards but does not prevent toetip regeneration. *Wound Repair Regen* 8:277–281.

Mohammad, K.S., et al. 1999. Bone growth is induced by nail transplantation in amputated proximal phalanges. *Calcif Tissue Int* 65:408–410.

Mohanty-Hejmadi, P., et al. 1992. Limbs generated at site of tail amputation in marbled balloon frog after vitamin A treatment. *Nature* 355:352–353.

Moment, G.B. 1949. On the relation between growth in length, the formation of new segments, and electric potential in an earthworm. *J Exp Zool* 112:1–12.

Mong, F.S.F. 1977. Histological and histochemical studies on the influence of nerves on minced muscle regeneration of triceps surae of the rat. *J Morphol* 151:451–462.

Monroy, A. 1941. Richerche sulla correnti elettriche derivabili dalla superficie del corpo di tritoni adulti normali e durante la rigenerazine degli arti e della coda. *Pubbl Staz Zool Napoli* 18:265–281.

Mont, M.A., et al. 2004. Use of bone morphogenetic proteins for musculoskeletal applications. *J Bone Joint Surg* 86-A(suppl 2):41–55.

Montesano, R., et al. 1983. *In vitro* rapid organization of endothelial cells into capillary-like networks is promoted by collagen matrices. *J Cell Biol* 97:1648–1652.

Moon, L.D., et al. 2001. Regeneration of CNS axons back to their target following treatment of adult rat brain with chondroitinase ABC. *Nat Neurosci* 4:465–466.

Mooney, D., et al. 1992. Switching from differentiation to growth in hepatocytes: Control by extracellular matrix. *J Cell Physiol* 151:497–505.

Moore, K.A., and I.R. Lemischka. 2006. Stem cells and their niches. *Science* 311:1880–1885.

Morais da Silva, S., et al. 2002. The newt ortholog of CD59 is implicated in proximodistal identity during amphibian limb regeneration. *Dev Cell* 3:547–555.

Morgan, T.H. 1900. Regeneration in planarians. *Arch f Entw-mech* 10:58–119.

Morgan, T.H. 1901. *Regeneration*. New York: Macmillan.

Morgan, T.H. 1902. Experimental studies of the internal factors of regeneration in the earthworm. *Arch f Entw-mech* 14:562–591.

Morrison, J.I., et al. 2006. Salamander limb regeneration involves the activation of a multipotent skeletal muscle satellite cell population. *J Cell Biol* 172:433–440.

Moussian, B., and A.E. Uv. 2005. An ancient control of epithelial barrier function and wound healing. *BioEssays* 27:987–990.

Moyer, E.K., et al. 1960. Numbers of fibers in regeneration after crushing the motor spinal nerve roots of young, mature and aged cats. *Am J Anat* 107:193–207.

Mufti, S.A. 1971. Role of peripheral nerves in amphibian tail regeneration. *Pakistan J Zool* 3:127–131.

Mufti, S.A. 1973. Tail regeneration following amputation in adult *Triturus viridescens*. *Pakistan J Zool* 5:31–49.

Mufti, S.A. 1977. Regeneration following denervation of minced gastrocnemius muscles in mice. *J Neurol Sci* 33:251–266.

Mufti, S.A., and S.B. Simpson. 1972. Tail regeneration following autotomy in the adult salamander *Desmognathus fuscus*. *J Morph* 136:297–312.

Mukouyama, Y.-S., et al. 2002. Sensory nerves determine the pattern of arterial differentiation and blood vessel branching in the skin. *Cell* 109:693–705.

Mullen, L.M., et al. 1996. Nerve dependency of regeneration: The role of Dlx and FGF signaling in amphibian limb regeneration. *Development* 122:3487–3497.

Müller, P., et al. 2005. Myocardial regeneration by endogenous adult progenitor cells. *J Mol Cell Cardiol* 39:377–387.

Muller, T.L., et al. 1999. Regeneration in higher vertebrates: Limb buds and digit tips. *Semin Dev Biol* 10:405–413.

Muneoka, K., and S.V. Bryant. 1982. Evidence that patterning mechanisms in developing and regenerating limbs are the same. *Nature* 298:369–371.

Muneoka, K., and D. Sassoon. 1992. Molecular aspects of regeneration in developing vertebrate limbs. *Dev Biol* 152:37–49.

Murray, P.D.F. 1936. *Bones*. Cambridge, UK: Cambridge Univ. Press.

Murry, C.E., et al. 2002. Cellular therapies for myocardial infarct repair. *Cold Springs Harbor Symp Quant Biol* LXVII:519–526.

Murry, C.E., et al. 2004. Haematopoietic stem cells do not transdifferentiate into cardiac myocytes inmyocardial infarcts. *Nature* 428:664 668.

Nakatomi, H., et al. 2002. Regeneration of hippocampal pyramidal neurons after ischemic brain injury by recruitment of endogenous neural progenitors. *Cell* 110:429–441.

Nardi, K.B., and D.L. Stocum. 1983. Surface properties of regenerating limb cells: Evidence for gradation along the proximodistal axis. *Differentiation* 25:27–31.

Nassonov, N.V. 1941. *Supernumerary formations developing after the implantation of cartilage beneath the skin of adult tailed amphibians* (Russian). Moscow: Izdatel. Akad. Nauk SSSR.

Naville, A. 1927. La perte du pouvoir régénérateur des Anoures étudiée par les homogreffes. *Rev Suisse de Zool* 34:269–284.

Nawata, T. 2001. Wound currents following amputation of tail tip in the Japanese newt, *Cynops pyrrhogaster. Zool Sci* 18:11–15.

Needham, A.E. 1952. *Regeneration and wound healing.* London: Methuen & Co.

Needham, A.E. 1960. Regeneration and growth. In: W.W. Nowinski, ed. *Fundamental aspects of normal and malignant growth.* Amsterdam: Elsevier. 588–663.

Nesmeyanova, T.N. 1977. *Experimental studies in regeneration of spinal neurons.* Washington, DC: Winston/Wiley.

Neufeld, D.A. 1980. Partial blastema formation after amputation in adult mice. *J Exp Zool* 212:31–36.

Neufeld, D.A. 1989. Epidermis, basement membrane, and connective tissue healing after amputation of mouse digits: Implications for mammalian appendage regeneration. *Anat Rec* 223:425–432.

Neufeld, D.A. 1996. Bone regeneration after amputation stimulated by basic fibroblast growth factor *in vitro. In Vitro Cell Dev Biol Anim* 32:63–65.

Neufeld, D.A., and F.A. Day. 1996. Perspective: A suggested role for basement membrane structures during newt limb regeneration. *Anat Rec* 246:155–161.

Nguyen, Q.T., et al. 2002. Pre-existing pathways promote precise projection patterns. *Nat Neurosci* 5:861–867.

Niazi, I.A., and S. Saxena. 1978. Abnormal hind limb regeneration in tadpoles of the toad, *Bufo andersoni,* exposed to excess vitamin A. *Fol Biol (Kraków)* 26:1–8.

Nicolas, S., et al. 1999. Two *Nkx-3*-related genes are expressed in the adult and regenerating central nervous system of the urodele *Pleurodeles waltl. Dev Genet* 24:319–328.

Nikitenko, M.F. 1957. The role of the central nervous system in regeneration of the shed tail in lizards (English translation). *Doklady Akad Nauk SSSR Biol Sci* 119:186–188.

Niu, M.C., et al. 1961. Ribonucleic acid-induced changes in mammalian cells. *Proc Natl Acad Sci USA* 47:1689–1700.

Norman, W.P., and A.J. Schmidt. 1967. The fine structure of tissues in the amputated-regenerating limb of the adult newt, *Diemictylus viridescens. J Morphol* 123:271–311.

Nosrat, A.C. 1998. Neurotrophic factors in the tongue: Expression patterns, biological activity, relation to innervation and studies of neurotrophin knockout mice. *Ann N Y Acad Sci* 855:28–49.

Nuccitelli, R. 2003. A role for endogenous electric fields in wound healing. *Curr Top Dev Biol* 58:1–26.

Nüesch, H. 1968. The role of the nervous system in insect morphogenesis and regeneration. *Ann Rev Entomol* 13:27–44.

Oakley, B. 1967. Altered temperature and taste responses from cross-regenerated sensory nerves in the rat's tongue. *J Physiol (Lond)* 188:353–371.

Oberpriller, J.O., and J.C. Oberpriller. 1971. Mitosis in the adult newt ventricle. *J Cell Biol* 49:560–563.

Oberpriller, J.O., and J.C. Oberpriller. 1974. Response of the adult newt ventricle to injury. *J Exp Zool* 187:249–260.

Oberpriller, J.O., and J.C. Oberpriller. 1991. Cell division in adult newt cardiac myocytes. In: J.O. Oberpriller, et al., eds. *The development and regenerative potential of cardiac muscle.* Chur, Switzerland: Harwood. 293–311.

Oberpriller, J.O., et al. 1988. Nuclear characteristics of cardiac myocytes following the proliferative response to mincing of the myocardium in the adult newt, *Notophthalmus viridescens*. *Cell Tissue Res* 253:619–624.

Oberpriller, J.O., et al., eds.: 1991. *The development and regenerative potential of cardiac muscle*. Chur, Switzerland: Harwood.

Odelberg, S.J. 2002. Inducing cellular dedifferentiation: A potential method for enhancing endogenous regeneration in mammals. *Semin Cell Dev Biol* 13:335–343.

Odelberg, S.J. 2005. Cellular plasticity in vertebrate regeneration. *Anat Rec* 287B:25–35.

Odelberg, S.J., et al. 2000. Dedifferentiation of mammalian myotubes induced by msx1. *Cell* 103:1099–1109.

Odland, G., and R. Ross. 1968. Human wound repair. I. Epidermal regeneration. *J Cell Biol* 39:135–151.

Ogden, D.A., and H.S. Mickelm. 1976. The fate of serially transplanted bone marrow populations from young and old donors. *Transplantation* 22:287–293.

Oh, Y.-S., et al. 2004. Scar formation after ischemic myocardial injury in MRL mice. *Cardiovasc Pathol* 13:203–206.

Ojingwa, J.C., and R.R. Isseroff. 2002. Electrical stimulation of wound healing. *Prog Dermatol* 36:1–12.

Olmsted, J.M.D. 1920. The results of cutting the seventh cranial nerve in *Ameiurus nebulosus* (Leseur). *J Exp Zool* 31:369–401.

Orechowitsch, W.N., and N.W. Bromley. 1934. Die histolisierende Eigenschaften des Regenerationsblastems. *Biol Zbl* 54:524–535.

Orlic, D. 2004. The strength of plasticity: Stem calls for cardiac repair. *Int J Cardiol* 95(suppl 1):S16–S19.

Orlic, D., et al. 2001a. Bone marrow cells regenerate infarcted myocardium. *Nature* 410:701–705.

Orlic, D., et al. 2001b. Mobilized bone marrow cells repair the infarcted heart, improving function and survival. *Proc Natl Acad Sci USA* 98:10344–10349.

Orlic, D., et al. 2002. Stem cells for myocardial regeneration. *Circ Res* 91:1092–1102.

Overturf K., et al. 1997. Serial transplantation reveals the stem-cell-like regenerative potential of adult mouse hepatocytes. *Am J Pathol* 151:1273–1280.

Palermo, A.T., et al. 2005. Bone marrow contribution to skeletal muscle: A physiological response to stress. *Dev Biol* 279:336–344.

Parenteau, N.L., et al. 2000. Skin. In: R.P. Lanza, et al., eds. *Principles of tissue engineering*. 2nd ed. San Diego: Academic Press. 879–890.

Park, C.M., and M.J. Hollenberg. 1989. Basic fibroblast growth factor induces retinal regeneration in vivo. *Dev Biol* 134:201–205.

Parker, G.H., and V.L. Paine. 1934. Progressive nerve degeneration and its rate in the lateral-line nerve of the catfish. *Am J Anat* 54:1–25.

Parmacek, M.S., and J.A. Epstein. 2005. Pursuing cardiac progenitors: Regeneration redux. *Cell* 120:295–298.

Patton, B.L. 2000. Laminins of the neuromuscular system. *Microsc Res Tech* 51:247–261.

Peadon, A.M., and M. Singer. 1966. The blood vessels of the regenerating limb of the adult newt, *Triturus*. *J Morphol* 118:79–89.

Penn, M.S., et al. 2004. Role of stem cell homing in myocardial regeneration. *Int J Cardiol* 95(suppl 1):S23–S25.

Penzlin, H. 1964. Die Bedeutung des Nervensystems für die Regeneration bei den Insekten. *Arch f Entw-mech* 155:152–161.

Perrone, C.E., et al. 1995. Collagen and stretch modulate autocrine secretion of insulin-like growth factor-1 and insulin-like growth factor binding proteins from differentiated skeletal muscle cells. *J Biol Chem* 270:2099–2106.

Pescitelli, M.J., and D.L. Stocum. 1980. The origin of skeletal structures during intercalary regeneration in larval *Ambystoma* limbs. *Dev Biol* 79:255–275.

Pettigrew, R.K., and W.F. Windle. 1976. Factors in recovery from spinal cord injury. *Exp Neurol* 53:815–829.

Phillips, G.D., and D.R. Knighton. 1990. Angiogenic activity in damaged skeletal muscle. *Proc Soc Exp Biol Med* 193:197–202.

Phillips, G.D., et al. 1987. Survival of myogenic cells in freely grafted rat rectus femoris and extensor digitorum longus muscles. *Am J Anat* 180:365–372.

Phillips, G.D., et al. 1991. An angiogenic extract from skeletal muscle stimulates monocyte and endothelial cell chemotaxis *in vitro*. *Proc Soc Exp Biol Med* 197:458–464.

Piatt, J. 1957. Studies on the problem of nerve pattern. III. Innervation of the regenerated forelimb in *Amblystoma*. *J Exp Zool* 136:229–248.

Pittack, C., et al. 1991. Basic fibroblast growth factor induces retinal pigment epithelium to generate neural retina in vitro. *Development* 113:577–588.

Plaghki, L. 1985. Régénération et myogenèse du muscle strié. *J Physiol Paris* 80:51–110.

Plant, G.W., et al. 2001. Inhibitory proteoglycan immunoreactivity is higher at the caudal than the rostral Schwann cell graft-transected spinal cord interface. *Mol Cell Neurosci* 17:471–487.

Plopper, G.E., et al. 1995. Convergence of integrin and growth factor receptor signaling pathways within focal adhesion complex. *Mol Biol Cell* 6:1349–1365.

Polejaiev, L.W. 1936. Sur la restauration de la capacité régénérative chez les anoures. *Arch d'Anat Microsc* 32:439–463.

Poležajew, L.W. 1945. Chemical methods for restoring the regenerative capacity of limbs in tadpoles. *Compt Rend Acad Sci URSS* 48:216–220.

Poležajew, L.W., and W.N. Faworina. 1935. Über die Rolle des Epithels in den anfänglichen Entwicklungsstadien einer Regenerationsanlage der Extremität beim Axolotl. *Arch f Entw-mech* 133:701–727.

Polezhaev, L.V. 1933a. Concerning the renewal of regenerative capacity in tailless amphibians (Russian). *Biol Zhur* 2:357–367.

Polezhaev, L.V. 1933b. Concerning processes of resorption, proliferation and relations of tissues during regeneration of limbs in axolotls (Russian). *Biol Zhur* 2:368–386.

Polezhaev, L.V. 1936. The role of the epithelium in regeneration and normal ontogenesis of limbs in amphibian. *Zool Zhur* 15:277–291.

Polezhaev, L.V. 1945a. Chemical methods of restoration of regenerative capacity of limbs in tadpoles (Russian). *Dokl Akad Nauk SSSR* 48:232–236.

Polezhaev, L.V. 1945b. *The fundamentals of vertebrate developmental mechanics* (Russian). Moscow: Izdatel. Akad. Nauk SSSR.

Polezhayev, L.V. 1946. The loss and restoration of regenerative capacity in the limbs of tailless amphibia. Biol Rev 21:141–147.

Polezhaev, L.V. 1946. Morphological data on regenerative capacity in tadpole limbs as restored by chemical agents. *Compt Rend Acad Sci URSS* 554:281–284.

Polezhaev, L.V. 1950. Some principles in the study of regeneration (Russian). *Zhur Obsch Biol* 11:253–273.

Polezhaev, L.V. 1951. The filling-in of bony defects of the skull in mice (Russian). *Doklad Akad Nauk SSSR* 77:525–528.

Polezhaev, L.V. 1972a. *Organ regeneration in animals*. Springfield, IL: Charles C. Thomas.

Polezhaev, L.V. 1972b. *Loss and restoration of regenerative capacity in tissues and organs of animals* (translation of a Russian monograph published in 1968). Cambridge, MA: Harvard Univ. Press.

Polezhaev, L.V. 1977a. Restoration of regenerative capacity suppressed by roentgen irradiation (regeneration of limbs in amphibians) (Russian). *Uspekh Sovrem Biol* 84:96–112.

Polezhaev, L.V. 1977b. *Regeneration by means of induction* (Russian). Moscow: Meditsina.

Polezhaev, L.V. 1979. The morphogenetic potency of the regeneration blastema (Russian). *Uspekh Sovrem Biol* 9:277–292.

Polezhaev, L.V. 1980. Regeneration of digits in children (Russian). *Chirurgia* 12:76–77.

Polezhaev, L.V. 1982. The repair of defects of the skull by regenerating bone (Russian). *Voprosi Neirochirurg* 2:53–57.

Polezhaev, L.V., and G.P. Ramenskaya. 1950. Regeneration of limbs in fire-bellied toads, elicited by products of the hydrolyzation of cartilage (Russian). *Dokl Akad Nauk SSSR* 70:141–144.

Polezhaev, L.V., et al. 1965. *Stimulation of regeneration of cardiac muscle* (Russian). Moscow: Izdatel. Nauka.

Polezhayev, L.V. 1946. The loss and restoration of regenerative capacity in the limbs of tailless amphibia. *Biol Rev* 21:141–147.

Pollack, S.R. 2001. Streaming potentials in bone. In: S.C. Cowin, ed. *Bone mechanics handbook*. 2nd ed. Boca Raton, FL: CRC Press. 24-1–24-22.

Pollin, M.M., et al. 1991. The effect of age on motor neurone death following axotomy in the mouse. *Development* 112:83–89.

Poole, B. 1966. The stimulus to hypertrophic growth. *Adv Morphogen* 5:93–129.

Popiela, H. 1976. Muscle satellite cells in urodele amphibians: Facilitated identification of satellite cells using ruthenium red staining. *J Exp Zool* 198:57–64.

Poss, K.D., et al. 2002. Heart regeneration in zebrafish. *Science* 298:2188–2190.

Prehn, R.T. 1970. Immunosurveillence, regeneration, and oncogenesis. *Prog Exp Tumor Res* 14:1–24.

Profyris, C., et al. 2004. Degenerative and regenerative mechanisms governing spinal cord injury. *Neurobiol Dis* 15:415–436.

Przibram, H. 1909. *Experimental-Zoologie. 2. Regeneration*. Leipzig: Franz Deuticke.

Purves, D., and J.W. Lichtman. 1985. *Principles of neural development*. Sunderland, MA: Sinauer Associates.

Quaini, F., et al. 2002. Chimerism of the transplanted heart. *N Engl J Med* 346:5–15.

Quinones, J.L., et al. 2002. Extracellular matrix remodeling and metalloproteinase involvement during intestine regeneration in the sea cucumber *Holothuria glaberrima*. *Dev Biol* 250:181–197.

Rafferty, N.S., and R. Smith. 1976. Analysis of cell populations of normal and injured mouse lens epithelium. I. Cell cycle. *Anat Rec* 186:105–114.

Rafii, S., et al. 1995. Characterization of hematopoietic cells arising on the textured surface of left ventricular assist devices. *Ann Thorac Surg* 60:1627–1632.

Rafii, S., et al. 2002. Contribution of marrow-derived progenitors to vascular and cardiac regeneration. *Semin Cell Dev Biol* 13:61–67.

Rageh, M.A.E., et al. 2002. Vasculature in pre-blastema and nerve-dependent blastema stages of regenerating forelimbs of the adult newt, *Notophthalmus viridescens*. *J Exp Zool* 292:255–266.

Raisman, G. 2004. Olfactory ensheathing cells and repair of brain and spinal cord injuries. *Clon Stem Cells* 6:364–368.

Ramer, L.M., et al. 2005. Setting the stage for functional repair of spinal cord injuries: A cast of thousands. *Spinal Cord* 43:134–161.

Ramon-Cueto, A., et al. 1998. Long-distance axonal regeneration in the transected adult rat spinal cord is promoted by olfactory ensheathing glia transplants. *J Neurosci* 18:3803–3815.

Rasmussen, H.S. 2000. Matrix metalloproteinase inhibitors in angiogenesis-mediated disorders with special emphasis on cancer. In: S.A. Mousa, ed. *Angiogenesis inhibitors and stimulators.* Georgetown, TX: Landes Bioscience. 124–133.

Rattan, S.I.S., and A. Derventzi. 1991. Altered cellular responsiveness during ageing. *BioEssays* 13:601–606.

Rawlinson, S.C.F., et al. 1991. Loading-related increases in prostaglandin production in cores of adult canine cancellous bone *in vitro:* A role for prostaglandin in adaptive remodeling? *J Bone Miner Res* 6:1345–1357.

Raya, A., et al. 2003. Activation of Notch signaling pathway precedes heart regeneration in zebrafish. *Proc Natl Acad Sci USA* 100:11889–11895.

Raya, A., et al. 2004. The zebrafish as a model of heart regeneration. *Clon Stem Cells* 6:345–351.

Reddien, P.W., et al. 2005. Identification of genes needed for regeneration, stem cell function, and tissue homeostasis by systematic gene perturbation in planaria. *Dev Cell* 8:635–649.

Reginelli, A.D., et al. 1995. Digit tip regeneration correlates with regions of Msx1 (Hox 7) expression in fetal and newborn mice. *Development* 121:1065–1076.

Reis, R.L., and J.S. Román, eds. 2005. *Biodegradable systems in tissue engineering and regenerative medicine.* Boca Raton, FL: CRC Press.

Renault V., et al. 2000. Skeletal muscle regeneration and the mitotic clock. *Exp Gerontol* 35:711–719.

Renault V., et al. 2002. Regenerative potential of human skeletal muscle during aging. *Aging Cell* 1:132–139.

Repesh, L.A., and J.C. Oberpriller. 1978. Scanning electron microscopy of epidermal cell migration in wound healing during limb regeneration in the adult newt, *Notophthalmus viridescens. Am J Anat* 151:539–556.

Repesh, L.A., and J.C. Oberpriller. 1980. Ultrastructural studies on migrating epidermal cells during the wound healing stage of regeneration in the adult newt, *Notophthalmus viridescens. Am J Anat* 159:187–208.

Reyer, R.W. 1954. Regeneration of the lens in the amphibian eye. *Q Rev Biol* 29:1–46.

Reyer, R.W. 1956. Lens regeneration from homoplastic and heteroplastic implants of dorsal iris into the eye chamber of *Triturus viridescens* and *Amblystoma punctatum. J Exp Zool* 133:145–190.

Reyer, R.W. 1962. Regeneration in the amphibian eye. In: D. Rudnick, ed. *Regeneration.* New York: Ronald Press. 211–265.

Reyer, R.W. 1977. The amphibian eye: Development and regeneration. In: F. Crescitelli, ed. *Handbook of sensory physiology. Vol. VII/5. The visual system of vertebrates.* Berlin: Springer Verlag. 309–390.

Reyer, R.W., et al. 1973. Stimulation of lens regeneration from the newt dorsal iris when implanted into the blastema of the regenerating limb. *Dev Biol* 32:258–281.

Richards, C.M., et al. 1975. Regeneration of digits and forelimbs in the Kenyan reed frog, *Hyperolius viridiflavus ferniquei. J Morphol* 146:431–445.

Richards, C.M., et al. 1977. A scanning electron microscopic study of differentiation of the digital pad in regenerating digits of the Kenyan reed frog, *Hyperolius viridiflavus ferniquei*. *J Morphol* 153:387–396.

Richardson, P.M., et al. 1980. Axons from CNS neurons regenerate into PNS grafts. *Nature* 284:264–265.

Ricklefs, R.E., and C.E. Finch. 1995. *Aging: A natural history*. New York: Scientific America Library.

Risau, W. 1997. Mechanisms of angiogenesis. *Nature* 386:671–674.

Roberts, A.B. 2000. Transforming growth factor-β. In: E. Canalis, ed. *Skeletal growth factors*. Philadelphia: Lippincott Williams & Wilkins. 221–232.

Roberts, P., et al. 1997. The host environment determines strain-specific differences in the timing of skeletal muscle regeneration: Cross-transplantation studies between SJL/J and BALB/c mice. *J Anat* 191:585–594.

Robinson, K.R., and M.R. Messerli. 2003. Left/right, up/down: The role of endogenous electrical fields as directional signals in development, repair and invasion. *BioEssays* 25:759–766.

Rogal, I.G. 1951. Stimulation of differentiation during limb regeneration in fire-bellied toads (Russian). *Dokl Akad Nauk SSSR* 81:953–956.

Rogers, S.L. 1982. Muscle spindle formation and differentiation in regenerating rat muscle grafts. *Dev Biol* 94:265–283.

Rogers, S.L., and B.M. Carlson. 1981. A quantitative assessment of muscle spindle formation in reinnervated and non-reinnervated grafts of the rat extensor digitorum longus muscle. *Neuroscience* 6:87–94.

Roguski, H. 1957. Wpływwrdzenia kręgowego na regeneracjęogona larw płazow ogoniastych i bezogonowych. *Folia Biol (Warsaw)* 5:249–266.

Rohovsky, S., and P.A. D'Amore. 1997. Growth factors and angiogenesis in wound healing. In: T.R. Ziegler et al., eds. *Growth factors and wound healing*. New York: Springer. 8–26.

Romanes, G.J. 1946. Motor localization and the effects of nerve injury on the ventral horn cells of the spinal cord. *J Anat* 80:117–131.

Romanova, L.K. 1971. *Experimental and clinical regeneration of the lungs* (Russian). Moscow: Izdatel. Meditsina.

Romero, R., and D. Bueno. 2001. Disto-proximal regional determination and intercalary regeneration in planarians, revealed by retinoic acid induced disruption of regeneration. *Int J Dev Biol* 45:669–673.

Rosania, G.R., et al. 2000. Myoseverin, a microtubule-binding molecule with novel cellular effects. *Nat Biotechnol* 18:304–308.

Rose, F.C., and S.M. Rose. 1947. Limb regeneration in frogs stimulated by heat shock. *Anat Rec* 99:653 (abstract).

Rose, F.R.A.J., et al. 2004. Delivery systems for bone growth factors: The new players in skeletal regeneration. *J Pharm Pharmacol* 56:415–427.

Rose, S.M. 1942. A method for inducing limb regeneration in adult anura. *Proc Soc Exp Biol Med* 49:408–410.

Rose, S.M. 1944. Methods of initiating limb regeneration in adult Anura. *J Exp Zool* 95:149–170.

Rose, S.M. 1945. The effect of NaCl in stimulating regeneration of limbs of frogs. *J Morphol* 77:119–139.

Rose, S.M. 1962. Tissue-arc control of regeneration in the amphibian limb. In: D. Rudnick, ed. *Regeneration*. New York: Ronald Press. 153–176.

Rose, S.M. 1970. *Regeneration: Key to understanding normal and abnormal growth and development.* New York: Appleton-Century-Crofts.

Rudnicki, M.A. 2003. Marrow to muscle, fission vs. fusion. *Nat Med* 9:1461–1462.

Rudolph, K.L., et al. 2000. Inhibition of experimental liver cirrhosis in mice by telomerase gene delivery. *Science* 287:1253–1258.

Ruegg, M.A. 1996. Agrin, laminin β2 (s-laminin) and ARIA: Their role in neuromuscular development. *Curr Opin Neurobiol* 6:97–103.

Rumyantsev, P.P. 1961. Evidence of regeneration of significant parts of myocardial fibers of frogs after trauma (Russian). *Arkh Anat Gist Embriol* 40:65–74.

Rumyantsev, P.P. 1973. Post-injury DNA synthesis, mitosis and ultrastructural reorganization of adult frog cardiomyocytes: An electron microscopic-autoradiographic study. *Z Zellforsch* 139:431–450.

Rumyantsev, P.P. 1977. Interrelations of the proliferation and differentiation processes during cardiac myogenesis and regeneration. *Int Rev Cytol* 51:187–273.

Rumyantsev, P.P. 1991. *Growth and hyperplasia of cardiac muscle cells.* London: Harwood Academic Publication.

Rumyantsev, P.P., and A.M. Kassem. 1976. Cumulative indices of DNA synthesizing myocytes in different compartments of the working myocardium and conductive system of the rat's heart muscle following extensive left ventricle infarction. *Virchows Arch B* 20:329–342.

Ruoslahti, E. 1991. Integrins. *J Clin Invest* 887:1–5.

Rzehak, K., and M. Singer. 1966. The number of nerve fibers in the limb of the mouse and its relation to regenerative capacity. *Anat Rec* 155:537–540.

Sadeh, M. 1988. Effects of aging on skeletal muscle regeneration. *J Neurol Sci* 87:67–74.

Sage, C., et al. 2005. Proliferation of functional hair cells in vivo in the absence of the retinoblastoma protein. *Science* 307:1114–1118.

Sakaguchi, D.S., et al. 1997. Basic fibroblast growth factor (FGF-2) induced transdifferentiation of retinal pigment epithelium: Generation of neurons and glia. *Dev Dyn* 209:387–398.

Saló, E., and J. Baguñà. 1989. Regeneration and pattern formation in planarians. II. Local origin and role of cell movements in blastema formation. *Development* 107:69–76.

Saltzman, W.M. 2000. Cell interactions with polymers. In: R.P. Lanza, et al., eds. *Principles of tissue engineering.* 2nd ed. San Diego: Academic Press. 221–235.

Samarova, V. 1940. Limb potency of the skin of the axolotl (Russian). *Byull Exp Biol Med* 10:228–232.

Samchukov, M.L., et al. 2001. *Craniofacial distraction osteogenesis.* St. Louis: Mosby.

Sanes, J.R. 1994. The extracellular matrix. In: A.E. Engel and C. Franzini-Armstrong, eds. *Myology.* 2nd ed. New York: McGraw-Hill. 242–260.

Sanes, J.R. 2003. The basement membrane/basal lamina of skeletal muscle. *J Biol Chem* 278:12601–12604.

Sanes, J.R., et al. 1978. Reinnervation of muscle fiber basal lamina after removal of muscle fibers. *J Cell Biol* 78:176–198.

Sauer, H., et al. 2004. Development of the cardiovascular system in embryoid bodies derived from embryonic stem cells. In: S. Sell, ed. *Stem cell handbook.* Totowa, NJ: Humana Press. 229–238.

Saunders, N.R., et al. 1998. Development of walking, swimming and neuronal connections after complete spinal cord transection in the neonatal opossum, *Monodelphis domestica. J Neurosci* 18:339–355.

Savard, P., et al. 1988. Position dependent expression of a homeobox gene transcript in relation to amphibian limb regeneration. *EMBO J* 7:4275–4282.

Sawamoto, K., et al. 2006. New neurons follow the flow of cerebrospinal fluid in the adult brain. *Science* 311:629–632.

Schächinger, V., et al. 2004. Transplantation of progenitor cells and regeneration enhancement in acute myocardial infarction. *J Am Coll Cardiol* 44:1690–1699.

Schaller, H.C. 1973. Isolation and characterization of a low-molecular-weight substance activating head and bud formation in hydra. *J Embryol Exp Morphol* 29:27–38.

Scharf, A. 1961. Experiments on regenerating rat digits. *Growth* 25:7–23.

Scharf, A. 1963. Reorganization of cornified nail-like outgrowths related with the wound healing process of the amputation sites of young rat digits. *Growth* 27:255–269.

Scherer, S.S., and J.L. Salzer. 2001. Axon-Schwann cell interactions during peripheral nerve degeneration and regeneration. In: K.R. Jessen and W.D. Richardson, eds. *Glial cell development*. 2nd ed. Oxford: Oxford Univ. Press. 299–330.

Scheuing, M.R., and M. Singer. 1957. The effects of microquantities of beryllium ion on the regenerating forelimb of the adult newt, *Triturus*. *J Exp Zool* 136:301–326.

Schmalbruch, H. 1984. Motoneuron death after sciatic nerve section in newborn rats. *J Comp Neurol* 224:252–258.

Schmid, V. 1992. Transdifferentiation in medusae. *Int Rev Cytol* 142:213–261.

Schmid, V., and S. Reber-Müller. 1995. Transdifferentiation of isolated striated muscle of jellyfish *in vitro*: The initiation process. *Semin Cell Biol* 6:109–116.

Schmidt, A.J. 1966. The molecular basis of regeneration: Enzymes. *Ill Monogr Med Sci* 6:1–78.

Schmidt, A.J. 1968. *Cellular biology of vertebrate regeneration and repair*. Chicago: Univ. of Chicago Press.

Schmidt, C.E., and J.B. Leach. 2003. Neural tissue engineering: Strategies for repair and regeneration. *Annu Rev Biomed Eng* 5:293–347.

Schmidt, J.T., et al. 1978. Expansion of the half retinal projection to the tectum in goldfish: An electrophysiological and anatomical study. *J Comp Neurol* 177:257–278.

Schnapp, E., and E.M. Tanaka. 2005. Quantitative evaluation of morpholino-mediated protein knockdown of GFP, MSX1, and PAX7 during tail regeneration in *Ambystoma mexicanum*. *Dev Dyn* 232:162–170.

Schnapp, E., et al. 2005. Hedgehog signaling controls dorsoventral patterning, blastema cell proliferation and cartilage induction during axolotl tail regeneration. *Development* 132:3243–3253.

Schnell, L., and M.E. Schwab. 1990. Axonal regeneration in the rat spinal cord produced by an antibody against myelin-associated neurite growth inhibitors. *Nature* 343:269–272.

Schotté, O.E. 1926. Système nerveux et régénération chez le Triton. *Rev Suisse Zool* 33:1–211.

Schotté, O.E., and E.G. Butler. 1941. Morphological effects of denervation and amputation of limbs in urodele larvae. *J Exp Zool* 87:279–322.

Schotté, O.E., and E.G. Butler. 1944. Phases in regeneration of the urodele limb and their dependence upon the nervous system. *J Exp Zool* 97:95–121.

Schotté, O.E., and M. Harland. 1943. Amputation level and regeneration in limbs of late *Rana clamitans* tadpoles. *J Morphol* 73:329–363.

Schotté, O.E., and C.B. Smith. 1961. Effects of ACTH and of cortisone upon amputational wound healing processes in mice digits. *J Exp Zool* 146:209–230.

Schotté, O.E., and J.F. Wilber. 1958. Effects of adrenal transplants upon forelimb regeneration in normal and hypophysectomized adult frogs. *J Embryol Exp Morphol* 6:247–269.

Schotté, O.E., et al. 1941. Effects of transplanted blastemas on amputated nerveless limbs of urodele larvae. *Proc Soc Exp Biol Med* 48:500–503.

Schultz, E., and B.H. Lipton. 1982. Skeletal muscle satellite cells: Changes in proliferation potential as a function of age. *Mech Ageing Dev* 20:377–383.

Schultz, G., et al. 1991. EGF and TGF-α in wound healing and repair. *J Cell Biochem* 45:346–352.

Schwab, M.E. 2004. Nogo and axon regeneration. *Curr Opin Neurobiol* 14:118–124.

Seale, P., et al. 2000. Pax-7 is required for the specification of myogenic satellite cells. *Cell* 102:777–786.

Seeherman, H., and J.M. Wozney. 2005. Delivery of bone morphogenetic proteins for orthopedic tissue regeneration. *Cytokine Growth Factor Rev* 16:329–345.

Selzer, M.E. 2003. Promotion of axonal regeneration in the injured CNS. *Lancet Neurol* 2:157–166.

Shavlakadze, T., et al. 2005. Reconciling data from transgenic mice that overexpress IGF-I specifically in skeletal muscle. *Growth Horm IGF Res* 15:4–18.

Shen, Q., et al. 2004. Endothelial cells stimulate self-renewal and expand neurogenesis of neural stem cells. *Science* 304:1338–1340.

Sherwood, R.I., et al. 2004. Isolation of adult mouse myogenic progenitors: Functional heterogeneity of cells within and engrafting skeletal muscle. *Cell* 119:543–554.

Shimizu, I., et al. 1990. Anatomical and functional recovery following spinal cord transection in the chick embryo. *J Neurobiol* 21:918–937.

Sicard, R.E. 1985. Leukocytic and immunological influence on regeneration of amphibian forelimbs. In: R.E. Sicard, ed. *Regulation of vertebrate limb regeneration*. New York: Oxford Univ. Press. 128–145.

Sidman, R.L., and M. Singer. 1951. Stimulation of forelimb regeneration in the newt, *Triturus viridescens*, by a sensory nerve supply isolated from the central nervous system. *Am J Physiol* 165:257–260.

Sidorova, V.F. 1976. *Age and the restorative capacity of mammalian organs* (Russian). Moscow: Meditzina.

Sidorova, V.F. 1978. *The postnatal growth and restoration of internal organs in vertebrates* (Translation of a Russian monograph published in 1969). Littleton, MS: PSG Publishing Co.

Sieber-Blum, M., et al. 2004. Pluripotential neural crest stem cells in the adult hair follicle. *Dev Dyn* 231:258–269.

Simon, H.-G., and C.J. Tabin. 1993. Analysis of *Hox-4.5* and *Hox-3.6* expression during newt limb regeneration: Differential regulation of paralogous *Hox* genes suggest different roles for members of different *Hox* clusters. *Development* 117:1397–1407.

Simon, H.-G., et al. 1997. A novel family of T-box genes in urodele amphibian limb development and regeneration: Candidate genes involved in vertebrate forelimb/hindlimb patterning. *Development* 124:1355–1366.

Simpson, S.B. 1961. Induction of limb regeneration in the lizard, *Lygosoma laterale*, by augmentation of the nerve supply. *Proc Soc Exp Biol Med* 107:108–111.

Simpson, S.B. 1964. Analysis of tail regeneration in the lizard, *Lygosoma laterale*. I. Initiation of regeneration and cartilage differentiation: The role of ependyma. *J Morphol* 114:425–435.

Simpson, S.B., and P.G. Cox. 1967. Vertebrate regeneration: Culture *in vitro*. *Science* 157:1330–1332.

Singer, A.J., and R.A.F. Clark. 1999. Cutaneous wound healing. *N Engl J Med* 341:738–746.

Singer, M. 1943. The nervous system and regeneration of the forelimb of adult *Triturus*. II. The role of the sensory supply. *J Exp Zool* 92:297–315.

Singer, M. 1952. The influence of the nerve in regeneration of the amphibian extremity. *Q Rev Biol* 27:169–200.

Singer, M. 1954. Induction of regeneration of the forelimb of the postmetamorphic frog by augmentation of the nerve supply. *J Exp Zool* 126:419–471.

Singer, M. 1959. The influence of nerves on regeneration. In: C.S. Thornton, ed. *Regeneration in vertebrates*. Chicago: Univ. Chicago Press. 59–80.

Singer, M. 1961. Induction of regeneration of body parts in the lizard, *Anolis*. *Proc Soc Exp Biol Med* 107:106–108.

Singer, M. 1965. A theory of the trophic nervous control of amphibian limb regeneration, including a re-evaluation of quantitative nerve requirements. In: V. Kiortsis and H.A.L. Trampusch, eds. *Regeneration in animals and related problems*. Amsterdam: North-Holland Publication Co. 20–32.

Singer, M. 1978. On the nature of the neurotrophic phenomenon in urodele limb regeneration. *Am Zool* 18:829–841.

Singer, M., and M.M. Salpeter. 1961. Regeneration in vertebrates: The role of the wound epithelium. In: M.X. Zarrow, ed. *Growth in living systems*. New York: Basic Books. 277–311.

Singer, M., et al. 1957. The influence of denervation upon trauma-induced regenerates of the forelimb of the postmetamorphic frog. *J Exp Zool* 136:35–51.

Singer, M., et al. 1967. The relation between the caliber of the axon and the trophic activity of nerves in limb regeneration. *J Exp Zool* 166:89–98.

Singer, M., et al. 1987. Open finger tip healing and replacement after distal amputation in Rhesus monkey with comparison to limb regeneration in lower vertebrates. *Anat Embryol* 177:29–36.

Sisken, B.F. 1988. Effects of electromagnetic fields on nerve regeneration. In A.A. Marino, ed. *Modern bioelectricity*. New York: Marcel Dekker, Inc. 497–527.

Sisken, B.F., et al. 1979. A comparison of the effects of direct current, nerve growth factor, and direct current plus nerve growth factor on amputated rat limbs. In: C.T. Brighton et al., eds. *Electrical properties of bone and cartilage*. New York: Grune & Stratton. 267–287.

Skuk, D., and J.P. Tremblay. 2003. Myoblast transplantation: The current status of a potential therapeutic tool for myopathies. *J Muscle Res Cell Motil* 24:285–300.

Slack, J.M.W., et al. 2004. Cellular and molecular mechanisms of regeneration in *Xenopus*. *Philos Trans R Soc Lond B* 359:745–751.

Smith, S.D. 1967. Induction and partial limb regeneration in *Rana pipiens* by galvanic stimulation. *Anat Rec* 158:89–97.

Smith, S.D. 1974. Effects of electrode placement on stimulation of adult frog limb regeneration. *Ann N Y Acad Sci* 238:500–507.

Smith, S.D., and G.L. Crawford. 1969. Initiation of regeneration in adult *Rana pipiens* by injection of homologous liver nuclear RNP. *Oncology* 23:299–307.

Snow, M.H. 1977. Myogenic cell formation in regenerating rat skeletal muscle injured by mincing. II. An autoradiographic study. *Anat Rec* 188:201–217.

Song, B., et al. 2002. Electrical cues regulate the orientation and frequency of cell division and the rate of wound healing *in vivo*. *Proc Natl Acad Sci USA* 99:13577–13582.

Soukup, T., and M. Novotová. 2000. Ultrastructure and innervation of regenerated intrafusal muscle fibres in heterochronous isografts of fast rat muscle. *Acta Neuropathol* 100:435–444.

Soukup, T., and L.-E. Thornell. 1997. Expression of myosin heavy chain isoforms in regenerated muscle spindle fibres after muscle grafting in young and adult rats–Plasticity of intrafusal satellite cells. *Differentiation* 62:179–186.

Soyfer, V.N. 1994. *Lysenko and the tragedy of soviet science.* New Brunswick, NJ: Rutgers Univ. Press.

Spallanzani, A. 1769. *An essay on animal reproductions.* London: T. Becket and P. A. de Hondt.

Speer, D.P. 1981. The pathogenesis of amputation stump overgrowth. *Clin Orthop Relat Res* 159:294–307.

Spemann, H. 1938. *Embryonic development and induction.* New Haven, CT: Yale Univ. Press.

Spencer, T., et al. 2003. New roles for old proteins in adult CNS axonal regeneration. *Curr Opin Neurobiol* 13:133–139.

Sperry, R.W. 1943. Effect of 180° rotation of the retinal field on visuomotor coordination. *J Exp Zool* 92:263–279.

Sperry, R.W. 1944. Optic nerve regeneration with return of vision in anurans. *J Neurophysiol* 7:57–69.

Sperry, R.W. 1963. Chemoaffinity in the orderly growth of nerve fiber patterns and connections. *Proc Natl Acad Sci USA* 50:703–710.

Sta Iglesia, D.D., and J.W. Vanable. 1998. Endogenous lateral electrical fields around bovine corneal lesions are necessary for and can enhance normal rates of wound healing. *Wound Repair Regen* 6:531–542.

Sta Iglesia, D.D., et al. 1996. Electric field strength and epithelialization in the newt (*Notophthalmus viridescens*). *J Exp Zool* 274:56–62.

Stahl, B., et al. 1990. Biochemical characterization of a putative axonal guidance molecule of the chick visual system. *Neuron* 5:735–743.

Stark, D.R., et al. 1998. *Hedgehog* family member is expressed throughout regenerating and developing limbs. *Dev Dyn* 212:352–363.

Steele, R.E. 2002. Developmental signaling in *Hydra:* What does it take to build a "simple" animal. *Dev Biol* 248:199–219.

Steen, T.P. 1968. Stability of chondrocyte differentiation and contribution of muscle to cartilage during limb regeneration in the axolotl *(Siredon mexicanum). J Exp Zool* 167:49–77.

Steen, T.P. 1970. Origin and differentiative capacities of cells in the blastema of the regenerating salamander limb. *Am Zool* 10:119–132.

Steen, T.P., and C.S. Thornton. 1963. Tissue interaction in amputated aneurogenic limbs of *Ambystoma* larvae. *J Exp Zool* 154:207–221.

Sternlicht, M.P., and Z. Werb. 2001. How matrix metalloproteinases regulate cell behavior. *Annu Rev Cell Dev Biol* 17:463–516.

Stevens, M.M., and J.H. George. 2005. Exploring and engineering the cell surface interface. *Science* 310:1135–1138.

Stevenson, T.J., et al. 2006. Tissue inhibitor of metalloproteinase 1 regulates matrix metalloproteinase activity during newt limb regeneration. *Dev Dyn* 235:606–616.

Stewart, P.A., and M.J. Wiley. 1981. Developing nervous tissue induces formation of blood-brain characteristics in invading endothelial cells: A study using quail-chick transplantation chimeras. *Dev Biol* 84:183–192.

Stock, U.A., and J.P. Vacanti. 2001. Tissue engineering: Current state and prospects. *Annu Rev Med* 52:443–451.

Stocker, E., and W.D. Heine. 1971. Regeneration of liver parenchyma under normal and pathological conditions. *Beitr z Pathol* 144:400–408.

Stocum, D.L. 1975. Regulation after proximal or distal transposition of limb regeneration blastemas and determination of the proximal boundary of the regenerate. *Dev Biol* 45:112–136.

Stocum, D.L. 1978a. Regeneration of symmetrical hindlimbs in larval salamanders. *Science* 200:790–793.

Stocum, D.L. 1978b. Organization of the morphogenetic field in regenerating amphibian limbs. *Am Zool* 18:883–896.

Stocum, D.L. 1995. *Wound repair, regeneration and artificial tissues.* Austin, TX: R.G. Landes Co.

Stocum, D.L. 2005. Amphibian regeneration and stem cells. In: E. Heber-Katz, ed. *Regeneration: Stem cells and beyond.* Berlin: Springer. 1–70.

Stocum, D.L., and S.D. Thoms. 1984. Retinoic acid-induced pattern completion in regenerating double anterior limbs of urodeles. *J Exp Zool* 232:207–215.

Stone, L.S. 1954. Further experiments on lens regeneration in eyes of the adult newt *Triturus v. viridescens. Anat Rec* 120:599–624.

Straube, W.L., et al. 2004. Plasticity and reprogramming of differentiated cells in amphibian regeneration: Partial purification of a serum factor that triggers cell cycle re-entry in differentiated muscle cells. *Clon Stem Cells* 6:333–344.

Strauer, B.E., et al. 2002. Repair of infarcted myocardium by autologous intracoronary mononuclear bone marrow cell transplantation in humans. *Circulation* 106:1913–1918.

Strebkov, V.S. 1966. Experiment on the application of the method of destruction according to L. V. Polezhaev with the goal of regeneration of bone in the region of defects in the skull (Russian). *Conditions of regeneration of organs and tissues in animals* (Russian). Moscow, 277–281.

Strey, C.W., et al. 2003. The proinflammatory mediators C3a and C5a are essential for liver regeneration. *J Exp Med* 198:913–923.

Stroeva, O.G., and V.I. Mitashov. 1983. Retinal pigment epithelium: Proliferation and differentiation during development and regeneration. *Int Rev Cytol* 83:221–293.

Studitsky, A.N. 1953. Types of new formations of cells from living substance in processes of histogenesis and regeneration (Russian). *Zhur Obshch Biol* 14:177–197.

Studitsky, A.N. 1959. *The experimental surgery of muscles* (Russian). Moscow: Izdatel. Akad. Nauk SSSR.

Studitsky, A.N. 1963. The role of neurotrophic influences upon the restitution of structure and function of regenerating muscles. In: E. Gutmann and P. Hník, eds. *The effect of use and disuse on neuromuscular functions.* Prague: Publishing House of Czechoslovak Acad. Sci. 71–82.

Studitsky, A.N. 1973. Regenerative morphogenesis and laws of the plastic activity of tissues (Russian). In: A.P. Avtsin, ed. *Dedifferentiation in the process of regeneration* (Russian). Moscow: Izdatel Moskovskovo Universiteta, 72–86.

Suhonen, J.O., et al. 1996. Differentiation of adult hippocampus-derived progenitors into olfactory neurons in vivo. *Nature* 383:624–627.

Sunderland, S. 1978. *Nerves and nerve injuries.* Edinburgh, UK: Churchill Livingstone.

Sussman, M.A., and P. Anversa. 2004. Myocardial aging and senescence: Where have the stem cells gone? *Annu Rev Physiol* 66:29–48.

Suttie, J.M., and P.F. Fennessy. 1985. Regrowth of amputated velvet antlers with and without innervation. *J Exp Zool* 234:359–366.

Suzuki, M., et al. 2005. Nerve-dependent and -independent events in blastema formation during *Xenopus* froglet limb regeneration. *Dev Biol* 286:361–375.

Svendgaard, N.A., et al. 1975. Regenerative properties of central monoamine neurons. *Adv Anat Embryol Cell Biol* 51:1–76.

Swift, M.E., et al. 1999. Impaired wound repair and delayed angiogenesis in aged mice. *Lab Invest* 79:1479–1487.

Syfestad, G.T., and M.R. Urist. 1982. Bone aging. *Clin Orthop Relet Res* 162:288–297.

Taban, C. 1955. Quelques problèmes de régénération chez les Urodèles. *Rev Suisse Zool* 62:387–468.

Taban, C. 1971. Tentatives d'induction de la régénération d'organes chez les mammifères. I. Results d'amputation et greffe de tissue nerveux chez l'embryon de lapin. *Rev Suisse Zool* 78:1252–1269.

Taipale, J., and J. Keski-Oja. 1997. Growth factors in the extracellular matrix. *FASEB J* 11:51–59.

Takahashi, T., et al. 1999. Ischemia- and cytokine-induced mobilization of bone marrow-derived endothelial cells for neovascularization. *Nat Med* 5:434–438.

Tanaka, E.M., et al. 1997. Newt myotubes reenter the cell cycle by phosphorylation of the retinoblastoma protein. *J Cell Biol* 136:155–165.

Tanaka, E.M., et al. 1999. Thrombin regulates S-phase re-entry by cultured newt myotubes. *Curr Biol* 9:792–799.

Tanaka, K., and H.deF. Webster. 1991. Myelinated fiber regeneration after crush injury is retarded in sciatic nerves of aging mice. *J Comp Neurol* 308:180–187.

Tank, P.W. 1978. The occurrence of supernumerary limbs following blastemal transplantation in the regenerating forelimb of the axolotl, *Ambystoma mexicanum*. *Dev Biol* 62:143–161.

Tank, P.W., et al. 1976. A staging system for forelimb regeneration in the axolotl, *Ambystoma mexicanum*. *J Morphol* 150:117–128.

Tartar, V. 1961. *The biology of stentor.* New York: Pergamon Press.

Tassava, R.A., and W.D. McCullough. 1978. Neural control of cell cycle events in regenerating salamander limbs. *Am Zool* 18:843–854.

Tassava, R.A., and A.L. Mescher. 1975. The roles of injury, nerves and the wound epidermis during the initiation of amphibian limb regeneration. *Differentiation* 4:23–24.

Tassava, R.A., and C.L. Olsen-Winner. 2003. Responses to amputation of denervated *Ambystoma* limbs containing aneurogenic grafts. *J Exp Zool* 297A:64–79.

Tassava, R.A., et al. 1996. Extracellular matrix protein turnover during salamander limb regeneration. *Wound Repair Regen* 4:75–81.

Tatsumi, R., et al. 1998. HGF/SF is present in normal adult skeletal muscle and is capable of activating satellite cells. *Dev Biol* 194:114–128.

Tatsumi, R., et al. 2001. Mechanical stretch induces activation of skeletal muscle satellite cells *in vitro*. *Exp Cell Res* 267:107–114.

Tatsumi, R., et al. 2002. Release of hepatocyte growth factor from mechanically stretched skeletal muscle satellite cells and role of pH and nitric oxide. *Mol Cell Biol* 13:2909–2918.

Taub, R. 2004. Liver regeneration: From myth to mechanism. *Nat Rev Mol Cell Biol* 5:836–847.

Temenoff, J.S., et al. 2004. Biodegradable scaffolds. In: V.M. Goldberg and A.I. Caplan, eds. *Orthopedic tissue engineering.* New York: Marcel Dekker. 77–103.

Terada, N., et al. 2002. Bone marrow cells adopt the phenotype of other cells by spontaneous cell fusion. *Nature* 416:542–545.

Theise, N.D., et al. 2000. Liver from bone marrow in humans. *Hepatology* 32:11–16.

Thiele, J., et al. 2004. Regeneration of heart muscle tissue: Quantification of chimeric cardiomyocytes and endothelial cells following transplantation. *Histol Histopathol* 19:201–209.

Thitoff, A.R., et al. 2003. Unique expression patterns of the retinoblastoma *(Rb)* gene in intact and lens regeneration-undergoing newt eyes. *Anat Rec* 271A:185–188.

Thompson, W.J., et al. 1990. The origin and selective innervation of early muscle fiber types in the rat. *J Neurobiol* 21:212–222.

Thornton, C.S. 1938. The histogenesis of muscle in the regenerating forelimb of larval *Amblystoma punctatum*. *J Morphol* 62:17–47.

Thornton, C.S. 1956. Epidermal modifications in regenerating and in non-regenerating limbs of anuran larvae. *J Exp Zool* 131:373–393.

Thornton, C.S. 1957. The effect of apical cap removal on limb regeneration in *Amblystoma* larvae. *J Exp Zool* 134:357–382.

Thornton, C.S. 1962. Influence of head skin on limb regeneration in urodele amphibians. *J Exp Zool* 150:5–15.

Thornton, C.S., and M.T. Thornton. 1970. Recuperation of regeneration in denervated limbs of *Ambystoma* larvae. *J Exp Zool* 173:293–302.

Ting, S.B., et al. 2005. A homolog of *Drosophila grainy head* is essential for epidermal integrity in mice. *Science* 308:411–413.

Tochimoto, T. 1997. Regeneration in the Japanese giant salamander, *Andrias japonicus,* in captivity (Japanese). *J Japan Assoc Zool Aqua* 38:65–68.

Todd, T.J. 1823. On the process of reproduction of the members of the aquatic salamander. *Q J Sci Lit Arts* 16:84–96.

Tokin, B.P. 1959. *Regeneration and somatic embryogenesis* (Russian). Leningrad: Izdatel. Leningradskovo Univ., Leningrad.

Tokin, B.P., ed. 1972. *Asexual reproduction, somatic embryogenesis and regeneration* (Russian). Izdatel. Leningrad: Leningrad. Univ.

Tomlinson, B.L., et al. 1985. Pattern-deficient forelimb regeneration in adult bullfrogs. *J Exp Zool* 236:313–326.

Tonge, D.A., and P.G. Leclere. 2000. Directed axonal growth towards axolotl limb blastemas *in vitro*. *Neuroscience* 100:201–211.

Tonna, E.A. 1978. Electron microscopic study of bone surface changes during ageing: The loss of cellular control and biofeedback. *J Gerontol* 33:163–177.

Toole, B.P., and J. Gross. 1971. The extracellular matrix of the regenerating limb: Synthesis and removal of hyaluronate prior to differentiation. *Dev Biol* 25:57–77.

Tornier, G. 1906. Kampf der Gewebe im Regeneration bei Begunstigung der Hautregeneration. *Arch f Entw-mech* 22:348–369.

Torok, M.A., et al. 1999. *Sonic hedgehog (shh)* expression in developing and regenerating axolotl limbs. *J Exp Zool* 284:197–206.

Tsai, E.C., and C.H. Tator. 2005. Neuroprotection and regeneration strategies for spinal cord repair. *Curr Pharm Des* 11:1211–1222.

Tsonis, P.A. 1996. *Limb regeneration*. Cambridge, UK: Cambridge Univ. Press.

Tsonis, P.A., et al. 2004. A novel role of the hedgehog pathway in lens regeneration. *Dev Biol* 267:450–461.

Umansky, E.E., and V.P. Kudokotsev. 1948. Concerning the restoration of regenerative capacity in the limbs of reptiles (Russian). *Dokl Akad Nauk SSSR* 61:757–760.

Umansky, E.E., and V.P. Kudokotsev. 1951. Stimulation of the regenerative process of reptile limbs through the action of parathyroid hormone. *Dokl Akad Nauk SSSR* 77:533–536.

Urbanek, K., et al. 2003. Intense myocyte formation from cardiac stem cells in human cardiac hypertrophy. *Proc Natl Acad Sci USA* 100:10440–10445.

Urist, M.R. 1965. Bone: Formation by autoinduction. *Science* 150:893–899.

Urist, M.R. 1997. Bone: Formation by autoinduction. *J NIH Res* 9:43–50.

Urist, M.R., et al. 1979. Solubilized and insolubilized bone morphogenetic protein. *Proc Natl Acad Sci USA* 76:1828–1832.

Urist, M.R., et al. 1984. Purification of bovine bone morphogenetic protein by hydroxyapatite chromatography. *Proc Natl Acad Sci USA* 81:371–375.

van Zant, G., and Y. Liang. 2003. The role of stem cells in aging. *Exp Hematol* 31:659–672.

Vanable, J.W. 1989. Integumentary potentials and wound healing. In: R.B. Borgens, et al., eds. *Electric fields in vertebrate repair.* New York: Alan R. Liss. 171–224.

Vandenburgh, H.H. 1982. Dynamic mechanical orientation of skeletal myofibers in vitro. *Dev Biol* 93:438–443.

Vandenburgh, H.H. 1987. Motion into mass: How does tension stimulate muscle growth? *Med Sci Sports Exer* 19:S142–S149.

Vandenburgh, H.H., and S. Kaufman. 1979. In vitro model for stretch-induced hypertrophy of skeletal muscle. *Science* 203:265–268.

Vaughan, D.W. 1992. Effects of advancing age on peripheral nerve regeneration. *J Comp Neurol* 323:219–237.

Verdú, E., et al. 1995. The effect of aging on efferent nerve fibers regeneration in mice. *Brain Res* 696:76–82.

Viguie, C.A., et al. 1997. A quantitative study of the effects of long-term denervation on the extensor digitorum longus muscle of the rat. *Anat Rec* 248:346–354.

Vinarsky, V., et al. 2005. Normal newt limb regeneration requires matrix metalloproteinase function. *Dev Biol* 279:86–98.

Vinogradova, T.P., and G.I. Lavrishcheva. 1974. *The regeneration and transplantation of bone* (Russian). Moscow: Meditsina.

von Boxberg, Y., et al. 1993. Guidance and topographic stabilization of nasal chick retinal axons on target-derived components in vitro. *Neuron* 10:345–357.

Vorontsova, M.A. 1949. *Regeneration of organs in animals* (Russian). Moscow: Sovietskaya Nauka.

Vorontsova, M.A. 1953. *The restoration of lost organs in animals and humans* (Russian). Moscow: Sovietskaya Nauka.

Vorontsova, M.A., and L.D. Liosner. 1960. *Asexual propagation and regeneration.* New York: Pergamon Press.

Vracko, R. 1974. Basal lamina scaffold: Anatomy and significance for maintenance of orderly tissue structure. *Am J Pathol* 77:314–346.

Wagers, A.J., and I.M. Conboy. 2005. Cellular and molecular signatures of muscle regeneration: Current concepts and controversies in adult myogenesis. *Cell* 122:659–667.

Wallace, H. 1981. *Vertebrate limb regeneration.* Chichester, UK: John Wiley & Sons.

Wanek, N., et al. 1989. Evidence for regulation following amputation and tissue grafting in the developing mouse limb. *J Exp Zool* 249:55–61.

Wang, C., et al. 2004. Mechanical, cellular, and molecular factors interact to modulate circulating endothelial cell progenitors. *Am J Physiol Heart Circ Physiol* 286:H1985–H1993.

Wang, Z.M., et al. 1998. Adult opossums *(Didelphis virginiana)* demonstrate near normal locomotion after spinal cord transection as neonates. *Exp Neurol* 151:50–69.

Washabaugh, C.H., et al. 2004. Nonmuscle stem cells fail to significantly contribute to regeneration of normal muscle. *Gene Ther* 11:1724–1728.

Weis, J.S. 1972. The effects of nerve growth factor on bullfrog tadpoles *(Rana catesbiana)* after limb amputation. *J Exp Zool* 180:385–392.

Weis, J.S., and L.P. Bleier. 1973. The effect of hydrocortisone on limb regeneration in the bullfrog tadpole, *Rana catesbiana. J Embryol Exp Morphol* 29:65–71.

Weismann, A. 1892. *Das Keimplasma. Eine Theorie der Vererbung.* Jena, Germany: G. Fischer Verlag.

Weiss, P. 1925. Unabhängigkeit der Extremitätenregeneration vom Skelett (bei *Triton cristatus*). *Arch mikroskop Anat u Entw-mech* 104:359–394.

Weiss, P. 1933. Functional adaptation and the role of ground substances in development. *Am Nat* 57:322–340.

Weiss, P. 1939. *Principles of development.* New York: Henry Holt & Co.

Weiss, P. 1950. Perspectives in the field of morphogenesis. *Q Rev Biol* 25:177–198.

Weiss, P. 1958. Cell contact. *Int Rev Cytol* 7:391–423.

Weiss, P. 1961. The biological foundations of wound repair. *The Harvey Lecture Series 1959–1960.* Series 55. New York: Academic Press, 13–42.

Weiss, P., and J.L. Kavanau. 1957. A model of growth and growth control in mathematical terms. *J Gen Physiol* 41:1–47.

Weiss, P., and A.G. Matoltsy. 1959. Wound healing in chick embryos *in vivo* and *in vitro*. *Dev Biol* 1:302–326.

Weiss, P.A. 1934. In vitro experiments on the factors determining the course of the outgrowing nerve fiber. *J Exp Zool* 68:393–448.

Weissman, I.L. 2000. Stem cells: Units of development, units of regeneration, and units in evolution. *Cell* 100:157–168.

Werner, S., and R. Grose. 2003. Regulation of wound healing by growth factors and cytokines. *Physiol Rev* 83:835–870.

Werner, S., et al. 1994. The function of KGF in morphogenesis of epithelium and reepithelialization of wounds. *Science* 266:819–822.

Whitehead, G.G., et al. 2005 *fgf20* is essential for initiating zebrafish fin regeneration. *Science* 310:1957–1960.

Whittemore, S.R., et al. 1999. Mitogen and substrate differentially affect the lineage restriction of adult rat subventricular zone neural precursor cell populations. *Exp Cell Res* 252:75–95.

Willcox, B.J., and J.N. Scott. 2004. Growth-associated proteins and regeneration-induced gene expression in the aging neuron. *Mech Ageing Dev* 125:513–516.

Winter, G.D. 1972. Epidermal regeneration studied in the domestic pig. In: H.I. Maibach and D.T. Rovee, eds. *Epidermal wound healing.* Chicago: Year Book Medical Publishers. 71–112.

Wislocki, G.B., and M. Singer. 1946. The occurrence and the function of nerves in the growing antlers of deer. *J Comp Neurol* 85:1–19.

Wolfe, A.D., et al. 2004. Early regeneration genes: Building a molecular profile for shared expression in a cornea-lens transdifferentiation and hindlimb regeneration in *Xenopus laevis.* *Dev Dyn* 230:615–629.

Wolff, E., and T. Lender. 1962. Les néoblastes et les phénomènes d'induction et d'inhibition dans la régénération des planaires. *Ann Biol* 1:500–529.

Wolff, G. 1895. Entwicklungsphysiologische Studien. I. Die Regeneration der Urodelenlinse. *Arch f Entwmech* 1:380–390.

Wolff, J. 1892. *Das Gesetz der Transformation der Knochen.* Berlin: Hirschwald.

Wolpert, L. 1969. Positional information and the spatial pattern of cellular differentiation. *J Theor Biol* 25:1–47.

Wolpert, L. 1971. Positional information and pattern formation. *Curr Top Dev Biol* 6:183–224.

Wolpert, L., et al. 1974. Positional information and positional signaling in Hydra. *Am Zool* 14:647–663.

Womble, M.D. 1986. The clustering of acetylcholine receptors and formation of neuromuscular junctions in regenerating mammalian muscle grafts. *Am J Anat* 176:191–205.

Wozniak, A.C., et al. 2005. Signaling satellite-cell activation in skeletal muscle: Markers, models, stretch, and potential alternate pathways. *Muscle Nerve* 31:283–300.

Wright, M.R. 1947. Regeneration and degeneration experiments on lateral-line nerves and sense organs in anurans. *J Exp Zool* 105:221–257.

Yamada, T. 1967. Cellular and subcellular events in Wolffian lens regeneration. *Curr Top Dev Biol* 2:247–283.

Yamada, T. 1977. *Control mechanisms in cell-type conversion in newt lens regeneration.* Basel, Switzerland: Karger.

Yamanouchi, K., et al. 2000. Expression of myostatin gene in regenerating skeletal muscle of the rat and its localization. *Biochem Biophys Res Commun* 270:510–516.

Yancopoulos, G.D., et al. 2000. Vascular-specific growth factors and blood vessel formation. *Nature* 407:242–248.

Yang, E.V., et al. 2005. Effects of exogenous FGF-1 treatment on regeneration of the lens and the neural retina in the newt, *Notophthalmus viridescens. J Exp Zool* 303A:837–844.

Yannas, I.V. 2001. *Tissue and organ regeneration in adults.* New York: Springer Verlag.

Yasuda, I. 1974. Mechanical and electrical callus. *Ann N Y Acad Sci* 238:457–465.

Yasuda, I. 1977. Electrical callus and callus formation by electret. *Clin Orthopaed* 124: 53–56.

Ying, Q.-L., et al. 2002. Changing potency by spontaneous fusion. *Nature* 416:545–548.

Yntema, C.L. 1959a. Regeneration in sparsely innervated and aneurogenic forelimbs of *Amblystoma* larvae. *J Exp Zool* 140:101–1214.

Yntema, C.L. 1959b. Blastema formation in sparsely innervated and aneurogenic forelimbs of *Amblystoma* larvae. *J Exp Zool* 142:423–440.

Yokoyama, H., et al. 2000. Mesenchyme with *fgf-10* expression is responsible for regenerative capacity in *Xenopus* limb buds. *Dev Biol* 219:18–29.

Yokoyama, H., et al. 2001. FGF-10 stimulates limb regeneration ability in *Xenopus laevis. Dev Biol* 233:72–79.

Yoon, M.G. 1971. Reorganization of retinotectal projection following surgical operations on the optic tectum in goldfish. *Exp Neurol* 33:395–411.

Young, L.J.T., et al. 1971. The influence of host and tissue age on life span and growth rate of serially transplanted mouse mammary gland. *Exp Gerontol* 6:49–56.

Yu, D., and R. Auerback. 1999. Brain-specific differentiation of mouse yolk sac endothelial cells. *Dev Brain Res* 117:159–169.

Zalewski, A.A. 1969. Combined effects of testosterone and motor, sensory, or gustatory nerve regeneration on the regeneration of taste buds. *Exp Neurol* 24:285–297.

Zalewski, A.A. 1972. Regeneration of taste buds after transplantation of tongue and ganglia grafts to the anterior chamber of the eye. *Exp Neurol* 35:519–528.

Zamaraev, V.N. 1973. Is regenerative capacity lost with increasing age? (Russian). *Ontogenez* 4:539–548.

Zammaretti, P., and M. Jaconi. 2004. Cardiac tissue engineering: regeneration of the wounded heart. *Curr Opin Biotechnol* 15:430–434.

Zaucha, J.M., et al. 2001. Hematopoietic responses to stress conditions in young dogs compared to elderly dogs. *Blood* 98:322–327.

Zelená, J. 1957. The morphogenetic influence of innervation on the ontogenetic development of muscle-spindles. *J Embryol Exp Morphol* 5:283–292.

Zelená, J. 1964. Development, degeneration and regeneration of receptor organs. In: M. Singer and J.P. Schadé. *Mechanisms of neural regeneration: Progress in brain research.* Vol. 19. Amsterdam: Elsevier. 175–213.

Zelená, J., and M. Sobotková. 1971. Absence of muscle spindles in regenerated muscles of the rat. *Physiol Bohemoslov* 20:433–439.

Zeltner, T.B., et al. 1987. The postnatal development and growth of the human lung. I. Morphometry. *Respir Physiol* 67:247–267.

Zhao, M., et al. 1999. A small, physiological electric field orients cell division. *Proc Natl Acad Sci USA* 96:4942–4946.

Zhao, P., and E.P. Hoffman. 2004. Embryonic myogenesis pathways in muscle regeneration. *Dev Dyn* 229:380–392.

Zhao, S., et al. 1995. In vitro transdifferentiation of embryonic rat retinal pigment epithelium to neural retina. *Brain Res* 677:300–310.

Zhao, W., and D.A. Neufeld. 1995. Bone regrowth in young mice stimulated by nail organ. *J Exp Zool* 271:155–159.

Zhenevskaya, R.P. 1960. The influence of de-efferentation on the regeneration of skeletal muscle (Russian). *Arkh Anat Gist Embriol* 39:42–50.

Zhenevskaya, R.P. 1961. The restoration of muscle by the method of transplantation of minced muscle tissue under conditions of sensory denervation (Russian). *Arkh Anat Gist Embriol* 40:46–53.

Zhenevskaya, R.P. 1962. Experimental histologic investigation of striated muscle tissue. *Rev Can Biol* 21:457–470.

Zhenevskaya, R.P. 1974. *Neurotrophic regulation of the plastic activity of muscular tissue* (Russian). Moscow: Izdatel. Nauka.

Zheng, W., et al. 2004. Stretch induces upregulation of key tyrosine kinase receptors in microvascular cells. *Am J Physiol Heart Circ Physiol* 287:H2739–H2745.

Zhou, R., and I.B. Black. 2000. Development of neural maps: Molecular mechanisms. In: M.S. Gazzaniga, ed. *The new cognitive neurosciences*. Cambridge, MA: MIT Press. 213–221.

Zochodne, D.W. 2000. The microenvironment of injured and regenerating peripheral nerves. *Muscle Nerve Suppl* 9:S33–S38.

Zupanc, G.K.H. 2001. Adult neurogenesis and neuronal regeneration in the central nervous system of teleost fish. *Brain Behav Evol* 58:250–275.

Index

Printed and bound by CPI Group (UK) Ltd, Croydon, CR0 4YY

03/10/2024

01040316-0001